An Introduction to the Physics of Nuclei and Particles (Second Edition)

Online at: https://doi.org/10.1088/978-0-7503-6094-4

An Introduction to the Physics of Nuclei and Particles (Second Edition)

Richard A Dunlap

Department of Physics and Atmospheric Science, Dalhousie University, Halifax, Nova Scotia, Canada

IOP Publishing, Bristol, UK

ISBN   978-0-7503-6094-4 (ebook)
ISBN   978-0-7503-6092-0 (print)
ISBN   978-0-7503-6095-1 (myPrint)
ISBN   978-0-7503-6093-7 (mobi)

DOI   10.1088/978-0-7503-6094-4

Version: 20231101

IOP ebooks

British Library Cataloguing-in-Publication Data: A catalogue record for this book is available from the British Library.

Published by IOP Publishing, wholly owned by The Institute of Physics, London

IOP Publishing, No.2 The Distillery, Glassfields, Avon Street, Bristol, BS2 0GR, UK

US Office: IOP Publishing, Inc., 190 North Independence Mall West, Suite 601, Philadelphia, PA 19106, USA

for Ewa

Contents

Preface

This text has developed from a one semester fourth year undergraduate level course in nuclear and particle physics that is designed for physics majors. The course is also taught as an introductory course for graduate students who have not previously had an undergraduate course in this subject. The same material with some exclusions (primarily some of the material in chapters 9 and 10) has been used for a third year undergraduate level course. In general students should have a minimum background of one semester of electricity and magnetism and one semester of introductory quantum mechanics or modern physics. The present text contains about 25% more material that can be comfortably covered in a one semester course (about 14 weeks) and gives the instructor some freedom in selecting topics in the latter half of the book.

Since the publication of the first edition of *An Introduction to the Physics of Nuclei and Particles*, numerous significant advances have been made in both nuclear and particle physics. These advances are, perhaps, most significant in the field of particle physics with the discovery of the Higgs boson and a more complete understanding of neutrino oscillations and masses. In the field of nuclear physics, recent developments include progress in our understanding of phenomena such as double β-decay and α-clustering in nuclei. From a more practical standpoint, the development of safer, more efficient fission reactors along with advances in the quest to develop fusion power, are notable.

The aim of the second edition of this text is to bring the material in both the nuclear physics and particle physics of the book up to date. Revisions include expanding and updating material on nuclear models, β-decay and fission and fusion reactors. Part IV of the text is substantially revised and includes new chapters on the Higgs boson, proton decay and neutrino physics. A new appendix on the design and physics of particle accelerators is included and may contribute to course material at the instructor's discretion. As well, detailed solutions to all even numbered end-of-chapter problems are included as an appendix rather than as a separate student manual, as for the first edition.

Acknowledgments

I would like to acknowledge the contributions of numerous individuals in the development and preparation of this book. Most significantly I am indebted to the students who have taken my nuclear and particle physics course over the years. They have provided invaluable feedback on the ideas presented here and have tested countless homework problems. I am also grateful to David Kiang for the support and encouragement he has provided for my interest in teaching this course and for his numerous comments and suggestions on the first edition. Finally, I am grateful to the reviewers of both the first and second editions of this book for their valuable comments and suggestions.

Author biography

Richard A Dunlap

 Richard A Dunlap received a BS from Worcester Polytechnic Institute (Physics 1974), an A.M from Dartmouth College (Physics 1976) and a PhD from Clark University (Physics 1981). Since 1981 he has been a professor in the Department of Physics and Atmospheric Science at Dalhousie University and currently holds a position as Research Professor. He was Faculty of Science Killam Research Professor from 2001 to 2006 and Director of the Dalhousie University Institute for Research in Materials from 2009 to 2015. Professor Dunlap has published more than 300 refereed research papers and his research interests include nuclear spectroscopies, magnetic materials, quasicrystals, critical phenomena and advanced battery materials. He is author of fourteen previous books including; Experimental Physics: Modern Methods (Oxford 1988), The Golden Ratio and Fibonacci Numbers (World Scientific 1997), An Introduction to the Physics of Nuclei and Particles—1st edn (Brooks/Cole 2004), Sustainable Energy (Cengage, 1st edn 2015, 2nd edn 2019), Novel Microstructures for Solids (IOP/Morgan & Claypool 2018), Particle Physics (IOP/Morgan & Claypool 2018), The Mössbauer Effect (IOP/Morgan & Claypool 2019), Electrons in Solids—Contemporary Topics (IOP/Morgan & Claypool 2019), Energy from Nuclear Fusion (IOP Publishing 2021), Lasers and Their Application in the Cooling and Trapping of Atoms (IOP, 2nd edn 2023) and Transportation Technologies for a Sustainable Future (IOP 2023).

Part I

Introduction

IOP Publishing

An Introduction to the Physics of Nuclei and Particles
(Second Edition)

Richard A Dunlap

Chapter 1

Basic concepts

1.1 Introduction

Much of our understanding of physics comes from the application of fundamental theoretical concepts to the description of physical systems. This is particularly true in the fields of atomic physics and solid state physics. An important example is the calculation of the energy levels of the electron in a hydrogen atom (see chapter 5 for a more detailed discussion), where the three-dimensional time independent Schrödinger equation,

$$-\frac{\hbar^2}{2m}\nabla^2\psi + V(r)\psi = E\psi. \tag{1.1}$$

is solved. Here $\hbar$ is the reduced Planck constant, m is the electron mass, $V(r)$ is the potential energy, ψ is the electron wave function, and E is the energy. Using the appropriate form of the Coulomb potential for $V(r)$, the energy eigenvalues can be easily obtained. The success of this approach in predicting the experimentally observed values, relies on a good fundamental understanding of the electromagnetic interactions responsible for the determination of the potential. The difficulty in applying equation (1.1) to atoms with many electrons lies, not in our lack of understanding of the fundamental properties of the electromagnetic interaction, but in the complexity of the mathematics. The situation in solid state physics is very similar. However, in nuclear physics the form of $V(r)$ has not been uniquely determined and a phenomenological approach is usually adopted. Although considerable success has been achieved by using potentials based on empirical observation and on what seems sensible, there is much of a fundamental nature concerning nuclear interactions that still needs to be learned. A better appreciation of nuclear physics can be gained by some understanding of the interactions between particles that are responsible for determining nuclear properties. It is with this idea

doi:10.1088/978-0-7503-6094-4ch1

in mind that the first two chapters of this book cover, not only a discussion of some basic concepts in nuclear physics, but a brief overview of particle properties and interactions.

1.2 Terminology and definitions

An atom consists of a nucleus and the atomic electrons bound to it by the Coulomb interaction. The simplest nucleus, that of the ^{1}H atom, consists of a single proton. All other nuclei are bound systems consisting of some combination of Z protons and N neutrons. It is the purpose of Parts II and III of this book to describe the properties of such nuclei. Such properties include the size, shape, mass, stability, and electromagnetic moments of the nucleus as well as its cross section for various reactions. The present chapter covers some basic background materials that will be helpful before proceeding to Part II of the text. We begin with some basic definitions that are used in nuclear physics. Following these, we discuss the units that are commonly used to describe nuclear quantities and some sources for further information.

Atomic number (Z) is the number of protons in the nucleus of an atom and also the number of electrons in a neutral atom.

Nucleon is the general term used to refer to a neutron or a proton when it is part of a nucleus.

Atomic mass unit (u) is a unit of mass used for atoms, nuclei, or particles that is one-twelfth of the mass of a neutral atom of ^{12}C (six protons, six neutrons, and six electrons). The superscript in front of the elemental name indicates the total number of nucleons.

Atomic mass is the mass of a neutral atom (usually expressed in atomic mass units) and includes the masses of the protons, neutrons, and electrons as well as all binding energies.

Nuclear mass is the mass of the nucleus (usually expressed in atomic mass units) and includes the masses of the protons and neutrons as well as the nuclear binding energy but does not include the mass of the atomic electrons or electronic binding energy.

Mass number (A) is the total number of nucleons in a nucleus ($A = Z + N$) and is also the integer closest to the nuclear mass in atomic mass units.

Nuclide is a specific nuclear species specified by the values of N and Z.

Isotopes are members of a family of nuclides with a common value of Z.

Radioisotopes are members of a family of unstable nuclides with a common value of Z.

Isotones are members of a family of nuclides with a common value of N.

Isobars are members of a family of nuclides with a common value of A.

Natural abundance is the relative proportion (as a fraction or as percent) of a particular isotope of an element in a naturally occurring terrestrial sample of the element.

Atomic weight is the average atomic mass of naturally occurring isotopes of an element weighted by their natural abundance.

1.3 Units and dimensions

In general, the units used in this book are those that are in common use in the fields of nuclear and particle physics. In most cases these are derived from standard SI units, although a few comments are necessary here to define units that are particular to these fields. The quantities that are most commonly encountered are distance, area, time, energy, and mass. These are considered below.

Distance Nuclear dimensions, as we will see in chapter 3, are of the order of 10^{-15} m. This represents one femtometer and is abbreviated fm. In nuclear physics this unit is commonly called a fermi.

Area In SI, area units are multiples of m^2 and are most commonly encountered in nuclear and particle physics when discussing cross sections. Nuclear cross sections vary over a wide range of values but are frequently in the range of hundreds of fm^2. The unit barn is defined as 100 fm^2.

Time The time scale for nuclear processes is usually quite short. However, lifetimes for nuclear or particle decays can cover a very large range of values, from less than 10^{-25} s to more than 10^{39} s. In many cases long time periods may be expressed in years. Intermediate time periods may be expressed in minutes, hours, or days as appropriate.

Energy Energy is most commonly expressed in multiples of electron volts (eV), usually MeV (10^6 eV) in nuclear physics or GeV (10^9 eV) in particle physics. In some cases, keV is also used when energies are in this range. The eV is related to the conventional SI energy unit (the Joule) by 1 eV = 1.602 177 $\times 10^{-19}$ J.

Mass Atomic and nuclear masses are most commonly given in atomic mass units (u) as defined above. This unit is related to the conventional SI mass unit (the kilogram) as 1 u = 1.660 5402 $\times 10^{-27}$ kg. Masses of subatomic particles are sometimes given in atomic mass units, particularly when these masses are used in the context of nuclear properties. When dealing with the properties of the particles themselves, it is more common to express masses in units of MeV/c^2 or GeV/c^2 (or some multiple of these quantities). These mass units result from Einstein's equivalence of mass and energy expressed as

$$E = mc^2. \tag{1.2}$$

The conversion between atomic mass units and MeV/c^2 can be obtained by writing the energy equivalent of a 1 u mass from equation (1.2) using the speed of light in SI units as

$$E = 1.660\ 5402 \times 10^{-27}\ \text{kg} \times (2.997\ 925 \times 10^8\ \text{m/s})^2 = 1.505\ 35 \times 10^{-10}\ \text{J}.$$

Using the above conversion from Joules to MeV,

$$E = 1.505\ 35 \times 10^{-10}\ \text{J} \times 1.602\ 177 \times 10^{-13}\ \text{MeV/J} = 931.494\ \text{MeV}.$$

Thus,

$$1\ \text{u} = 931.494\ \text{MeV}/c^2.$$

This relationship provides a value for the speed of light squared in alternative units that is useful for many nuclear and particle physics calculations,

$$c^2 = 931.494 \text{ MeV/u}.$$

1.4 Sources of information

There are numerous textbooks on nuclear and particle physics that can provide supplemental information. Recommended books appear in the references at the end of the chapters. There are a number of sources for nuclear data such as atomic masses, decay properties, reaction cross sections, and particle masses. The *Table of Isotopes*, 8th edn, ed R B Firestone and V S Shirley (New York: Wiley 1998) is the standard reference for information about nuclear decays. There are a number of sites on the Internet that provide useful nuclear and particle data. Some of the publicly accessible sites that are particularly useful are as follows:

Japan Atomic Energy Research Institute (https://wwwndc.jaea.go.jp/jendl/j5/j5.html) has very convenient tables of nuclear data that include atomic masses for the nuclides.

Laboratoire National Henri Becquerel (www.lnhb.fr/nuclear-data/) gives detailed decay information for individual nuclides.

Live Chart of Nuclides nuclear structure and decay data (https://www-nds.iaea.org/relnsd/vcharthtml/VChartHTML.html) gives detailed graphs of energy levels of individual nuclides.

The Lund/LBNL Nuclear Data Search (http://nucleardata.nuclear.lu.se/toi/) contains simplified decay schemes as a function of mass number.

Particle Data Group (http://pdg.lbl.gov) provides up-to-date information about subatomic particles and includes all fundamental properties such as mass, charge, and spin.

Table of Nuclides (www.nndc.bnl.gov/nudat3) provides much of the information that is available in the *Table of Isotopes* book. Some decay information may not be as complete as that given in the book.

Web Elements (www.webelements.com) is aimed at providing chemical and physical data for the elements and includes a number of important nuclear properties as well.

IOP Publishing

An Introduction to the Physics of Nuclei and Particles
(Second Edition)

Richard A Dunlap

Chapter 2

Subatomic particles and their interactions

2.1 Classification of subatomic particles

Our initial discussion of nuclear properties will deal almost exclusively with neutrons and protons. In part III of the text, we will encounter electrons and neutrinos and in part IV we will examine the properties of a wide variety of subatomic particles. Although the details of the properties of subatomic particles, including the neutron and the proton, will be given in the latter portions of the text, a brief overview of the classification and some of the properties of these particles will be presented in this chapter to allow for a more complete discussion of the interactions that take place within the nucleus.

All subatomic particles can be categorized as either fermions or bosons depending on whether their spin is 1/2 integral or integral, respectively. The former particles can be described by Fermi–Dirac statistics and the latter by Bose–Einstein statistics. The general division of fermions and bosons into categories based on other particle properties is summarized in figure 2.1. It is seen from the figure that the fermions are either leptons or baryons. The bosons are categorized as either mesons or gauge bosons. As will be demonstrated in chapter 15, baryons and mesons share certain properties and are collectively referred to as *hadrons*. The basic properties of the particles of relevance to parts II and III of this book are summarized in table 2.1.

The properties of the nucleus depend on the properties of these particles as determined by the interactions that act on each class of particle and by the conservation laws that are applicable in each case. A selective interaction is, for example, the electrostatic interaction that acts only on objects that are electrically charged. An example of a conservation law is the conservation of charge; positively charged objects and negatively charged objects may interact but the total charge (sum of the positive and negative charges) will be the same before and after the interaction.

doi:10.1088/978-0-7503-6094-4ch2

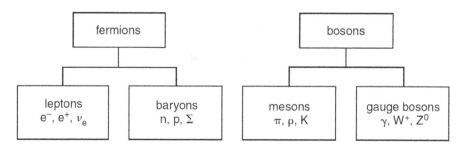

Figure 2.1. Classification of fermions and bosons. This diagram shows some examples of particles in each category.

Table 2.1. Properties of some subatomic particles that are of importance in nuclear physics.

Particle	Symbol	Classification	Lepton number	Baryon number	Charge (e)	Mass $(MeV\,/c^2)$	Lifetime (s)
Electron	e^-	lepton	+1	0	−1	0.511	∞
Positron	e^+	lepton	−1	0	+1	0.511	∞
Electron neutrino	ν_e	lepton	+1	0	0	~0	∞
Electron antineutrino	$\bar{\nu}_e$	lepton	−1	0	0	~0	∞
Proton	p	baryon	0	+1	+1	938.28	$>10^{39}$
Neutron	n	baryon	0	+1	0	939.57	898

The remainder of this chapter gives a brief overview of the properties of the various classes of particles that are of relevance to the properties of the nucleus. This includes a discussion of the interactions that apply to each type of particle and the conservation laws that must be considered.

2.2 Classification and ranges of interactions

There are four known forces in Nature as summarized in table 2.2. The table gives the relative strengths of these interactions and the range over which they act. The gravitational interaction is sufficiently weak that it plays no role in the behavior of nuclei. The electromagnetic interaction is long range and affects macroscopic objects as well as nuclei. The strong and weak interactions are short range and are only important on a size scale comparable to the dimension of the nucleus.

The electromagnetic interaction acts on objects that possess charge. Along similar lines the strong and weak interactions each act on certain types of particles. Although the weak interaction is very weak and very short range (compared with the strong interaction), it is of great importance to the behavior of nuclei because it acts upon leptons as well as hadrons. The strong interaction does not act on leptons, so the weak interaction is an important consideration for many processes that

Table 2.2. Properties of the four interactions in nature. The relative strength is normalized to unity for the strong interaction.

Interaction	Acts on	Strength	Range
Strong	Hadrons	1	10^{-15} m
Electromagnetic	Electric charges	10^{-2}	long ($1/r^2$)
Weak	Leptons and hadrons	10^{-5}	10^{-18} m
Gravity	Masses	10^{-39}	long ($1/r^2$)

Table 2.3. Quantities that are conserved (Y) and nonconserved (N) for the strong, electromagnetic, and weak interactions.

Quantity	Strong	Electromagnetic	Weak
Mass/energy	Y	Y	Y
Linear momentum	Y	Y	Y
Angular momentum	Y	Y	Y
Charge	Y	Y	Y
Parity	Y	Y	N
Lepton number	Y	Y	Y
Baryon number	Y	Y	Y
Lepton generation	Y	Y	Y
Isospin	Y	N	N
Strangeness	Y	Y	N
Charm	Y	Y	N
Bottom	Y	Y	N
Top	Y	Y	N

involve only leptons. In fact, chapters 15 and 16 will show that strong and weak interactions act differently on hadrons and that the weak interaction is significant even in many processes that involve only hadrons.

2.3 Conservation laws

A detailed analysis of any nuclear process requires a consideration of the applicable conservation laws. Certain quantities (for example, angular momentum) are believed to be conserved in all processes while other quantities (for example, parity) may or may not be conserved depending on the nature of the interactions that are present. Table 2.3 summarizes the quantities that are conserved for the strong, electromagnetic, and weak interactions. An important distinction between the strong interaction and the weak interaction is seen from the table. Certain quantities (for example, isospin, strangeness, etc) are not conserved in processes that are dominated by the weak interaction while these same quantities must be conserved in processes involving only the strong interaction. Details of these quantities will be discussed in part IV of the book. When the weak interaction is present, it is the non-conservation of these quantities that allows certain processes (such as β-decay) to occur.

Nuclear processes must satisfy all appropriate conservation laws. Mass/energy and angular momentum must be conserved in all nuclear processes as is the case for macroscopic processes. Of particular relevance to nuclear processes is the conservation of charge, lepton number, and baryon number. These quantities are shown for particles of interest in table 2.1. An investigation of reactions and decay processes as described in parts II and III of this book will demonstrate the conservation of these quantities. Further discussion on the conservation or nonconservation of certain quantities in reactions and decays of particles will be presented in part IV.

Suggestions for further reading

Dunlap R A 2018 *Particle Physics* (San Rafael, CA: Morgan & Claypool)
Krane K S 1988 *Introductory Nuclear Physics* (New York: Wiley)

Part II

Nuclear properties and models

IOP Publishing

An Introduction to the Physics of Nuclei and Particles
(Second Edition)

Richard A Dunlap

Chapter 3

Nuclear composition and size

3.1 Composition of the nucleus

The physical properties of a nucleus are determined by its configuration of nucleons. The most important factor in determining nuclear stability is the number of neutrons and protons in the nucleus. The behavior of a system of neutral neutrons and positively charged protons is determined by both the electromagnetic interaction and the strong interaction. In order for a nucleus to be formed, the Coulomb repulsion between the protons must be overcome by an attractive strong interaction that involves both the neutrons and the protons. A certain combination of neutrons and protons may form a stable nucleus, it may form an unstable nucleus, or it may not form a nucleus at all. One might expect that adding more neutrons would help to stabilize a nucleus because it increases the attractive strong interaction but does not influence the repulsive Coulomb interaction between protons. However, this is not entirely true. As we can see from the example in table 3.1, carbon forms two stable isotopes with $Z = 6$ and either $N = 6$ or $N = 7$. Fewer than six neutrons combined with six protons do not form a stable nucleus. This is, at least in part, due to an insufficient attractive strong interaction to overcome the repulsive Coulomb interaction. We also see that more than seven neutrons combined with six protons does not form a stable nucleus. In fact, it can be seen from the table that the more the number of neutrons differs from six or seven the greater the nuclear instability as indicated by a decreasing mean lifetime. Thus, it is clear that for a given number of protons either too few neutrons or too many neutrons will yield a stable nucleus. The reasons for these properties will become clear in chapters 4 and 5.

The known nuclides, both stable and unstable as a function of N and Z are illustrated in figure 3.1. This graph is referred to as a Segrè plot. The general features shown in this figure can be summarized as follows:

doi:10.1088/978-0-7503-6094-4ch3

Table 3.1. Properties of the known isotopes of carbon.

Nuclide	Atomic mass (u)	Stability	Natural abundance (%)	Lifetime
^{9}C	9.031 400 87	Unstable	0	182 ms
^{10}C	10.016 853 11	Unstable	0	27.8 s
^{11}C	11.011 433 818	Unstable	0	29.4 m
^{12}C	12.000 000 000	Stable	98.9	—
^{13}C	13.003 354 838	Stable	1.1	—
^{14}C	14.003 241 988	Unstable	0	8267 y
^{15}C	15.010 599 258	Unstable	0	3.533 s
^{16}C	16.014 701 243	Unstable	0	1.078 s
^{17}C	17.022 583 712	Unstable	0	278 ms

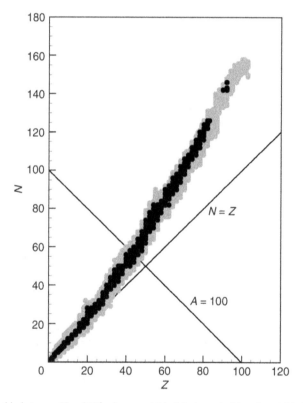

Figure 3.1. Relationship between N and Z for known stable (black symbols) and unstable (gray area) nuclides. Lines for $N = Z$ and a typical constant value of A are shown.

1. The unstable nuclides form a region on either side of the stable ones.
2. For small values of Z there is a tendency for $N = Z$.
3. For large values of Z, $N > Z$.

Table 3.2. Number of β-stable nuclei for various combinations of N and Z.

N	Z	A	Number known
Even	Odd	Odd	50
Odd	Even	Odd	55
Even	Even	Even	165
Odd	Odd	Even	4

It can be shown that the stability of a nucleus is closely related to whether N and/or Z is even or odd. In fact, as illustrated by the number of β-stable nuclides given in table 3.2, some very remarkable trends can be observed (β-stability will be discussed in detail in chapter 4). In this table, nuclides are defined as even–even, even–odd, odd–even, or odd–odd depending on whether N and Z are even or odd. The four situations defined above correspond to values of A that are even, odd, odd, and even, respectively. These data clearly show that even–even nuclei have a greater probability of being stable, followed by approximately equal probabilities for even–odd and odd–even nuclei, while almost no odd–odd nuclei are stable.

The models that will be presented in chapters 4 and 5 will describe the tendency of N to equal Z for small A, for N to exceed Z for large nuclei and the nuclear stability as a function of an even or odd number of nucleons. However, before proceeding to develop specific models to describe nuclear properties, it is important to understand a little about the size and structure of the nucleus as discussed in the remainder of this chapter.

3.2 Rutherford scattering

Much of what has been learned about the structure of the nucleus and the distribution of matter within the nucleus has come from studies of scattering cross sections. The simplest type of scattering that can be observed is Rutherford scattering (named after Ernest Rutherford (1871–1937)). This is the result of the nonrelativistic Coulomb scattering of point charges. Deviations of experimental results from the scattering cross sections predicted by the Rutherford formula are very important in understanding the structure of the nucleus. We begin, therefore, with the basic nonrelativistic derivation of the Rutherford scattering cross section. This simple derivation assumes that the nucleus is sufficiently massive that it does not recoil during the scattering process.

The geometry of the Rutherford scattering problem is illustrated in figure 3.2. The scattering particle (with charge ze) approaches the scatterer (with charge Ze) at an impact parameter defined in the figure as b and with an initial velocity v_0. At any time, the location of the scattering particle is given relative to the scatterer by the coordinates (r, ϕ) where the angle ϕ is measured relative to the line AB. For the point $\phi = 0$ the radial velocity of the particle is zero and the distance, r, is a minimum. Conservation of angular momentum requires that,

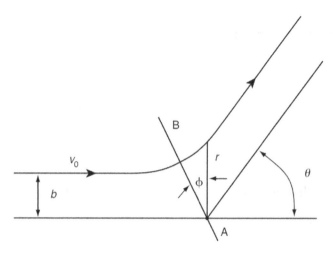

Figure 3.2. Geometry for the Rutherford scattering problem.

$$mv_0 b = mr^2 \frac{d\phi}{dt}. \tag{3.1}$$

The force acting on the particle due to the Coulomb interaction is given in terms of the linear momentum, p, as

$$\overrightarrow{F} = \frac{\overrightarrow{dp}}{dt} = \frac{Zze^2}{4\pi\varepsilon_0 r^3}\overrightarrow{r}. \tag{3.2}$$

The change in momentum over a time interval t_1 to t_2 is found by integrating equation (3.2),

$$\Delta\overrightarrow{p} = \int_{t_1}^{t_2} \overrightarrow{F} \, dt. \tag{3.3}$$

The time at which the particle is at $\phi = 0$ is defined as $t = 0$. The linear momentum of the particle at $t = -\infty$ along the direction AB is

$$p = -mv_0 \sin\frac{\theta}{2} \tag{3.4}$$

where θ is the scattering angle as defined in the figure. Over the time interval $t = -\infty$ to $+\infty$, equation (3.3) can be written in terms of the Coulomb force as

$$2mv_0 \sin\frac{\theta}{2} = \int_{-\infty}^{+\infty} \frac{Zze^2}{4\pi\varepsilon_0 r^2} \cos\phi \, dt.$$

A change of variables in the integration from t to ϕ can be accomplished using equation (3.1) to give,

$$\sin\frac{\theta}{2} = \frac{Zze^2}{8\pi\varepsilon_0 mv_0^2 b} \int_{-(\pi-\theta)/2}^{+(\pi-\theta)/2} \cos\phi \, d\phi. \tag{3.5}$$

Integration of equation (3.5) is straightforward and yields,

$$\tan\frac{\theta}{2} = \frac{Zze^2}{4\pi\varepsilon_0 m v_0^2 b}. \tag{3.6}$$

This expression shows that increasing the impact parameter decreases the scattering angle. Thus, all particles that are scattered by an angle greater than some value of θ, must have impact parameters less than some value of b. Incident particles with b less than a particular value define an area, or cross section, given by

$$\sigma = \pi b^2. \tag{3.7}$$

Thus, it can be said from equations (3.6) and (3.7) that the scattering cross section for scattering by an angle greater than θ is

$$\sigma = \pi \left[\frac{Zze^2 \cot\frac{\theta}{2}}{4\pi\varepsilon_0 m v_0^2} \right]^2. \tag{3.8}$$

Experiments typically measure the differential scattering cross section, that is the scattering into a differential solid angle, $d\Omega$. Some simple solid geometry gives,

$$d\Omega = 2\pi \sin\theta d\theta \tag{3.9}$$

where the differential angle as shown in figure 3.3 has been integrated to account for the axial symmetry of the problem. The differential scattering cross section is defined as

$$\frac{d\sigma}{d\Omega} = \frac{d\sigma}{d\theta}\frac{d\theta}{d\Omega}$$

or from equations (3.8) and (3.9)

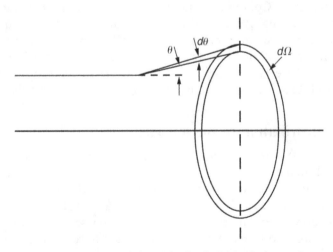

Figure 3.3. Relationship between $d\theta$ and $d\Omega$ for the Rutherford scattering problem.

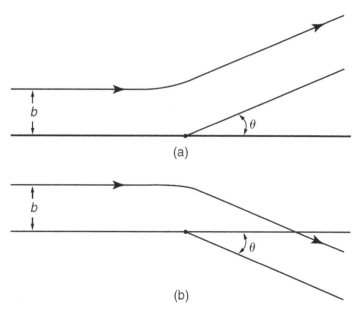

Figure 3.4. Particle trajectories for Rutherford scattering for (a) repulsive interaction and (b) attractive interaction.

$$\frac{d\sigma}{d\Omega} = \left[\frac{Zze^2}{8\pi\varepsilon_0 m v_0^2}\right]^2 \csc^4\frac{\theta}{2}. \tag{3.10}$$

This is known as the Rutherford differential scattering cross section and will be used in the next section for the analysis of nuclear charge distributions. Although our analysis was based entirely on classical arguments, it is a fortunate coincidence that a thorough quantum mechanical treatment yields the exact same answer. This is true for all scattering problems involving a $1/r$ potential. A final note of interest here concerns the sign of the Coulomb interaction. Throughout this derivation it has been implied that this is repulsive as indicated by figure 3.2. Because equation (3.10) shows that the cross section is proportional to $(Zz)^2$ the sign of Zz is irrelevant. Figure 3.4 shows that although the trajectory of the scattered particle is not exactly the same for repulsive and attractive scattering, the relationship between impact parameter b and scattering angle, θ, is independent of the sign of the interaction.

3.3 Charge distribution of the nucleus

The distribution of charge in the nucleus is determined by the distribution of protons, $\rho_p(r)$. Initial scattering experiments conducted by Rutherford to study the structure of the atom used the scattering of α-particles (^{4}He nuclei) and observed the general size of the nucleus. Since the α-particles are nuclei and have their own structure, it is now more common to utilize electron scattering to observe these phenomena. In order to properly analyze the results of scattering experiments it is important to include two factors that have, thus far, been omitted—relativistic effects and the spatial distribution of charges within the nucleus. Relativistic effects can be included in the derivation given above and lead to a relativistic differential scattering cross section given by

$$\left(\frac{d\sigma}{d\Omega}\right)_{rel} = \left(\frac{d\sigma}{d\Omega}\right)_{nonrel}\left[1 - \frac{v_0^2}{c^2}\sin^2\frac{\theta}{2}\right].$$

Inclusion of the non-point charge nature of the nucleus is a substantially more complex problem but leads to some highly significant information about nuclear forces and structure. In general, the effect of a finite size nucleus is to reduce the scattered intensity at all angles (except zero) from the prediction for the scattering of point charges. The scattering of charges from a nucleus can be understood by drawing an analogy with the diffraction of light waves by a circular aperture. For any scattering angle, the total measured intensity is the result of interference effects between different components of the beam that are diffracted from different points within the aperture. This gives rise to the characteristic ring-like pattern. Similarly, for a beam of (say) electrons scattered from a finite size nucleus the interference (of the wave functions) of electrons scattered from different locations within the nuclear charge distribution gives rise to oscillations in the scattered intensity as a function of scattering angle. The nature of these oscillations depends on the wavelength—that is, energy—of the electrons as well as the spatial distribution of charges within the nucleus. The former quantity is known and can be controlled in the experiment. The latter quantity can be extracted from the experimental measurement of scattered intensity as a function of scattering angle. The above description can be quantified by writing the differential scattering cross section as

$$\frac{d\sigma}{d\Omega} = \left(\frac{d\sigma}{d\Omega}\right)_{Rutherford}|F(\theta)|^2 \qquad (3.11)$$

where the form factor $F(\theta)$ is a function that is characteristic of the nuclear charge distribution $\rho_p(r)$ but also depends on the energy of the scattering particles. Here we have assumed that the charge distribution of the nucleus is spherically symmetric. This is a reasonable approximation for most, but not all, nuclei (see chapter 6). The form factor can be related to $\rho_p(r)$ as

$$F(\theta) = \frac{1}{Ze}\int \rho_p(\vec{r})e^{i\delta(\theta,\,\vec{r})}dV \qquad (3.12)$$

where the integration is over the volume of the nucleus and δ is a phase factor given by

$$\delta = \frac{\Delta\vec{p}\cdot\vec{r}}{\hbar}.$$

The magnitude of the change in the momentum is given by equation (3.4) as

$$|\Delta\vec{p}| = 2p\sin\frac{\theta}{2}.$$

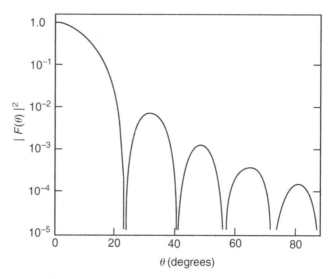

Figure 3.5. Square of the form factor for electrons with a de Broglie wavelength of half the nuclear radius.

Knowing $\rho_p(r)$ allows for the calculation of $F(\theta)$ as illustrated for a simple case in figure 3.5. This demonstrates the expected oscillatory nature of the form factor as a function of angle with $F(\theta)$ normalized to unity for $\theta = 0$. From an experimental standpoint, however, $\rho_p(r)$ is not known *a priori* and equation (3.12) may be substituted into equation (3.11) in order to relate $\rho_p(r)$ to the measured differential scattering cross section. Two approaches can be taken to analyze experimental data on the basis of this integral: a model independent numerical technique to obtain $\rho_p(r)$ and the assumption of a particular functional form for the radial dependence of $\rho_p(r)$ and the fitting of the experimental data to the above equations to obtain the numerical values of unknown parameters. The latter is not necessary for the analysis of scattering data but provides a convenient analytical expression for the nuclear charge distribution that can be implemented in nuclear model calculations. We will begin with a discussion of the former method, since it is more general.

Figure 3.6 shows some typical experimental data for scattering of electrons from a nucleus. This clearly shows the combined effects of the geometric factor predicted by the Rutherford formula and the oscillatory nature introduced by the form factor. Figure 3.7 shows the charge distribution that is extracted from these data. In general, such analyses of scattering data show similar features, a relatively flat region near the center and a gradual decrease in density towards the surface. A good representation of this shape is provided by the Woods–Saxon model,

$$\rho_p(r) = \frac{\rho_{p0}}{1 + \exp\left[\frac{r - R_0}{a}\right]}. \tag{3.13}$$

R_0 gives the mean nuclear radius (the radius at which the density drops to 1/2 of its maximum value) and a is related to the width of the surface region. The

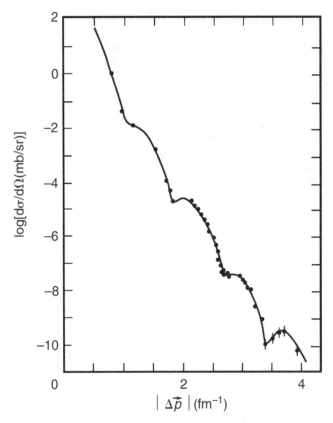

Figure 3.6. Measured differential scattering cross section for 450 MeV electrons incident on ^{58}Ni. The solid line is a fit using the charge distribution shown in figure 3.7. Reprinted with permission from Sick *et al* (1975), copyright (1975) by the American Physical Society.

coefficient ρ_{p0} in this function is given by normalizing the total integrated nuclear charge as

$$Z = \int \rho_p(r)dV.$$

An analysis of $\rho_p(r)$ for several nuclei is shown in figure 3.8. It is seen that the shape of this function is a reasonable approximation of the results of the model independent analysis shown above. The figure shows some important aspects of the nuclear charge distributions.

1. Larger nuclei have a larger mean diameter.
2. The surface region has a similar width in all nuclei.
3. The charge density at the center is greater in light nuclei than in heavy nuclei.

3.4 Mass distribution of the nucleus

The measurement of $\rho_p(r)$ is based on the distribution of charge and hence the distribution of protons in the nucleus. It is also important to know the distribution of

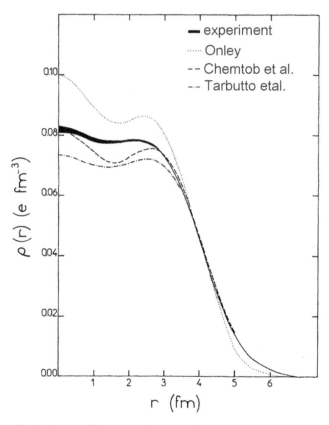

Figure 3.7. Charge distribution for ^{58}Ni extracted numerically from the measurements shown in figure 3.6 along with theoretical predictions. Theoretical predictions are from Onley (1969), Chemtob *et al* (1974) and Tarbutton and Davies (1968), as noted in the figure. Reprinted with permission from Sick *et al* (1975), copyright (1975) by the American Physical Society.

all nucleons (neutrons and protons) in the nucleus. Since neutrons are uncharged, Coulomb scattering experiments do not observe them. It is, however, fairly straightforward to infer the total nucleon density (or nuclear mass density), $\rho(r)$, from measurements of $\rho_p(r)$. A nucleus with N neutrons and Z protons has an overall neutron to proton ratio N/Z, and a total number of nucleons $A = N + Z$. It is usually assumed that the ratio of neutron density, $\rho_n(r)$, to proton density, of $\rho_p(r)$, is the same everywhere within the nucleus. This is written as

$$\frac{\rho_n(r)}{\rho_p(r)} = \frac{N}{Z}.$$

It is also known that the total density is the sum of the neutron and proton densities,

$$\rho(r) = \rho_p(r) + \rho_n(r).$$

These two equations can be easily solved to give,

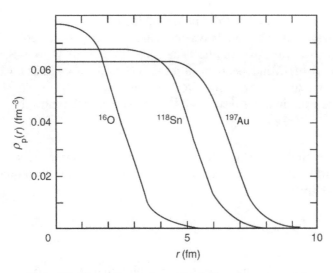

Figure 3.8. Woods–Saxon charge distributions for some nuclei.

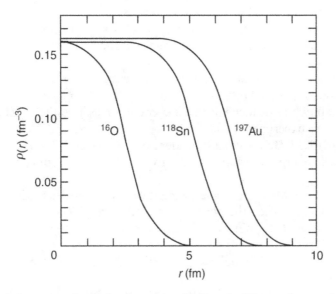

Figure 3.9. Woods–Saxon mass distributions for some nuclei. Note the differences between these distributions and the charge distributions for the same nuclei shown in figure 3.8.

$$\rho(r) = \rho_p(r)\left[1 + \frac{N}{Z}\right].$$

Using this equation, data of the form shown in figure 3.8 can be rescaled for different N and Z to give the total nucleon density, as illustrated in figure 3.9. It is significant here that the mean nuclear radii and the width of the surface regions has not changed from that shown in figure 3.8. It is also seen that all nuclei, regardless of total mass, have virtually the same nucleon density in their centers. This is a very important feature of nuclei and confirms that the strong interaction is very short ranged and is dominated by the interaction between nearest neighbor nucleons. This will be discussed further in the next chapter. The constant a in the Woods–Saxon equation is found to be virtually independent of total nuclear mass and has a value of 0.52 ± 0.01 fm. The mean nuclear radius, R_0, increases with total nuclear mass and an analysis of the experimental results leads to the empirical expression,

$$R_0 = 1.2 \times A^{1/3} \text{ (fm).} \tag{3.14}$$

This is what would be anticipated for nuclei with differing mass but the same density. This will play an important role in the development of the liquid drop model described in chapter 4.

Problems

3.1.
   (a) For the Coulomb scattering of a point charge $+ze$ from a heavy nucleus of charge $+Ze$, calculate the distance of closest approach for $b = 0$. Ignore the recoil of the heavy nucleus.
   (b) For $b = 0$ calculate the point of closest approach for an 8 MeV α-particle incident on an Au nucleus and compare with the expected nuclear radius.
3.2. For a nonrelativistic Coulomb scattering experiment, the number of particles scattered per unit time at a fixed scattering angle can be measured as a function of incident particle energy. Describe the expected results of such an experiment.
3.3. α-particles (^{4}He nuclei) are incident on Au nuclei. What is the minimum α-particle energy necessary for the two nuclei to come in contact?
3.4.
   (a) For an 8 MeV α-particle incident on an Au nucleus, what is the impact parameter when the particle is scattered at 90°?
   (b) What is the point of closest approach for the α-particle scattered at 90°?
   (c) What is the kinetic energy of the α-particle at the point of closest approach?
3.5. Calculate the density of matter at the center of a ^{208}Pb nucleus.
3.6.
   (a) Using the data shown in figure 3.9, estimate the width of the surface region of a nucleus; that is, the distance over which the density drops from 90% of its central value to 10% of its central value.
   (b) Using the result of part (a), estimate the value of a in equation (3.13).
3.7.
   (a) For 10 MeV α-particles incident on Au nuclei, calculate the total scattering cross section for scattering angles $\theta > 1°$, $\theta > 5°$, and $\theta > 20°$.
   (b) For the conditions given in part (a), calculate the differential scattering cross section for $\theta = 1°$, $\theta = 5°$, and $\theta = 20°$.

3.8.
  (a) For the scattering of 0.1 MeV electrons from ^{119}Sn nuclei, calculate the relative size of the relativistic correction to the differential scattering cross section for scattering angles of 20° and 90°.
  (b) Repeat part (a) for 1 MeV and 100 MeV electrons.

3.9.
  (a) Using figure 3.9, estimate the fraction of nucleons that are in the surface region of an ^{16}O nucleus.
  (b) Repeat part (a) for ^{118}Sn and ^{197}Au nuclei.

3.10.
  (a) For a 10 MeV α-particle incident on an ^{16}O nucleus with an impact parameter $b = 0$, calculate the recoil energy of the nucleus.
  (b) Repeat part (a) for 10 MeV α-particles incident on ^{118}Sn and ^{197}Au nuclei

References and suggestions for further reading

Chemtob M, Moniz E J and Rho M 1974 Deuteron electromagnetic structure at large momentum transfer *Phys. Rev.* C **10** 344–52

Cottingham W N and Greenwood D 2001 *An Introduction to Nuclear Physics* 2nd edn (Cambridge: Cambridge University Press)

Enge H A 1966 *Introduction to Nuclear Physics* (Reading, MA: Addison-Wesley)

Krane K S 1988 *Introductory Nuclear Physics* (New York: Wiley)

Onley D S 1969 Corrections to high-energy electron–nucleus scattering *Nucl. Phys.* A **118** 486–48

Sick I, Bellicard J B, Bernheim M, Frois B, Huet M, Leconte P H, Mougey J, Xuan-Ho P, Royer D and Turck S 1975 Shell structure of the ^{58}Ni charge density *Phys. Rev. Lett.* **35** 910–3

Tarbutton R M and Davies K T R 1968 Further studies of the Hartree–Fock approximation in finite nuclei *Nucl. Phys.* A **120** 1–24

IOP Publishing

An Introduction to the Physics of Nuclei and Particles
(Second Edition)

Richard A Dunlap

Chapter 4

Binding energy and the liquid drop model

4.1 Definition and properties of the nuclear binding energy

The nuclear binding energy, B, is the minimum energy required to separate a nucleus into its component neutrons and protons. The mass equivalent of the nuclear binding energy relates the nuclear mass (m_N) to the mass of the component neutrons (m_n) and protons (m_p),

$$m_N = Nm_n + Zm_p - \frac{B}{c^2}. \tag{4.1}$$

B is defined to be explicitly positive for a bound system, so from equation (4.1) it is seen that the binding energy of the nucleus decreases its total mass. For a given contribution to the mass from the neutrons and protons, the nuclear stability is increased as B increases, and this corresponds to a reduction in total mass. Nuclear masses are usually not measured directly. Instead, for a given nuclide the atomic mass of a neutral atom is generally measured. This can be related to the nuclear mass by

$$m = Nm_n + Z(m_p + m_e) - \frac{B}{c^2} - \frac{b}{c^2} \tag{4.2}$$

where b is the total binding energy associated with all the atomic electrons. In almost all cases b is sufficiently small that it can be ignored in calculations involving nuclear masses and binding energies. Since calculations of decay energies involve differences in atomic masses, the effects of the electronic binding energy tend to cancel out in any case. However, in cases where it is necessary to include this term in equation (4.2), it is often sufficient to use the results of a semiclassical calculation that gives,

$$b = 20.8 \times Z^{7/3} \text{ (eV)}.$$

doi:10.1088/978-0-7503-6094-4ch4

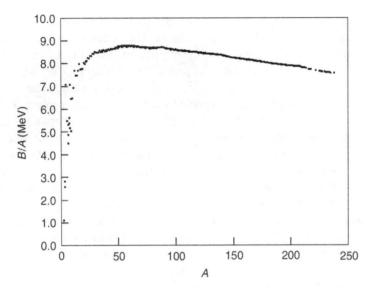

Figure 4.1. Binding energy per nucleon as a function of mass number, A.

Atomic masses are usually given in atomic mass units and are tabulated in appendix B. Tables in some references give the mass excess Δ (usually in MeV), which is defined in terms of the atomic mass m as

$$\Delta = (m - A)c^2.$$

The total nuclear binding energy shows an approximately linear relationship to A. This means that, to first order, each nucleon is bound to the nucleus with the same energy regardless of the size of the nucleus. This is consistent with the conclusions that can be drawn concerning the range of the strong interaction on the basis of nucleon density results, as discussed in chapter 3. However, on close inspection the binding energy shows some interesting features. These are most easily seen in a plot of B/A as a function of A, as shown in figure 4.1. If B is linear in A, then this quantity would be a constant. The general behavior as seen in the figure can be explained by the liquid drop model. This yields the so-called semiempirical mass formula, as discussed in the following section.

4.2 The liquid drop model

Much is known about nucleon–nucleon interactions from an experimental stand-point. Evidence indicates that this interaction is both strong and short range. It is also known that nucleon–nucleon interactions are spin dependent and that they have both central and noncentral components. Although a detailed description of these interactions is complex, much can be learned about their effects on the behavior of nuclei on the basis of some simple models. Two of these, the liquid drop model, which draws analogies with the behavior of liquids, and the shell model, which is based on the quantum mechanical behavior of neutrons and protons, are discussed in detail in this book.

The liquid drop model is a semiempirical model that describes the behavior of the nuclear binding energy as a function of the number of neutrons and protons. It is also an effective method of investigating the relative stability of various nuclides. This model is based by analogy with the properties of a drop of liquid, in part, on a consideration of fundamental electrostatics and, in part, on empirical observations. The analogy with a liquid drop is expected to be, at least to a point, valid, as the molecular interactions (van der Waals interactions) holding a drop of liquid together are short ranged. However, we should not be tempted to extend this analogy to the point of trying to understand any of the basic properties of the strong interaction by comparison with the interactions present in a liquid. The binding energy as predicted by the liquid drop model can be expressed as the sum of a number of terms due to different contributions. These are each discussed in detail below. It is assumed here that the nucleus is spherically symmetric which, in most cases, is a good approximation and that the nuclear volume is directly related to the number of nucleons.

Volume term This term results from the interaction of each nucleon with its nearest neighbor nucleons. We would expect this to be the dominant contribution to the overall interaction energy as the nucleon–nucleon interaction is very short ranged. Because the nucleus contains A nucleons (we do not distinguish between neutrons and protons here) the associated binding energy, which is proportional to the nuclear volume, is given by,

$$B_V = a_V A. \tag{4.3}$$

The coefficient a_V is a constant that can be determined by comparison with experimental data. The form of this term is consistent with the short-range nature of the strong interaction. In the case of long-range interactions, such as Coulomb interactions, then each nucleon would interact with each other nucleon and the resulting energy would be proportional to $A(A-1)$; see the discussion of Coulomb interactions below. In general, the form given by equation (4.3) is an overestimate of the binding energy since the other interactions that will be described below (except the pairing interaction) tend to destabilize the nucleus and, therefore, decrease the total binding energy.

Surface term All nucleons in the nucleus are not equivalent. Those on the surface are surrounded by a smaller number of nearest neighbor nucleons than those in the interior. This means that the nucleons on the surface are less tightly bound, and the binding energy given above must be decreased by a quantity that is proportional to the number of surface nucleons or to the surface area of the nucleus. For a spherical nucleus the surface area is related to the volume as $S \propto V^{2/3}$ so this contribution to the energy can be written

$$B_S = -a_S A^{2/3}$$

where the minus sign indicates that this decreases the overall nuclear stability. Again, the coefficient is determined by a comparison with experimental data.

Coulomb term The repulsive Coulomb interaction between the protons in the nucleus reduces the binding energy further. Basic electrostatics shows that the potential energy associated with a uniform spherical distribution of Z charges is proportional to $Z(Z-1)/r$, and using equation (3.14) this can be related to a term in the binding energy of the form

$$B_C = -a_C \frac{Z(Z-1)}{A^{1/3}}.$$

The coefficient a_C can be calculated analytically and has the value of 0.72 MeV. In the present discussion we will leave this as a free parameter in the analysis. One (of several) valid reason for this approach is the nonuniform density of charges near the edge of the nucleus, as was shown in figure 3.9.

Symmetry term As was illustrated in figure 3.1, there is a tendency for stable nuclei with small A to have $N = Z$. For large A, N tends to be somewhat greater than Z. On the basis of the first two terms discussed above there is no restriction placed on the relative number of protons and neutrons in the nucleus. The Coulomb term would favor nuclei with many neutrons and few protons. This is clearly not what is observed. The symmetry term favors $N = Z$ but becomes less important as A becomes large. The usual form of this term is

$$B_{\mathrm{sym}} = -a_{\mathrm{sym}} \frac{(A - 2Z)^2}{A}$$

where again a_{sym} is a free parameter in the analysis. Here it is important to note that $A-2Z = N-Z$ and is a measure of the deviation from the $N = Z$ line previously shown in figure 3.1.

Pairing term The behavior that was described in table 3.2 is taken into account by the pairing term, which is of the form,

$$B_p = -\frac{a_p}{A^{3/4}}.$$

The free parameter, a_p, takes on a positive value for odd–odd nuclei indicating a decrease in stability, a negative value of even–even nuclei indicating an increase in stability and zero for odd A nuclei.

The total binding energy per nucleon as determined by the sum of the above terms and per nucleon is expressed as

$$\frac{B}{A} = a_V - \frac{a_S}{A^{1/3}} - \frac{a_C Z(Z-1)}{A^{4/3}} - \frac{a_{\mathrm{sym}}(A - 2Z)^2}{A^2} - \frac{a_p}{A^{7/4}}.$$

This can be compared with the experimental results as indicated in figure 4.1 to obtain best fit values of the coefficients. The general shape of the curve is determined by the first four terms above. The relative importance of these terms is illustrated in figure 4.2. The value of a_p is determined from a consideration of fluctuations in B/A as A changes from odd to even. This analysis gives the best fit values as

$$a_V = 15.5 \text{ MeV}$$

$$a_S = 16.8 \text{ MeV}$$

$$a_C = 0.72 \text{ MeV}$$

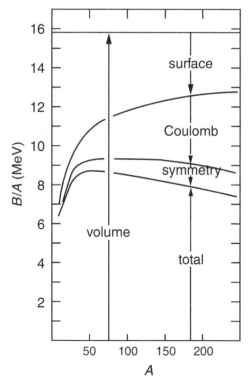

Figure 4.2. Relative importance of the various contributions to the binding energy predicted by the semiempirical mass formula.

$$a_{\text{sym}} = 23.2 \text{ MeV}$$

$$a_p = +34 \text{ MeV } N, Z = \text{odd–odd}$$

$$a_p = 0 \text{ MeV } A = \text{odd}$$

$$a_p = -34\text{MeV } N, Z = \text{even–even}.$$

Note that the value for a_C is in agreement with the analytical result calculated for a uniform charge distribution. As discussed in the next section, these values can be used to describe the stability of certain nuclides.

4.3 Beta stability

From the development given above the total atomic mass can be determined to be

$$m = (A - Z)m_{\text{n}} + Z(m_p + m_e) - \frac{a_V A}{c^2} + \frac{a_S A^{2/3}}{c^2}$$

$$+ \frac{a_C Z(Z - 1)}{A^{1/3} c^2} + \frac{a_{\text{sym}}(A - 2Z)^2}{A c^2} + \frac{a_p}{A^{3/4} c^2} \tag{4.4}$$

where N is written as $A-Z$ and the electronic binding energy is ignored. It is interesting to consider the Z dependence of this expression for a constant value of A. This is equivalent to examining the behavior of isobars. Collecting together powers of Z, it is seen that the above expression is a quadratic. The simplest case to consider is for nuclei with odd A since the pairing term is zero. An example of plotting m versus Z at constant A is shown in figure 4.3 for $A = 135$.

The data plotted here are experimentally measured atomic masses, but the parabolic shape of the curve as predicted by the semiempirical mass formula is evident. Minimizing the mass corresponds to maximizing nuclear stability. It is seen in the figure that the nuclide closest to the minimum in the figure will be stable; in this case it is ^{135}Ba with 56 protons. Nuclides with $A = 135$ and more or fewer than 56 protons are unstable and decay towards $Z = 56$. For a given (odd) A the minimum in the mass parabola can be calculated on the basis of the semiempirical mass formula. By setting $dm/dZ = 0$ equation (4.4) gives the value of Z, which minimizes the mass as

$$Z_{min} = \frac{(m_n - m_p - m_e)c^2 + a_C A^{-1/3} + 4a_{sym}}{2a_C A^{-1/3} + 8a_{sym}A^{-1}}.$$

The most stable nuclide is given by the integer value closest to Z_{min}.

The physical basis for the instability of nuclides with Z on either side of the minimum can be described as follows: nuclei with Z less than Z_{min} have too many neutrons (for the number of protons) and the instability can be explained by the

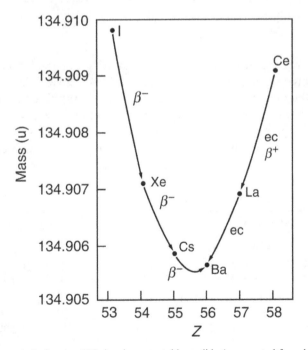

Figure 4.3. Mass parabola for $A = 135$ showing one stable nuclide (as expected for odd A) with $Z = 56$.

corresponding large value of the symmetry term in equation (4.4). Nuclei with Z greater than $Z_{\min}$ have too many protons (for the number of neutrons) and the instability can be explained by a combination of the symmetry and Coulomb terms in equation (4.4).

In the first case, a negative β-decay (β^- decay) process occurs. This is expressed as

$$\substack{A\\Z}X^N \rightarrow \substack{A\\Z+1}Y^{N-1} + e^- + \bar{\nu}_e \tag{4.5}$$

where an electron is emitted during the process. Here a nucleus of element X has been converted to a nucleus of element Y and we have written the corresponding values of Z as subscripts before the element name and the values of N as superscripts after the element name in order to help keep track of all the nucleons in the problem. This process is the result of the weak interaction (since neutrinos are involved, see chapter 2) and an inspection of equation (4.5) will show that charge, lepton number, and baryon number are conserved. In fact, the appearance of the antineutrino on the right-hand side of the equation is necessary to conserve lepton number. The basic process described by equation (4.5) is the conversion of a neutron to a proton,

$$n \rightarrow p + e^- + \bar{\nu}_e. \tag{4.6}$$

The second case corresponds to positive β-decay (β^+ decay) and is

$$\substack{A\\Z}X^N \rightarrow \substack{A\\Z-1}Y^{N+1} + e^+ + \nu_e$$

where a positron (i.e., antielectron) is emitted. Again, the neutrino is necessary to satisfy conservation of lepton number. This decay corresponds to the conversion of a proton to a neutron or

$$p \rightarrow n + e^+ + \nu_e. \tag{4.7}$$

These β-decay processes, as well as electron capture, which is equivalent to β^+ decay, will be considered in detail in chapter 9. Table 4.1 gives some of the relevant properties for the nuclides shown in figure 4.3. A general observation can be made concerning the lifetimes of the nuclides described in the table. As the nuclides decay towards ^{135}Ba from either side the lifetimes become longer. This is directly related to the decreasing difference in mass between the parent and daughter nucleus. This same feature is seen in subsequent decays described in this chapter.

Table 4.1. Properties of nuclides with $A = 135$ (ec = electron capture).

Element	N	Z	Mass (u)	Lifetime	Decay mode	Daughter	Decay energy (MeV)
I	82	53	134.909 823	9.6 h	β^-	^{135}Xe	2.51
Xe	81	54	134.907 130	13.1 h	β^-	^{135}Cs	1.16
Cs	80	55	134.905 885	4.3×10^6 y	β^-	^{135}Ba	0.21
Ba	79	56	134.905 665	Stable	—	—	—
La	78	57	134.906 953	28.0 h	ec	^{135}Ba	1.20
Ce	77	58	134.909 117	25.3 h	ec, β^+	^{135}La	2.02

On the basis of figure 4.3 it is expected that for a given value of odd A there should be uniquely one β stable nuclide. Since odd A can result from N and Z odd–even or even–odd we would expect approximately equal numbers of these two types of nuclei to exist. This is in agreement with table 3.2.

The situation for even A, which occurs for odd–odd or even–even nuclei, is much more complex because the pairing term is nonzero. In general, we expect two parabolas for m as a function of Z, one shifted up (the odd–odd parabola) and one shifted down (the even–even parabola), as illustrated in figure 4.4 for $A = 140$. The stable nucleus again occurs for the minimum value of mass and the figure shows that as nuclei with too few or too many protons decay by β-decay processes they alternate from the odd–odd parabola to the even–even parabola. For the case shown ^{140}Ce with $Z = 58$ lies very close to the minimum in m and represents the β stable nuclide with $A = 140$. Properties of the decay process shown in the figure are given in table 4.2.

Another situation for an even A nucleus ($A = 128$) is illustrated in figure 4.5. Here there are two stable nuclei on the even–even parabola, ^{128}Te and ^{128}Xe. ^{128}I can decay by either β^- decay to ^{128}Xe or by β^+ decay to ^{128}Te. This means that it becomes more stable either by converting a neutron to a proton or a proton to a neutron. This can be understood on the basis of the semiempirical mass formula because either process will change an odd–odd nucleus to an even–even nucleus. However, the real implications of this pairing behavior will become obvious in the next chapter. Relevant nuclear properties are given in table 4.3.

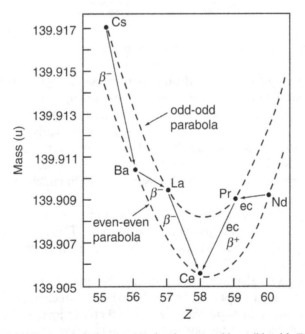

Figure 4.4. Mass parabola for $A = 140$ showing one stable nuclide with $Z = 58$.

Table 4.2. Properties of nuclides with $A = 140$ (ec = electron capture).

Element	N	Z	Mass (u)	Lifetime	Decay mode	Daughter	Decay energy (MeV)
Cs	85	55	139.917 338	95 s	β^-	^{140}Ba	6.29
Ba	84	56	139.910 518	18.3 d	β^-	^{140}La	1.03
La	83	57	139.909 471	58.0 h	β^-	^{140}Ce	3.76
Ce	82	58	139.905 433	Stable	–	–	–
Pr	81	59	139.909 071	4.9 m	ec, β^+	^{140}Ce	3.39
Nd	80	60	139.909 036	4.75 d	ec	^{140}Pr	0.22

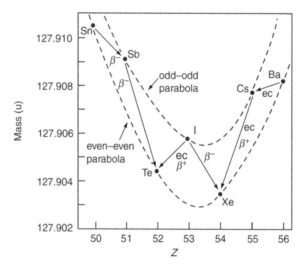

Figure 4.5. Mass parabola for $A = 128$ showing two stable nuclides with $Z = 52$ and 54.

A third possibility for an even A nucleus is shown in figure 4.6 ($A = 130$). Here there are three stable nuclei on the even–even parabola, ^{130}Te, ^{130}Xe and ^{130}Ba. In this case, ^{130}Cs can decay by either β^- or β^+ decay to ^{130}Xe or ^{130}Ba, respectively. ^{130}I decays by β^- decay to ^{130}Xe. Although it is energetically favorable for ^{130}I to decay to ^{130}Te, this process has not been observed. Relevant nuclear properties are given in table 4.4. The stability of nuclides such as ^{130}Te, which are at a local minimum on the mass curve will be discussed further in chapter 9 in the context of double β-decay.

From figures 4.4 through 4.6, it is seen that either one, two, or three even–even β-stable nuclei can occur for a given (even) value of A. This feature is responsible for the anomalously large number of known stable even–even nuclei as was indicated in table 3.1.

Table 3.1 indicated that there are four β-stable nuclei that are odd–odd. If we examine a situation as shown in figure 4.5 where a nuclide is situated near the minimum in the odd–odd parabola (^{128}I) it is difficult to imagine how any odd–odd nuclei could be stable, since the odd–odd nucleus can decay to the even–even nuclei

Table 4.3. Properties of nuclides with $A = 128$ (ec = electron capture).

Element	N	Z	Mass (u)	Lifetime	Decay mode	Daughter	Decay energy (MeV)
Sn	78	50	127.910 467	85 m	β^-	^{128}Sb	1.30
Sb	77	51	127.909 072	15.6 m	β^-	^{128}Te	4.29
Te	76	52	127.904 463	Stable	–	–	–
I	75	53	127.905 810	36 m	ec, β^+	^{128}Te	1.26
					β^-	^{128}Xe	2.12
Xe	74	54	127.903 531	Stable	–	–	–
Cs	73	55	127.907 762	5.5 m	ec, β^+	^{128}Xe	3.94
Ba	72	56	127.908 237	3.46 d	ec	^{128}Cs	0.44

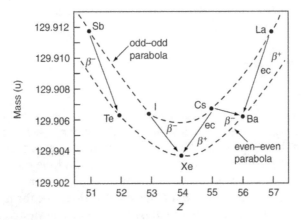

Figure 4.6. Mass parabola for $A = 130$ showing three stable nuclides with $Z = 52$, 54, and 56.

Table 4.4. Properties of nuclides with $A = 130$ (ec = electron capture).

Element	N	Z	Mass (u)	Lifetime	Decay mode	Daughter	Decay energy (MeV)
Sb	79	51	129.911 546	10 m	β^-	^{130}Te	4.95
Te	78	52	129.906 229	Stable	–	–	–
I	77	53	129.906 713	17.6 h	β^-	^{130}Xe	2.98
Xe	76	54	129.903 509	Stable	–	–	–
Cs	75	55	129.906 753	43.2 m	ec, β^+	^{130}Xe	3.02
					β^-	^{130}Ba	0.51
Ba	74	56	129.906 282	Stable	–	–	–
La	73	57	129.912 320	12.5 m	ec, β^+	^{130}Ba	5.62

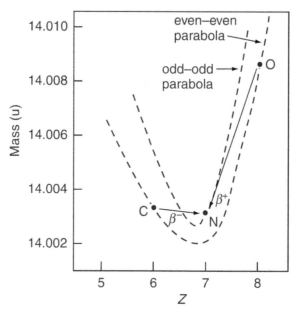

Figure 4.7. Mass parabola for $A = 14$ showing the existence of a stable odd–odd nuclide.

on either side. If we examine the shape of the mass parabola as given by equation (4.4), we learn that the parabola becomes narrower as A decreases. Thus, for small A we can have the situation as shown in figure 4.7 for $A = 14$. Here the sides of the parabolas are sufficiently steep that the minimum in the odd–odd parabola lies below the adjacent points on the even–even parabola and the odd–odd nucleus ^{14}N is the stable $A = 14$ nuclide. Other situations can exist, for example ^{2}H, where β-decay cannot occur because the daughter nucleus does not form a bound state.

The predictions of the semiempirical mass formula can be viewed in the context of the distribution of stable nuclides as shown in figure 3.1. In this figure, each particular point in N–Z space corresponds to a particular value of A. Constant A isobars are represented in the figure by lines parallel to the $A = 100$ line shown. Thus, in a three-dimensional plot with figure 3.1 as the x–y plane and mass plotted on the z-axis, mass parabolas for constant A lines would form a parabolic surface with the minimum following the stability line in the N–Z plane. This is referred to as the β-stability valley.

4.4 Nucleon separation energies

A measure of nuclear stability that will be discussed in some detail in the next chapter is the energy required to remove one nucleon from the nucleus. This is not the same for neutrons and protons. We can define two quantities, S_n and S_p, the neutron separation energy and the proton separation energy, respectively.

The removal of a neutron from a nucleus corresponds to the process,

$$^{A}_{Z}X^{N} \rightarrow {}^{A-1}_{Z}X^{N-1} + \text{n} \tag{4.8}$$

where the number of protons (and hence the identity of the element) has not changed but the number of neutrons (and hence the value of A) has decreased by one. The energy required to produce this process (assuming that it is endothermic and does not occur spontaneously) is given in terms of the masses of the components in equation (4.8) as

$$S_n = \left[m\left({}^{A-1}_{Z}X^{N-1} \right) + m_n - m\left({}^{A}_{Z}X^{N} \right) \right] c^2. \tag{4.9}$$

Here there is no change in the number of atomic electrons, therefore atomic masses (rather than nuclear masses) can be used because the electron masses will cancel out.

The removal of a proton from the nucleus corresponds to the process,

$$ {}^{A}_{Z}X^{N} \rightarrow {}^{A-1}_{Z-1}Y^{N} + p $$

and the separation energy is given by

$$S_p = \left[m\left({}^{A-1}_{Z-1}Y^{N} \right) + m_p + m_e - m\left({}^{A}_{Z}X^{N} \right) \right] c^2$$

where the electron mass is included to allow for the use of atomic masses.

Table 4.5 gives the nucleon separation energies for some light nuclei. It is clear that S_n and S_p can, in some cases, be substantially different. On the basis of the shell model discussed in the next chapter, this behavior is readily expected. These nucleon separation processes can be related to the predictions of the semiempirical mass formula. Figure 4.8 shows a typical example for neutron separation from a moderately heavy nucleus. In this example the separation of a neutron from an odd A nucleus is considered, however, an analogous diagram can be constructed for the even A case. No specific Z values are indicated but the horizontal axis does

Table 4.5. Binding energies and nucleon separation energies for some light nuclei.

Nuclide	N	Z	B/A (MeV)	S_n (MeV)	S_p (MeV)
^{2}H	1	1	1.11	2.22	–
^{3}He	1	2	2.57	–	5.49
^{4}He	2	2	7.08	20.58	19.81
^{6}Li	3	3	5.33	5.66	4.59
^{7}Li	4	3	5.61	7.25	9.98
^{9}Be	5	4	6.46	1.67	16.89
^{10}B	5	5	6.48	8.44	6.59
^{11}B	6	5	6.93	11.45	11.22
^{12}C	6	6	7.68	18.72	15.96
^{13}C	7	6	7.47	4.95	17.53
^{14}N	7	7	7.48	10.55	7.55
^{15}N	8	7	7.70	10.83	10.21
^{16}O	8	8	7.98	15.66	12.13
^{17}O	9	8	7.75	4.14	13.78
^{18}O	10	8	7.77	8.04	15.94

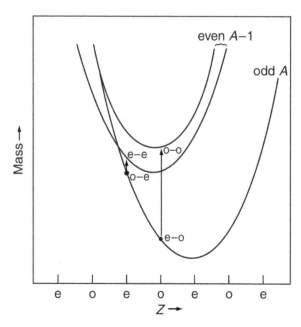

Figure 4.8. Representation of neutron separation energies. Arrows show the transitions from the odd A parabola to the even $A-1$ parabola for odd–even and even–odd starting nuclei.

illustrate that Z alternates between even (e) and odd (o) values. Mass parabolas for A and $A-1$ are plotted on the same graph. The mass parabolas for $A-1$ are shifted upward from their actual values by the mass of one neutron in order to properly account for the energetics of equation (4.9). On the average (that is, the pairing term ignored) the mass of the most stable nuclide with $(A-1)$ nucleons plus the mass of one nucleon is about 8 MeV greater than the mass of the most stable nuclide with A nucleons. This is just the mass associated with the average binding energy of one nucleon. In general, this means that neutron separation is an endothermic process and does not occur spontaneously. Neutron separation results in a transition from the odd A parabola to one of the even A parabolas. Since there is no change in Z the transition is represented by a vertical line, as shown in the figure. If the initial nucleus was an odd–even nucleus, then the final nucleus is an even–even nucleus; if the initial nucleus is even–odd then the final one is odd–odd. This is shown in the figure for both initial even and odd values of Z. It is clear that the energy involved in these two cases is substantially different and this is consistent with the indications in table 4.5 and the subsequent discussion in chapter 5.

For proton separation the situation is somewhat more complex as there is a change in the value of Z. This is illustrated in figure 4.9 for transitions from odd–even to odd–odd and from even–odd to even–even nuclei. It is clear from the figures that neutron separation energies can, in some cases, be much smaller than proton separation energies and this is consistent with the data in table 4.5.

In a few cases very neutron-rich nuclei can decay spontaneously from their excited states by neutron emission. This is most commonly observed in fission by-products

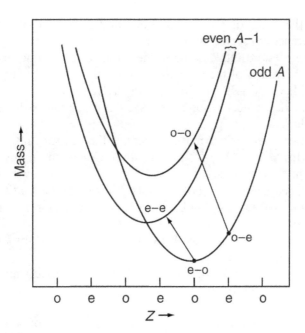

Figure 4.9. Representation of proton separation energies. Arrows show the transitions from the odd A parabola to the even $A-1$ parabola for even–odd and odd–even starting nuclei.

(see chapter 12). Neutron emission is also known from excited states (see chapters 6 and 11) in nuclei with small A where the mass parabolas are very steep.

Problems

4.1. Using the semiempirical mass formula and the values of the parameters given in the text, calculate the value of Z for the minimum in the mass parabola for $A = 91$ and $A = 123$. Discuss these results in the context of the known stability for nuclei with these values of A.

4.2.
   (a) Use the semiempirical mass formula to calculate the binding energy per nucleon for ^{3}H, ^{4}He, ^{64}Cu and ^{119}Sn.
   (b) Use measured atomic masses to calculate the binding energy per nucleon for the three nuclides in part (a).
   (c) Discuss the results of parts (a) and (b) and comment on the validity of the liquid drop model.

4.3. Derive an expression for the internal (Coulomb) energy of a uniformly charged sphere of radius r and total charge $+Ze$. Compare this with the form of the Coulomb term in the semiempirical mass formula.

4.4.
   (a) Using measured atomic masses calculate the binding energies of ^{13}C and ^{13}N. These nuclei are referred to as mirror nuclei.

(b) On the basis of the liquid drop model, describe the reasons for the differences observed in part (a).

(c) From these results determine the radius of a ^{13}C nucleus.

4.5. Using the values of the mass excess for ^{4}He, ^{18}F, and ^{198}Pt, of +2.42 MeV, +0.872 MeV and −29.91 MeV, respectively, calculate the atomic masses of these nuclides. Check your answers by comparison with tabulated atomic mass values.

4.6. Using measured atomic masses calculate the neutron and proton separation energies for ^{235}U, ^{236}U, ^{235}Np and ^{236}Np. Comment on any trends that you observe in these results.

4.7. Use the semiempirical mass formula to estimate the relative importance of the volume, surface, Coulomb, and symmetry terms for nuclei with $A = 3, 19, 99$, and 201.

4.8. Use the semiempirical mass formula to estimate N/Z for stable nuclei with $A = 3, 19, 99$, and 201.

4.9. Identify all stable odd–odd nuclides. Sketch and discuss mass diagrams (see figure 4.7) for each of these cases.

4.10.

(a) Rewrite the semiemprical mass formula in terms of A and N.

(b) Use the results of part (a) to derive an expression for the two-proton separation energy (that is, the energy required to remove two protons from the nucleus).

(c) Why is it more convenient to deal with the two-proton separation energy in this case than the one-proton separation energy?

4.11.

(a) Plot the atomic masses of all nuclides given in appendix B with $A = 64$.

(b) Use this figure to estimate the value of a_p in the semiempirical mass formula and compare it with the accepted value given in the text.

Suggestions for further reading

Burcham W E and Jobes M 1995 *Nuclear and Particle Physics* (Harlow, Essex: Pearson)

Krane K S 1988 *Introductory Nuclear Physics* (New York: Wiley)

Martin B R 2006 *Nuclear and Particle Physics—An Introduction* (Chichester, West Sussex: Wiley)

Segrè E 1977 *Nuclei and Particles—An Introduction to Nuclear and Subnuclear Physics* 2nd edn (Reading, MA: Benjamin/Cummings)

Williams W S C 1991 *Nuclear and Particle Physics* (Oxford: Oxford University Press)

IOP Publishing

An Introduction to the Physics of Nuclei and Particles
(Second Edition)

Richard A Dunlap

Chapter 5

The shell model

5.1 Overview of atomic structure

Previous chapters stressed the importance of neutron and proton pairing. This is analogous in some ways to the atomic situation, as we are dealing with a system of bound fermions. However, there are important differences, and these are summarized in table 5.1. In the atomic case, we need to consider only the electromagnetic force, which is well known, but in the nuclear case the strong interaction must also be included and is expected to be the dominant contribution to the nuclear potential. In order to understand the similarities and differences between the behavior of atoms and nuclei, we begin with a brief look at some of the characteristics of electrons in an atom.

Consider a single electron in the Coulomb potential of the positive charge of the nucleus, $+eZ$. This system is referred to as a hydrogen-like atom and the energy levels are illustrated in figure 5.1(a). These energy levels are characterized by a single quantum number n. The actual situation for a many-electron atom is substantially more complex. Principally, the electrostatic interactions affecting an electron are due not only to the charge of $+eZ$ on the nucleus, but also the charges of the other $(Z-1)$ atomic electrons. This gives rise to the splitting of the energy levels, as illustrated in figure 5.1(b). Additional fine structure splitting of the energy levels (not shown in the figure) results from the magnetic coupling between the spins and orbital angular momenta of the electrons. Mathematically, the problem of calculating the energy levels in a multielectron atom is complex and methods are described in most quantum mechanics texts. Each energy level is defined in terms of the principal quantum number n and the quantum number associated with the orbital angular momentum of the level, l. According to conventional spectroscopic notation the letters s, p, d, f, g, h,... correspond to the values of $l = 0, 1, 2, 3, 4, 5,....$ The degeneracy of each energy level is given by the product of the spin degeneracy, 2,

doi:10.1088/978-0-7503-6094-4ch5

Table 5.1. Similarities and differences in the modeling of atomic electrons and nucleons.

Property	Electrons	Nucleons
Type of particles	Fermions	Fermions
Identity of particles	Electrons	Neutrons and protons
Charges	All charged	Some charged
Occupancy considerations	Pauli principle	Pauli principle
Interactions	Electromagnetic	Strong and electromagnetic

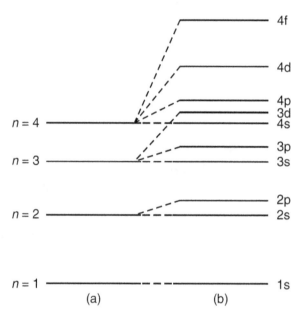

Figure 5.1. Atomic energy levels for $n = 1$–4 in (a) a hydrogen-like atom and (b) a many-electron atom. Note that the energy axis is not to scale.

and the orbital angular momentum degeneracy, $2l + 1$. These degeneracies result from the allowed values of the z-component of the spin, $m_s = -1/2, +1/2$ and the z-component of the orbital angular momentum, $m_l = -l - l + 1 \ldots l - 1, l$. The total degeneracy, $2(2l + 1)$ gives the electron occupancy of each level as indicated in table 5.2.

As figure 5.1 illustrates there are clusters of energy levels with gaps between them. These clusters of energy levels are referred to as shells and are given the designations K-shell, L-shell, M-shell, etc, with increasing energy. Atoms with filled electronic shells are particularly stable and this is manifested as an anomalously large ionization energy. On the other hand, atoms with one electron more than necessary to fill a shell have very low ionization energies as this final electron is very weakly bound. Figure 5.2 shows the measured ionization energy as a function of atomic number, Z. Peaks in the ionization energy clearly correspond to filled electron shells

Table 5.2. Occupancy of atomic energy levels in order of increasing energy.

Shell	n	l	Notation	Degeneracy $2(2l+1)$	Accumulated occupancy
K	1	0	1s	2	2
L	2	0	2s	2	4
L	2	1	2p	6	10
M	3	0	3s	2	12
M	3	1	3p	6	18
N	4	0	4s	2	20
N	3	2	3d	10	30
N	4	1	4p	6	36
O	5	0	5s	2	38
O	4	2	4d	0	48
O	5	1	5p	6	54
P	6	0	6s	2	56
P	4	3	4f	14	70
P	5	2	5d	10	80
P	6	1	6p	6	86

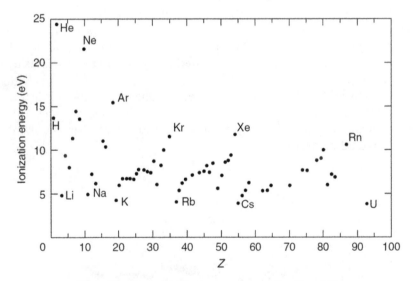

Figure 5.2. Electron ionization energies as a function of Z.

as indicated in the table. The number of electrons corresponding to these particular configurations are referred to as magic numbers, in this case the atomic magic numbers. Table 5.2 and figure 5.2 show that the atomic magic numbers are 2, 10, 18, 36, 54, 86.... Although the problem of calculating the energy levels of multielectron atoms is mathematically complex, the theoretically predicted atomic magic numbers are in agreement with the experimental observations.

5.2 Evidence for nuclear shell structure

There are indications of a shell structure, and corresponding magic numbers, associated with the nuclear energy levels that are analogous to the electronic case. The most straightforward evidence for this kind of behavior might be a consideration of nucleon separation energy as a function of the number of nucleons. This kind of approach for nuclei is somewhat complicated as there are two kinds of nucleons and, as shown in the last chapter, their separation energies can be quite different. Thus, it is not apparent whether A, N, or Z or some combination of these would be most relevant in determining nuclear stability. Some experimental observations that support nuclear shell structure are discussed below.

Binding energy Since the derivation of the semiempirical mass formula did not include any information about the shell structure of the nucleus, it is expected that deviations of experimental data from these model predictions can be indicative of shell effects. Figure 5.3 shows the difference between the liquid drop model predictions and actual measurements of B/A as a function of A. Clear deviations from the model prediction can be seen for certain values of A. Since stable nuclei have a reasonably well-defined relationship between A, N, and Z these anomalies correspond to specific N and/or Z values, as indicated in the figure. There is evidence from the figure that there is excess binding energy for nuclei with N or $Z = 28, 50, 82$, and 126 indicating particular stability for these nuclei. For light nuclei the behavior is illustrated in figure 5.4. The binding energy per nucleon is shown for nuclei with $N = Z$. The general increase in B/A as predicted by the semiempirical mass formula is seen, but slightly higher values are indicated for $N = 2$ and 8. The pairing effects are also obvious.

Nuclear radius Deviations in the nuclear radius from the simple behavior predicted by equation (3.14) are indicative of the nuclear shell structure. Figure 5.5 illustrates this behavior and shows the existence of magic numbers for 20, 28, 50, and 82 neutrons.

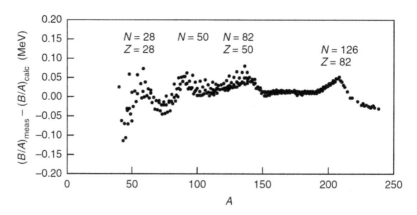

Figure 5.3. Differences between the measured binding energy per nucleon and the value predicted by the liquid drop model as a function of A.

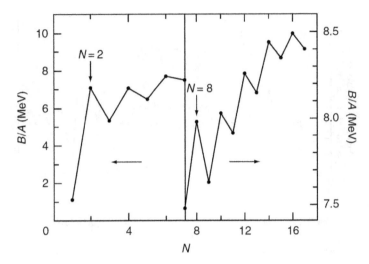

Figure 5.4. Binding energy per nucleon for light nuclei with $N = Z$.

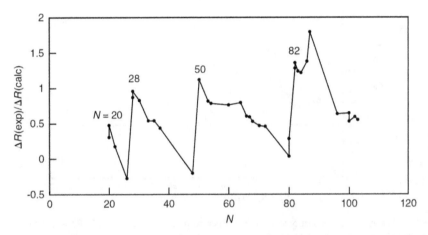

Figure 5.5. Change in the measured nuclear radius for a change in neutron number $\Delta N = 2$ normalized to the change predicted by equation (3.14). Data from Engfer *et al* (1974), Heilig and Steudel (1974) and Shera *et al* (1976).

Number of stable nuclides Although isotopes of an element all have the same value of Z, several values of N are possible. Thus, several stable nuclides with the same N (or isotones) can exist. A larger number of stable nuclides are possible for certain values of N, as illustrated in figure 5.6. This indicates that nuclei with $N = 20, 28, 50,$ and 82 have particular stability.

Neutron absorption cross section The cross section for fast neutron absorption, s, as a function of N is shown in figure 5.7. If a specific nucleon configuration is particularly stable then we expect that there will be a low probability of absorbing an additional neutron, and hence a low cross section. The figure shows that anomalously low cross sections occur for $N = 50, 82,$ and 126.

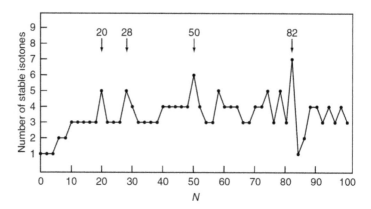

Figure 5.6. Number of stable isotones as a function of N (even).

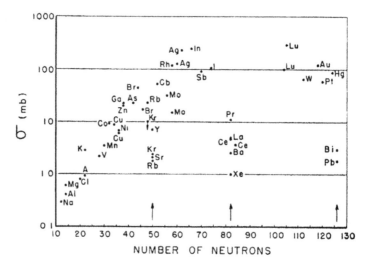

Figure 5.7. Absorption cross sections for 1 MeV neutrons. Magic numbers 50, 82 and 126 are shown. Reprinted with permission from Hughes and Sherman (1950). Copyright 1958 by the American Physical Society.

First excited state energies Nuclei possess excited states, and these can be populated during various decay processes or reactions. Excited nuclear states can be viewed much in the same way as excited atomic electron states, by the occupation of a normally unoccupied higher energy level by one of the nucleons. Details of excited states will be discussed in chapter 6. If a particular nuclear configuration is particularly stable, then it is expected that a larger amount of energy would be required to introduce a transition to an excited state. This is the case for nuclei with filled shells and is shown in figure 5.8. Clear peaks are seen for N or $Z = 20, 28, 50, 82,$ and 126. The nuclei with $Z = 20$ (Ca) are of particular interest. Figure 5.9 shows the excited state energies for even–even Ca. Clear peaks are seen for $N = 20$ and $N = 28$. In fact, these two nuclei have both N and Z magic and are referred to as doubly magic nuclei.

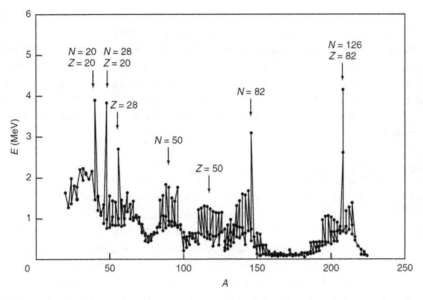

Figure 5.8. Energies of the first excited 2^+ states of even–even nuclei as a function of A and Z. The relevance of the first excited 2^+ state is discussed in chapter 6.

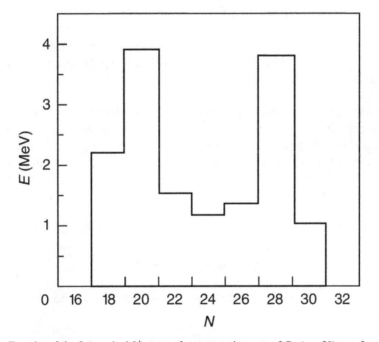

Figure 5.9. Energies of the first excited 2^+ states of even–even isotopes of Ca ($z = 20$) as a function of N.

Other nuclear properties that show characteristic behavior as a function of nucleon number will be discussed in subsequent chapters. Overall, observations of properties related to nuclear stability can be summarized in terms of the so-called nuclear magic numbers, 2, 8, 20, 28, 50, 82, and 126, and indicate that nuclei with N or Z equal to a magic number show particular stability.

It would be desirable to develop a model that, along the lines of the atomic situation described above, would correctly predict these nuclear magic numbers. The use of an appropriate potential in the Schrödinger equation that could predict these magic numbers would provide evidence that our understanding of the nature of the strong interaction is valid. The remainder of this chapter is devoted to the development of such a model.

5.3 The infinite square well potential

Unlike the atomic case, we do not have a simple analytical form of the potential for the nuclear interaction. However, some simple models can be surprisingly effective for describing the behavior of nuclei. We will begin with a consideration of the energy levels for neutrons, which is slightly simpler, and will discuss the situation for protons (which are subject to Coulomb interactions) in somewhat less detail later. To some extent the solution to this problem is a matter of refinement. The procedure is as follows:

1. Approximate the mutual interaction between nucleons by a single particle potential, called the nuclear potential.
2. Solve the Schrödinger equation for the energy eigenstates.
3. Use these results to determine the corresponding magic numbers and compare these with the experimental numbers given above.
4. If the calculated results do not agree with the experiment, revise the form of the potential and try again.

The solution obtained in this manner is sometimes referred to as the single particle model solution for the nucleus as it considers the behavior of each nucleon in a fixed nuclear potential. We begin with the simplest possible potential that could give reasonable results; the infinite spherical square well. This potential will account for the fact that the nucleons are well bound within the nucleus, that the nucleon density is relatively constant within the nucleus, and that this density is more or less independent of the total nuclear mass.

The time independent Schrödinger equation in three dimensions is expressed as

$$-\frac{\hbar^2}{2m}\nabla^2\psi + V(r)\psi = E\psi. \tag{5.1}$$

For a spherically symmetric potential we assume that the solution is separable and can be written as

$$\psi(r, \theta, \phi) = R(r)\Theta(\theta)\Phi(\phi). \tag{5.2}$$

The radially dependent potential only affects the solution for $R(r)$ meaning that the form of the Θ and Φ functions can be determined independently of $V(r)$. Substituting equation (5.2) into (5.1) gives the differential equation in Φ as

$$\frac{d^2\Phi}{d\phi^2} + m_l^2\Phi = 0$$

where m_l is an integer that takes on values $m_l = 0, \pm1, \pm2, \ldots$. The solutions to this equation are of the form,

$$\Phi_{m_l}(\phi) = \frac{1}{\sqrt{2\pi}}e^{im_l\phi}. \tag{5.3}$$

The Schrödinger equation for Θ is of the form,

$$\frac{1}{\sin\theta}\frac{d}{d\theta}\left[\sin\theta\frac{d\Theta}{d\theta}\right] + \left[l(l+1) - \frac{m_l^2}{\sin^2\theta}\right]\Theta = 0 \tag{5.4}$$

where $l = 0, 1, 2, \ldots$ and m_l must take on integer values, $0, \pm1, \pm2, \ldots \pm l$. The solution to equation (5.4) is a polynomial in $\sin\theta$ or $\cos\theta$ (an associated Legendre polynomial) and can be combined with equation (5.3) to give the angular dependence in terms of the spherical harmonics,

$$Y_{lm_l}(\theta, \phi) = \Theta_{lm_l}(\theta)\Phi_{m_l}(\phi).$$

Typical solutions are given as a function of l and m_l in table 5.3.

The energy eigenvalues are obtained from the radial part of the Schrödinger equation,

$$-\frac{\hbar^2}{2m}\left[\frac{d^2R}{dr^2} + \frac{2}{r}\frac{dR}{dr}\right] + \left[V(r) + \frac{l(l+1)\hbar^2}{2mr^2}\right]R = ER \tag{5.5}$$

where $l = 0, 1, 2, 3, \ldots$.

An infinite spherical square well of radius a is defined as

$$V(r) = 0 \quad r < R_0$$

$$V(r) = \infty \quad r \geqslant R_0.$$

Solutions to equation (5.5) with $V = 0$ are the spherical Bessel functions, $j_l(kr)$, as given in table 5.4 where $k = \sqrt{2mE/\hbar^2}$. The energy eigenvalues are found from these

Table 5.3. Example of spherical harmonics for small values of l.

l	m_l	$Y_{lm_l}(\theta, \phi)$
0	0	$(1/4\pi)^{1/2}$
1	0	$(3/4\pi)^{1/2} \cos\theta$
1	±1	$\mp(3/8\pi)^{1/2} \sin\theta\, e^{\pm i\phi}$
2	0	$(5/16\pi)^{1/2} (3\cos^2\theta - 1)$
2	±1	$\mp(15/8\pi)^{1/2} \sin\theta \cos\theta\, e^{\pm i\phi}$
2	±2	$(15/32\pi)^{1/2} \sin^2\theta\, e^{\pm i\phi}$

Table 5.4. Some examples of spherical Bessel functions.

l	$j_l(kr)$
0	sin $kr/(kr)$
1	[sin $kr/(kr)2$] – [cos $kr/(kr)$]
2	[3sin $kr/(kr)3$] – [3cos $kr/(kr)2$] – [sin $kr/(kr)$]

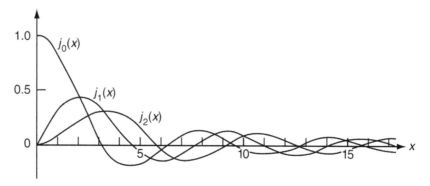

Figure 5.10. Spherical Bessel functions for l = 0, 1, and 2.

solutions by the application of the boundary condition $j_l(kR_0) = 0$. Examples of spherical Bessel functions are shown in figure 5.10. The boundary condition is satisfied for specific values of kR_0, that is, the zeros of the spherical Bessel function, $z_n(l)$. From the definition of k it is easily seen that the energy eigenvalues are given in terms of the $z_n(l)$ as

$$E = \frac{\hbar^2 z_n(l)^2}{2R_0^2 m} = E_0 \frac{z_n(l)^2}{\pi^2} \tag{5.6}$$

where E_0 is defined as $E_0 = 2\hbar^2\pi^2/2R_0^2 m$. Values of the $z_n(l)$ can be tabulated using the expressions given in table 5.4 and some of these are summarized in table 5.5. For a given value of l there is a series of zeros, as given in the table, and these can be seen as the zero crossing points of the spherical Bessel functions in figure 5.10. These different $z_n(l)$ correspond to different values of the quantum number, n. Tabulating calculated values of E in increasing order from each of the $z_n(l)$, allows the energy levels to be obtained as a function of n and l and these are given in table 5.6 The spectroscopic notation as shown in the table has been adapted from the atomic nomenclature discussed before where the l values 0, 1, 2, 3, 4,... are designated s, p, d, f, g,... It is important to note that n is not the principal quantum number, as it is in the case of atomic energy levels. Rather, n is a number that counts the number of levels with a particular value of l. Analogous to the atomic case each of the nuclear states is degenerate in m_l where the degeneracy is given by $(2l + 1)$, and also in spin, where the degeneracy is 2 (i.e., $m_s = \pm 1/2$). An inspection of the energy of the various

Table 5.5. The first four zeros of the spherical Bessel functions as a function of l.

l	$z_n(l)$
0	3.1416, 6.2832, 9.4248, 12.5664,...
1	4.4934, 7.7253, 10.9041, 14.0662,...
2	5.7635, 9.0950, 12.3229, 15.5146,...
3	6.9879, 10.4171, 13.6980, 16.9236,...
4	8.1826, 11.7049, 15.0397, 18.3013,...
5	9.3558, 12.9665, 16.3547, 19.6532,...
6	10.5128, 14.2074, 17.6480, 20.9835,...
7	11.6570, 15.4313, 18.9230, 22.2953,...

Table 5.6. Energies and occupancy of nucleon states for the infinite square well potential.

n	l	Notation	E/E_0	Occupancy	Accumulated occupancy
1	0	1s	1.00	2	2
1	1	1p	2.05	6	8
1	2	1d	3.37	10	18
2	0	2s	4.00	2	20
1	3	1f	4.96	14	34
2	1	2p	6.04	6	40
1	4	1g	6.78	18	58
2	2	2d	8.38	10	68
1	5	1h	8.88	22	90
3	0	3s	9.00	2	92
2	3	2f	10.99	14	106
1	6	1i	11.20	26	132
3	1	3p	12.05	6	138
1	7	1j	13.76	30	168

levels, the neutron-level occupancy and accumulative number of neutrons as given in the table 5.6 show that it is difficult to see clearly defined shells. It is clear, however, that aside from 2, 8, and 20, the accumulated occupancies given in the table cannot correspond to the observed nuclear magic numbers.

5.4 Other forms of the nuclear potential

Because the nuclear model as described above does not yield the correct magic numbers, it might seem reasonable to modify the nuclear potential. Because we do not expect infinite potential walls at the edges of the nucleus, it would be reasonable to consider a finite square well. The implementation of this model follows along the lines of the infinite well, and as long as the potential remains spherically symmetric, the energy eigenvalues are determined from a consideration of the radial equation. In this case the nucleon wave functions will penetrate the side of the well and the boundary conditions for the wave functions must be satisfied at $r = R_0$. In general, this has the effect of uniformly lowering the energy levels from those calculated for the infinite square well but does not help to predict the correct magic numbers.

From a physical standpoint the most reasonable nuclear potential would be one that looked like the total nucleon density in the nucleus, as shown in figure 3.9. This can be accomplished by using a potential in the Schrödinger equation that is proportional to the Woods–Saxon function given by equation (3.13). In practice, other analytical forms for the potential are mathematically easier to deal with. One that is often used is a finite square well with exponential sides. The details of the shape of the potential have relatively little influence on the success of this model. The energy levels for a nuclear potential described by a rounded square well are shown in figure 5.11. Compared to the infinite square well described in table 5.6, some minor changes in the ordering and spacing of the energy levels occur, but the model still fails to predict the correct magic numbers.

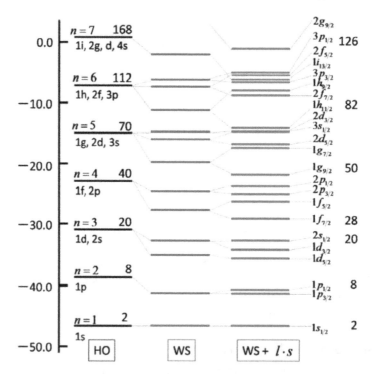

Figure 5.11. Energy levels predicted by the nuclear shell model. From left: energy levels for a harmonic oscillator potential, energy levels for a Woods–Saxon potential and energy levels for a Woods–Saxon potential with spin–orbit coupling. The spectroscopic notation for each energy level is shown, along with the resulting magic numbers. From Ito and Ikeda (2014), copyright IOP Publishing. Reproduced with permission. All rights reserved.

5.5 Spin–orbit coupling

It is clear that the details of the potential used in the Schrödinger equation have only a minor influence on the ordering of the nuclear energy levels and that any reasonable potential will probably be unable to explain the experimentally observed magic numbers. The proper description of experimental data, therefore, requires the introduction of an additional factor. It is well known that the correct description of atomic energy levels requires a consideration of the interaction between the electron's spin and its orbital angular momentum; the so-called spin–orbit inter-action. In a classical sense this interaction can be thought of as a magnetic interaction between the electron's magnetic dipole moment and the atom's Coulomb field. Although the spin–orbit interaction is crucial to the proper description of atomic energy levels, its possible introduction into the nuclear problem requires a consideration of the energy scales involved. The typical spacing of electronic energy levels is of the order of a few eV; the typical spacing of nuclear energy levels is closer to an MeV. The spin–orbit interaction as applied to atomic energy levels would be insufficient to have any appreciable effect on nuclear energy levels. Scattering experiments suggest the importance of this interaction in nuclei, so it is reasonable to hypothesize a strong nuclear spin–orbit coupling. The nuclear shell model including strong spin–orbit coupling was developed in 1949 by Eugene Paul Wigner (1902–1995), Maria Goeppert Mayer (1906–1972) and J Hans D Jensen (1907–1973). The success of the model described below on the basis of this hypothesis provides evidence for the existence of such an interaction. Thus, the potential in the Schrödinger equation is modified as

$$V(r) \rightarrow V(r) + f(r)\vec{l} \cdot \vec{s} \qquad (5.7)$$

where $f(r)$ is a function of position and $\vec{l}$ and $\vec{s}$ are the nucleon spin and orbital angular momenta. In the interior of the nucleus where the nucleon density is relatively constant, the spin–orbit interaction cancels out. It is in the edge region of the nucleus where there is a nucleon density gradient that the spin–orbit interaction is important. Thus, it is often customary to express the function $f(r)$ in the above as

$$f(r) \propto \frac{1}{r}\frac{dV(r)}{dr}.$$

In the case where there is a strong spin–orbit coupling the energy levels are determined by the total angular momentum, $\vec{j}$, as given by

$$\vec{j} = \vec{l} + \vec{s}. \qquad (5.8)$$

For nucleons $s = 1/2$ and the strong spin–orbit coupling imposes the condition on the expectation values for $\vec{j}$,

$$j = l + 1/2 \text{ or } j = l - 1/2 \qquad (5.9)$$

except when $l = 0$, in which case only $j = 1/2$ is allowed. Thus, $\vec{l}$ and $\vec{s}$ add vectorially (either parallel or antiparallel) to give the total $\vec{j}$. Each energy level is then designated by three quantum numbers n, l, and j and is written in spectroscopic notation as nl_j. This means that each l state is split into two j substates with the two allowed values of j (except for the $l = 0$ s-states, which remain unsplit). The magnitude of the splitting is proportional to $2l + 1$ and $f(r)$. This can be easily shown as follows: The expectation values for l^2, s^2, and j^2 are given by,

$$\langle l^2 \rangle = l(l + 1)\hbar^2$$

$$\langle s^2 \rangle = s(s + 1)\hbar^2$$

$$\langle j^2 \rangle = j(j + 1)\hbar^2.$$

From equation (5.8), the square of j is found using the cosine rule as

$$j^2 = l^2 + s^2 + 2\vec{l} \cdot \vec{s}. \tag{5.10}$$

This can be readily solved for $\vec{l} \cdot \vec{s}$ as

$$\vec{l} \cdot \vec{s} = \frac{1}{2}(j^2 - l^2 - s^2).$$

The expectation value, determined from equation (5.10), is

$$\langle \vec{l} \cdot \vec{s} \rangle = \frac{\hbar^2}{2}[j(j + 1) - l(l + 1) - s(s + 1)]. \tag{5.11}$$

The difference in energy between the two j states can be found from equation (5.7) by using $s = 1/2$ and substituting the two values of j from equation (5.9) into equation (5.11). This gives a change in energy of

$$\Delta V \propto 2l + 1.$$

Experimental evidence has shown that the sign of $f(r)$ is explicitly negative so that the energy of the state with $j = l - 1/2$ is higher than the energy of the state with $j = l + 1/2$. The splitting of the energy levels including the spin–orbit coupling is shown in figure 5.11.

In order to determine the occupancy of each level it is necessary to properly establish the degeneracy. The degeneracy of the unsplit l levels was given by $2l + 1$ for the allowed values of m_l and 2 from the allowed values of m_s, giving a total degeneracy of $2(2l + 1)$. In the spin–orbit split case, l_z and s_z do not commute with the Hamiltonian, meaning that m_l and m_s are no longer 'good' quantum numbers. Because j_z does commute with the Hamiltonian, the relevant 'good' quantum number is m_j. The degeneracy of the j states is therefore given by $2j + 1$. As an example, consider the 1g state ($n = 1$, $l = 4$). This has a degeneracy of $2(2l + 1) = 18$. In the presence of spin–orbit coupling, this splits into two states with $j = 9/2$ and $j = 7/2$ (in order of increasing energy). These have degeneracies of $2j + 1 = 10$ and $2j + 1 = 8$, respectively, giving the total degeneracy of the initial l state.

As illustrated in this figure the splitting of l states resulting from the spin–orbit interaction causes a reordering of energy levels and introduces well-defined gaps. These gaps define the various energy shells and as shown in the figure, correspond to occupancies that match the observed nuclear magic numbers.

5.6 Nuclear energy levels

The nuclear shell model that includes a strong spin–orbit coupling has been successful in predicting the correct magic numbers. It is, however, interesting to see how this model can predict the actual energies of the nucleons in their energy levels. A diagram, as illustrated in figure 5.11, shows the energy levels of the neutrons in a specific nucleus. If we are interested in predicting the behavior of a parameter such as separation energy as a function of N or A it is necessary to calculate the energy level diagram for nuclei of different N (or A) and to understand how these energy levels are occupied. In principle it is easy to see how this will be reflected in the calculation, since the nuclear potential depends on the size of the nucleus. Thus, although the sequence of energy levels is more or less as indicated by figure 5.11, the actual value of the energy corresponding to each of these levels is a function of the number of nucleons. For a given number of nucleons, the energy of the highest occupied level (the Fermi energy) is found by filling the energy levels with the required number of nucleons as determined by the degeneracy of each state. An example is illustrated in figure 5.12. The Fermi energy, as determined by the energy of the last nucleon, follows the trends that are observed for the neutron separation energy of various nuclei as a function of A.

It is interesting to note that the spacing of the nuclear energy levels is a sensitive function of the size of the nucleus. Figure 5.12 shows that although there are clear anomalies in the Fermi energy for nuclei with filled shells, the overall change in the Fermi energy between very light nuclei and very heavy nuclei is relatively small. Since heavy nuclei have more nucleons and hence more occupied states, the average spacing between the states below the Fermi energy must decrease as nuclei become heavier. Analogous to the terminology used in solid state physics the number of occupied states per unit energy is referred to as the density of states. This increase in the density of states for heavier (or larger) nuclei is apparent in the simple infinite square well model. As equation (5.6) shows, for a given value of $z_n(l)$, the energy decreases with increasing R_0. This will be considered in more detail in the discussion of excited states in the next chapter.

A final point to consider here is the question of the energy levels of the protons. The above discussion is valid for neutrons, but it is clear that the presence of Coulomb interactions must alter the form of the potential seen by the protons. There are basically three effects introduced by the proton charge on the potential seen by these charged particles:

1. A nonzero potential outside the nucleus that follows a $1/r$ dependence as expected outside a spherical distribution of charges;
2. A curvature in the bottom of the well as is expected inside a uniform charge distribution;

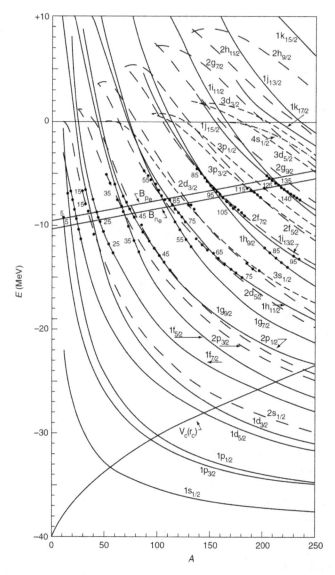

Figure 5.12. Nuclear energy levels for β stable odd A nuclei as predicted by the shell model. The energy of the odd neutron is shown by the closed circle. Reprinted with permission from Green (1956), copyright (1956) by the American Physical Society.

3. A decrease in the depth of the potential well, which accounts for the decreased binding energy for protons as a result of their Coulomb repulsion.

These features are illustrated in figure 5.13. The first two characteristics have relatively little effect on the structure of the energy levels; however, this third feature tends to stretch the proton energy levels apart in energy. This is shown in figure 5.14.

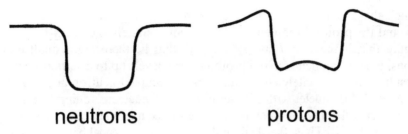

Figure 5.13. Comparison of potential well shapes for (a) neutrons and (b) protons.

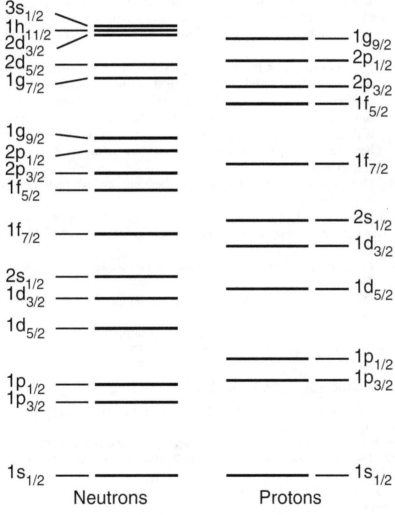

Figure 5.14. Comparison of nuclear energy levels for (a) neutrons and (b) protons.

Although there is not a strict equality, there is a tendency for the neutron Fermi energy and the proton Fermi energy to be approximately the same. It is clear from the figure that if the Fermi energy is small, that is, there are a small number of nucleons, then filling neutron and proton energy levels up to a certain Fermi energy will result in approximately equal numbers of occupied neutron and proton states; hence $N \approx Z$. For nuclei with a large number of nucleons, filling levels up to the Fermi energy will allow for a greater number of occupied neutron states than proton states; hence $N > Z$. Thus, the shell model provides a physical basis for the behavior previously illustrated in figure 3.1 and for the inclusion of the symmetry term in the liquid drop model.

Problems

5.1. For a neutron in a three-dimensional infinite spherical well calculate the actual values of the energies of the 1s, 1p, 1d, and 2s states (in MeV). Construct a plot along the lines of figure 5.12 illustrating the results.

5.2.

(a) Find an expression for the energy levels of a three-dimensional harmonic oscillator. This should be described in most introductory quantum mechanics texts. Do not actually solve the Schrödinger equation for this problem. The three-dimensional harmonic oscillator potential is widely used as an approximation for the shell model potential, not because it is particularly accurate, but because it is simple and convenient.

(b) Draw an energy level diagram for these energy levels and determine the degeneracy of each level.

(c) Use this model to predict the magic numbers for the system.

5.3. Draw the ground state nucleon configuration for ^{19}F, ^{33}S, ^{55}Mn, and ^{91}Zr.

5.4.

(a) From measured masses calculate the binding energy per nucleon for ^{38}K, ^{40}Ca, and ^{42}Sc.

(b) Discuss these results in terms of the semiempirical mass formula and the shell model.

5.5. Using an analogy with a particle in an infinite square well, explain the general trends that are observed in figure 5.12.

References and suggestions for further reading

Cottingham W N and Greenwood D 2001 *An Introduction to Nuclear Physics* 2nd edn (Cambridge: Cambridge University Press)

D'Arienzo M 2013 Emission of β^+ particles via internal pair production in the $0^+ \rightarrow 0^+$ transition of ^{90}Zr: Historical background and current applications in nuclear medicine imaging *Atoms* **1** 2–12

Das A and Ferbel T 2003 *Introduction to Nuclear and Particle Physics* 2nd edn (Singapore: World Scientific)

Enge H A 1966 *Introduction to Nuclear Physics* (Reading, MA: Addison-Wesley)

Engfer R, Schneuwly H, Vuilleumier J L, Walter H K and Zehnder A 1974 Charge-distribution parameters, isotope shifts, isomer shifts, and magnetic hyperfine constants from muonic atoms *At. Data Nucl. Data Tables* **14** 509–97

Green A E S 1956 Approximate analytical wave functions for the nuclear independent-particle model *Phys. Rev.* **104** 1617–24

Heilig K and Steudel A 1974 Changes in mean-square nuclear charge radii from optical isotope shifts *At. Data Nucl. Data Tables* **14** 613–38

Hughes D J and Sherman D 1950 Fast neutron cross sections and nuclear shells *Phys. Rev.* **78** 632–3

Ito M and Ikeda K 2014 Unified studies of chemical bonding structures and resonant scattering in light neutron-excess systems, 10,12Be *Rept. Prog. Phys.* **77** 096301

Preston M A 1962 *Physics of the Nucleus* (Reading, MA: Addison-Wesley)

Shera E B *et al* 1976 Systematics of nuclear charge distributions in Fe, Co, Ni, Cu, and Zn deduced from muonic x-ray measurements *Phys. Rev. C* **14** 731–47

IOP Publishing

An Introduction to the Physics of Nuclei and Particles
(Second Edition)

Richard A Dunlap

Chapter 6

Properties of the nucleus

6.1 Ground state spin and parity

In addition to properly predicting the nuclear magic numbers, the shell model is effective at describing other aspects of nuclear properties. Among these are nuclear spin and parity. These properties can be determined experimentally by looking at, for example, decay processes, transitions between nuclear states, and interactions of nuclei with magnetic fields. The spin of a nucleus is calculated from the shell model results given in the last chapter by considering the angular momenta of the individual nucleons. The total nuclear spin, denoted as $\vec{I}$, is the vector sum of the angular momenta of the individual nucleons, $\vec{j}$. Here uppercase characters are used to denote the overall property of the nucleus, while lowercase is used as in chapter 5 to denote the properties of individual nucleons. Thus, we write,

$$\vec{I} = \sum \vec{j_i}$$

It is important to consider how the individual spins align vectorially. This is analogous to the problem in solid state physics where the relative orientation of the electron spins in the various energy levels must be determined in order to calculate magnetic moments. In nuclear physics the situation is actually somewhat simpler because of the strong pairing tendency for nucleons. Following the assumptions of the extreme single-particle shell model, we note that like nucleons will pair with j antiparallel yielding a net zero contribution to the overall nuclear spin and it will be the spin of the unpaired neutron, j_n, and/or unpaired proton, j_p, that will determine the nuclear spin. To understand how to determine the total nuclear spin, we consider the ground state properties of some specific types of nuclei.

doi:10.1088/978-0-7503-6094-4ch6

Even–even nuclei In the case of an even number of neutrons and an even number of protons, the spin contribution from both the neutrons and protons will be zero yielding a net zero I for the nucleus. This has been found to be true experimentally.

Even–odd and odd–even nuclei These are perhaps the most interesting in the context of the shell model and can be described by the so-called single-nucleon model. The even species of nucleon will contribute zero to the overall I and the contribution of the odd species of nucleon will be determined by the spin of the single unpaired nucleon. Thus, in the case of an odd neutron $I = j_n$ and in the case of an odd proton $I = j_p$.

Odd–odd nuclei These are the most complex to deal with because both j_n and j_p are nonzero. Thus, the total nuclear spin is given by the vector sum of these two quantities,

$$\vec{I} = \vec{j_n} + \vec{j_p}. \tag{6.1}$$

In general, the vector relationship between these two quantities is not known, but equation (6.1) allows us to place certain limits on the total nuclear spin as,

$$|j_n - j_p| \leqslant I \leqslant j_n + j_p.$$

The above rules can be used to predict (or at least understand) the behavior of particular nuclei. Before looking at some examples, we consider the question of parity. The parity of a nuclear state is given by the parity of the nuclear wave function. This can be either even (denoted +) or odd (denoted −) depending on the mathematical properties of the wave function. An even wave function is symmetric and has the property,

$$\psi(\vec{r}) = \psi(-\vec{r})$$

and an odd wave function is antisymmetric and has the property,

$$\psi(\vec{r}) = -\psi(-\vec{r}).$$

For the single-nucleon model the parity of the overall wave function is given by the parity of the wave function of the unpaired nucleon. The symmetry of the wave function for a given value of l defines the parity as

$$\pi = (-1)^l$$

where $l = 0, 1, 2, 3,...$ correspond to s, p, d, f,... states. Thus s-states have even parity, p-states have odd parity, d-states have even parity, and so on.

In standard spectroscopic notation the occupancy of each state is given by a superscript. As an example, consider the odd A nuclei ^{15}O and ^{17}O. Both nuclides have a filled proton shell ($Z = 8$). ^{15}O has $N = 7$ while ^{17}O has $N = 9$. The state occupancy for ^{15}O and ^{17}O is given in table 6.1 and is determined by sequentially filling states in order of increasing energy. Figure 5.11 provides some guidance in this respect. In both cases the proton states contribute zero to I and the net nuclear angular momentum is the result of the properties of the one unpaired neutron. For

Table 6.1. Nucleon configurations for ^{15}O and ^{17}O.

Nuclide	Proton state	Neutron state	I^{π}
^{15}O	$(1s_{1/2})^2(1p_{3/2})^4(1p_{1/2})^2$	$(1s_{1/2})^2(1p_{3/2})^4(1p_{1/2})^1$	$1/2^-$
^{17}O	$(1s_{1/2})^2(1p_{3/2})^4(1p_{1/2})^2$	$(1s_{1/2})^2(1p_{3/2})^4(1p_{1/2})^2 (1d_{5/2})^1$	$5/2^+$

Table 6.2. Nucleon configurations for ^{27}Si and ^{29}Si.

Nuclide	Proton state	Neutron state	I^{π}
^{27}Si	$(1s_{1/2})^2(1p_{3/2})^4(1p_{1/2})^2 (1d_{5/2})^6$	$(1s_{1/2})^2(1p_{3/2})^4(1p_{1/2})^2 (1d_{5/2})^5$	$5/2^+$
^{29}Si	$(1s_{1/2})^2(1p_{3/2})^4(1p_{1/2})^2 (1d_{5/2})^6$	$(1s_{1/2})^2(1p_{3/2})^4(1p_{1/2})^2 (1d_{5/2})^6(2s_{1/2})^1$	$1/2^+$

^{15}O the unpaired neutron is in a $1p_{1/2}$ state giving $I = j_n = 1/2$ and an odd parity as appropriate for a p-state wave function. This state is sometimes referred to as $I^{\pi} = 1/2^-$. For ^{17}O the unpaired neutron is in a $1d_{5/2}$ state giving $I = j_n = 5/2$ and even parity, as appropriate for a d-state wave function. In many cases, particularly heavier nuclides, the spectroscopic notation is given by only the outermost level or levels (since inner filled levels do not contribute to the net nuclear spin or parity). Thus, the neutron states for ^{15}O would be given as $(1p_{1/2})^1$ and for ^{17}O as $(1d_{5/2})^1$.

A slightly more complex example is provided by two isotopes of Si, as shown in table 6.2. Since both ^{27}Si and ^{29}Si have $Z = 14$ there is no contribution to the nuclear spin from the protons. The neutron configuration of ^{27}Si is $(1d_{5/2})^5$ indicating an unpaired d-state neutron giving a spin 5/2 and even parity. For ^{29}Si a strict interpretation of figure 5.11 would provide the incorrect spin and parity of $3/2^+$ (for the unpaired neutron). The correct interpretation of these states is given in figure 5.12 where it is seen that the curves for $2s_{1/2}$ and $1d_{3/2}$ cross and for light nuclei the $2s_{1/2}$ state is filled before the $1d_{3/2}$. This gives the correct spin and parity of $1/2^+$. A careful examination of this figure shows that several such cases exist and provides a warning for over-interpreting a single energy level diagram such as figure 5.11.

In general, it is seen from a comparison of experimental results and calculations based on the single-nucleon model that the ground state of even–even nuclei is properly described as 0^+. There is remarkable agreement between model and experiment for the ground state properties of odd A nuclei, provided a careful consideration of the ordering of the energy levels is taken. However, the model fails to provide any simple systematic method of predicting the properties of odd–odd nuclei.

6.2 Excited nuclear states

Nuclei can exist in excited states and these states are typically populated by decays from other unstable nuclides or by interactions with other particles. Since the typical spacing of excited states in a nucleus is of the order of an MeV, thermal excitations play a negligible role (unlike the case for electronic energy levels). Here we discuss excited states in the context of the shell model. Further information about excited states will be given in later chapters.

The properties of many excited nuclear states can be described by nonground state distributions of nucleons in the energy levels. A reliable test for the accuracy of this kind of model is a direct comparison with experimentally determined nuclear spins and parities. The simplest description that is applicable to many excited nuclear states is the single-particle model where an excited state is described by the excitation of a single nucleon. This approach is most applicable to nuclei that have one unpaired nucleon outside filled neutron and proton shells. This is easiest to see in an example. Here we consider ^{41}Ca, which has a filled proton shell and one unpaired $1f_{7/2}$ neutron as indicated in figure 6.1. The properties of the ground state and the energy of the first few excited states of ^{41}Ca are illustrated in figure 6.2. Most of these states can be described as single-particle states as illustrated in figure 6.3. For each single-particle state a single neutron is excited into a higher energy state resulting in an unpaired neutron in one of the states. This unpaired neutron readily gives the resulting overall nuclear spin and parity. The state designation as given in figure 6.3 provides the relevant nuclear properties. In cases where a nucleon from a filled level is excited to a higher level resulting in an unpaired nucleon in a level that had previously been filled, this missing nucleon can be thought of as a 'hole.' The spectroscopic notation can then be used to denote the state in terms of this missing

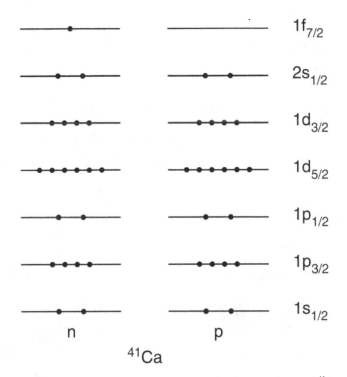

Figure 6.1. Neutron and proton configuration for the ground state of ^{41}Ca.

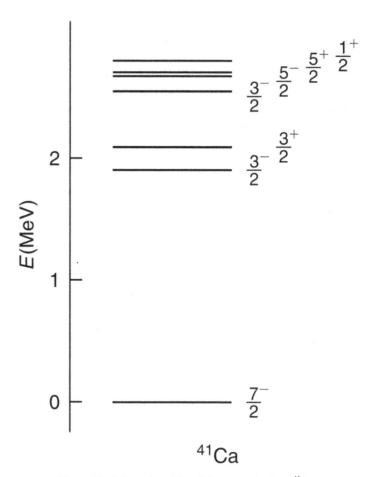

Figure 6.2. Spins and parities of the energy levels of ^{41}Ca.

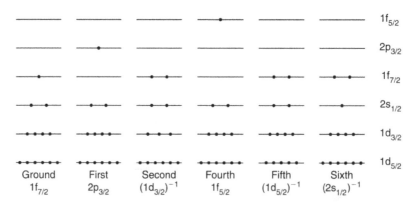

Figure 6.3. Neutron configurations for the ground and first few excited states of ^{41}Ca.

nucleon. It is therefore customary to refer to the second excited state of ^{41}Ca as $(1d_{3/2})^{-1}$ rather than $(1d_{3/2})^3$. Similar designations apply for the fifth and sixth excited states.

The shell model therefore provides a reliable means of determining the properties of excited states, in particular in the cases where excited state properties are accurately described by single-particle states. In some cases, an excited state cannot be described by the excitation of a single nucleon and two or more nucleons are involved giving rise to two or more unpaired nucleons. This may involve either neutrons or protons or both. The description of these states is analogous to the description of odd–odd nuclei and clearly defined rules cannot be described.

6.3 Mirror nuclei

Mirror nuclei are pairs of nuclei that have the same values of A but the values of N and Z interchanged. Some examples are ^{3}H—^{3}He, ^{7}Li—^{7}Be, ^{9}Be—^{9}B, ^{11}B—^{11}C, ^{13}C—^{13}N,.... The shell model would predict very similar properties for these pairs of nuclides and an experimental investigation of excited states in mirror nuclei gives some interesting insight. For one nuclide of each pair, for example, ^{13}C, the nuclear properties should be determined by a single unpaired neutron, for the other nuclide, for example, ^{13}N, the properties are determined by a single unpaired proton. If we ignore the presence of Coulomb interactions, then these cases are indistinguishable. Thus, differences between the properties of mirror nuclei are an indication of the importance of Coulomb interactions and similarities are suggestive of the importance of the strong interaction. The first few excited states of ^{13}C and ^{13}N are compared in figure 6.4. The spins and parities of the energy levels shown in figure 6.4 are the same, indicating that for the first few energy levels (at least) the single-particle excitations are the same for the unpaired neutron in ^{13}C and the unpaired proton in ^{13}N. The typical downward shift of energy levels for ^{13}N compared to ^{13}C is consistent with Coulomb effects, as illustrated in figure 5.14.

6.4 Electromagnetic moments of the nucleus

The measurement of the electromagnetic moments of nuclei provides important information about nuclear structure, nuclear charge distributions, and nuclear shapes, as well as insight into the properties of the neutron and proton themselves. The electromagnetic moments of a nucleus result from the distribution of charges and currents within the nucleus. The electric monopole moment results from the total electric charge and is merely $+eZ$. For symmetry reasons the electric dipole moment of the nucleus (as well as all other static multipole moments with odd parity; for example, magnetic monopole, magnetic quadrupole, or electric octupole) must vanish. The electric quadrupole moment exists for charge distributions that are nonspherical. In previous chapters we assumed that nuclei are spherical. However, this is not always the case, and a measure of the nuclear electric quadrupole moment is an indication of nonspherical nuclear symmetry. The magnetic multipole moment that is of interest is the magnetic dipole moment. This results from current distributions in the nucleus; recall the classical magnetic dipole consisting of a

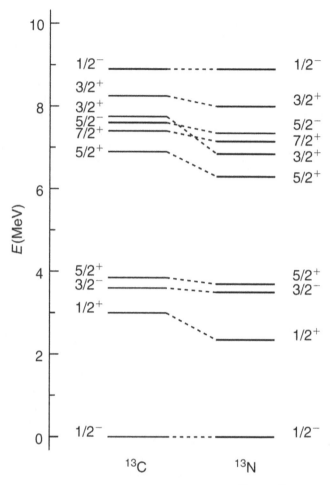

Figure 6.4. Excited states of the mirror nuclei ^{13}C and ^{13}N.

Table 6.3. Characteristics of the lowest order electric and magnetic multipole moments of the nucleus.

Moment	Electric	Magnetic
Monopole	Net nuclear charge	0
Dipole	~ 0	Due to currents
Quadrupole	Due to nonspherical $\rho_p(r)$	~ 0

single current loop. Table 6.3 summarizes information about the lower order moments of nuclei. The electric quadrupole moment and the magnetic dipole moment of the nucleus are of most relevance to the models described in previous chapters and these are discussed in detail below.

6.5 Electric quadrupole moments

The interaction energy of a charge distribution $\rho(r)$ and an external electric potential $\phi(r)$ can be written as

$$U = \int \rho(\vec{r})\phi(\vec{r})d^3\vec{r}. \tag{6.2}$$

If $\phi(r)$ is a slowly varying function of r then it can be expanded in a Taylor series about $r = 0$ as

$$\phi(\vec{r}) = \phi(0) + (\vec{r}\cdot\nabla\phi)_{r=0} + \frac{1}{2}\sum_{i,j}x_ix_j\frac{\partial^2\phi}{\partial x_i\partial x_j}\bigg|_{r=0} + \ldots$$

where the x_i, and x_j are the spatial coordinates x, y, and z. Substituting this into equation (6.2) gives the interaction energy as

$$U = \phi(0)\int\rho(\vec{r})d^3\vec{r} + \left(\int\rho(\vec{r})\vec{r}d^3\vec{r}\right)\cdot\nabla\phi + \frac{1}{2}\sum_{i,j}\frac{\partial^2\phi}{\partial x_i\partial x_j}\int\rho(\vec{r})x_ix_jd^3\vec{r} + \ldots \tag{6.3}$$

The first term on the right-hand side is the monopole term and defines the electric monopole moment (or total charge) as

$$eZ = \int\rho(\vec{r})d^3\vec{r}$$

The second term on the right-hand side gives the electric dipole moment, P, as

$$P = \int\rho(\vec{r})\vec{r}d^3\vec{r}.$$

The third term on the right-hand side of equation (6.3) comes from the electric quadrupole moment of the nucleus and, in the case where there is axial symmetry along the z-axis, this gives the electric quadrupole moment, Q, as

$$Q = \int\rho(\vec{r})[3z^2 - r^2]d^3\vec{r}. \tag{6.4}$$

Some measured electric quadrupole moments as a function of Z are illustrated in figure 6.5. An interesting feature of these data is the crossing from positive to negative values of Q for Z equal to a magic number. This feature is also observed for N equal to a magic number. In the context of the shell model, it is encouraging to see that nuclei with filled shells have near zero Q and are, therefore, spherically symmetric. The meaning of nonzero Q can be readily understood from an inspection of equation (6.4). Prolate nuclei are those that are elongated along the z-axis and will have $Q > 0$. This is seen from the equation by writing $r^2 = x^2 + y^2 + z^2$ and observing that when one integrates over the volume of the nucleus, $3z^2$ will, on the average be greater than r^2. Oblate nuclei are compressed along the z-axis and, as equation (6.4) indicates, will have $Q < 0$. In a qualitative sense, the shell model can explain positive Q values for nuclei with Z slightly less than a magic number in terms of the existence of proton hole states. Similarly, a negative Q for Z greater than a magic number

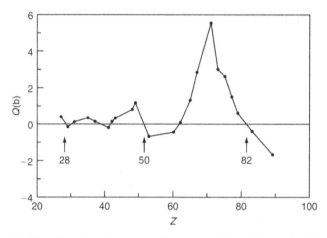

Figure 6.5. Electric quadrupole moments for some odd A nuclei as a function of Z.

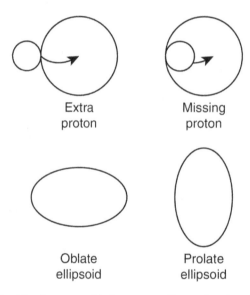

Figure 6.6. Relationship of proton and hole states to oblate and prolate elliptical distortion.

corresponds to unpaired proton states. Figure 6.6 shows that a nucleus with one less proton than a filled shell is approximated by a prolate ellipsoid and yields a positive quadrupole moment. The figure also shows how a nucleus with one proton more than a spherical closed shell can be approximated by an oblate ellipsoid and will give rise to a negative Q. Although the qualitative predictions of the shell model can explain the sign of the quadrupole moments, this model is rather ineffective at predicting the magnitude of Q. Figure 6.5 shows that nuclei without closed shells can have quite large nonspherical deformations. In the extreme cases, this corresponds to $\Delta R/R$ of about 30% and this substantially exceeds expectations based on the single-nucleon shell model. The question of neutron number is also difficult to deal with.

Since the electric quadrupole moment as given by equation (6.4) involves the distribution of charges, the role of unpaired neutrons in determining Q is not obvious. These features of electric quadrupole moments of nuclei are discussed in section 6.7 in terms of the collective model.

6.6 Magnetic dipole moments

In the context of the shell model, the magnetic dipole moment of a nucleus is related to the properties of unpaired nucleons. even–even nuclei are straightforward to deal with as the net nuclear spin is zero and the corresponding net dipole moment is also zero. This is observed experimentally. odd–odd nuclei are difficult to understand because of the uncertainty in the relationship of the angular momentum of the odd neutron and that of the odd proton. Therefore, the most interesting cases for comparison with experimental observations are the odd A nuclei. Here it is necessary to consider the properties of the single unpaired nucleon. The nuclear dipole moment results from the total angular momentum of this nucleon. The components of the total angular momentum, that is, the spin and the orbital angular moments, can both contribute to the total dipole moment and the way in which these two components add must be considered in detail. Without worrying too much about the difference between even–odd and odd–even nuclei we begin with a fairly general discussion of a 'generic' unpaired nucleon.

The nuclear magnetic dipole moment that arises from the orbital angular momentum, l, of an unpaired nucleon is given by

$$\vec{\mu_l} = \frac{g_l \mu_N \vec{l}}{\hbar} \tag{6.5}$$

where μ_N is the nuclear magneton and is defined as $e/2m_p$ (m_p = proton mass). g_l is a quantity analogous to the Landé g-factor used in atomic physics and relates the magnitude of the dipole moment to the orbital angular momentum in units of $\hbar$. An analogous expression can be written for the spin component:

$$\vec{\mu_s} = \frac{g_s \mu_N \vec{s}}{\hbar}. \tag{6.6}$$

It is important to note that, although the vector sum of $\vec{l}$ and $\vec{s}$ is along the direction of $\vec{j}$ (equation (5.8)), the vector sum of the spin and orbital components of $\vec{\mu}$ as given by equations (6.5) and (6.6) is not along the direction of $\vec{j}$. The direction of $\vec{j}$ is well defined, while the directions of $\vec{l}$ and $\vec{s}$ precess around $\vec{j}$. Similarly, the direction of $\vec{\mu}$ precesses around $\vec{j}$ and it is the j component of $\vec{\mu}$, defined as μ, that is a measurable quantity. From equations (6.5) and (6.6) we define this component of μ as

$$\mu = \frac{g_l \mu_N}{\hbar} \frac{\vec{l} \cdot \vec{j}}{j} + \frac{g_s \mu_N}{\hbar} \frac{\vec{s} \cdot \vec{j}}{j}. \tag{6.7}$$

The first dot product on the right-hand side of this equation is found from the scalar product,

$$s^2 = l^2 + j^2 - 2\vec{l}\cdot\vec{j}$$

to be

$$\frac{\langle \vec{l}\cdot\vec{j} \rangle}{\langle j \rangle} = \hbar\frac{l(l + 1) - s(s + 1) + j(j + 1)}{2[j(j + 1)]^{1/2}}. \tag{6.8}$$

Similarly, the second dot product is obtained from,

$$l^2 = s^2 + j^2 - 2\vec{s}\cdot\vec{j}$$

to be

$$\frac{\langle \vec{s}\cdot\vec{j} \rangle}{\langle j \rangle} = \hbar\frac{s(s + 1) - l(l + 1) + j(j + 1)}{2[j(j + 1)]^{1/2}}. \tag{6.9}$$

Combining equations (6.7), (6.8), and (6.9) gives

$$\mu = \frac{g_l\mu_N}{2}\left[\frac{l(l + 1) - s(s + 1) + j(j + 1)}{2[j(j + 1)]^{1/2}}\right] + \frac{g_s\mu_N}{2}\left[\frac{s(s + 1) - l(l + 1) + j(j + 1)}{2[j(j + 1)]^{1/2}}\right]. \tag{6.10}$$

Experimentally, it is the component of μ along the direction of the magnetic field that is measured. This is obtained by multiplying equation (6.10) by a factor $j/[j(j + 1)]^{1/2}$. After some simplification this yields

$$\mu = \frac{\mu_N}{2}\left[(g_l + g_s)j + (g_l - g_s)\frac{l(l + 1) - s(s + 1)}{j + 1}\right]. \tag{6.11}$$

For protons and neutrons $s = 1/2$ and, for a nuclear state defined by the properties of a single unpaired nucleon, the minimum and maximum allowed values of j are given by equation (5.9); that is $j = l \pm 1/2$. Substituting these two values for j into equation (6.11) yields

$$\mu = \frac{\mu_N}{2}[g_s + (2j - 1)g_l] \quad \text{for } j = l + 1/2 \tag{6.12}$$

and

$$\mu = \frac{\mu_N}{2}\frac{j}{j + 1}[-g_s + (2j + 3)g_l] \quad \text{for } j = l - 1/2. \tag{6.13}$$

It is now possible to apply the shell model to the prediction of nuclear magnetic dipole moments of specific nuclei with odd A. To do this we need to know the l and j values of the unpaired nucleon. This will tell us if we are dealing with the $j = l + 1/2$ or the $j = l - 1/2$ case. We also need to know the appropriate values for g_l and g_s. If we are considering even–odd nuclei, then the values of g_l and g_s should be those appropriate for the unpaired proton. If we are considering odd–even nuclei, then the

g_l and g_s values should be those appropriate for the unpaired neutron. Since the neutron is uncharged, its value of g_l is, by definition, zero. If the dipole moment in equation (6.5) is measured in units of the nuclear magnetons, then the value of g_l for the proton is 1. Experimentally determined values of g_s for a free neutron and a free proton are given in table 6.4. The nonzero g_s value for the uncharged neutron results from the fact that hadrons are not fundamental particles but have internal structure. This is discussed further in chapter 15.

Equations (6.12) and (6.13) can readily be applied to the two cases for nuclei with odd A; those with odd N (an unpaired neutron) and those with odd Z (an unpaired proton). Values of μ in nuclear magnetons can be calculated as a function of j of the unpaired nucleon, for both neutron states and proton states and for both alignments of l and s. These values determine the so-called Schmidt lines as a function of j as illustrated in figures 6.7 and 6.8. The data illustrated in the figures show that the model provides somewhat less than ideal agreement with experiment. In fact, the Schmidt lines seem to be limiting values, although, in general, data points lie closer to the correct line. A number of possible explanations have been suggested for the lack of agreement observed here. Two possibilities as described below are of particular interest.

In many cases nuclear states may not be pure states but may be a mixture of states resulting from different configurations of neutrons and protons in the various energy

Table 6.4. g-Factors for the neutron and proton.

Nucleon	g_l	g_s
Neutron	0	−3.8261
Proton	1	5.5856

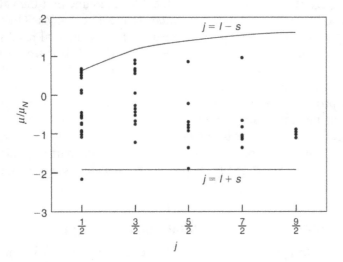

Figure 6.7. Magnetic dipole moments and Schmidt lines for odd N nuclei (neutron states).

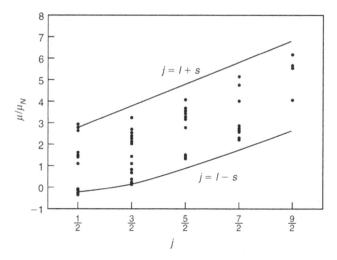

Figure 6.8. Magnetic dipole moments and Schmidt lines for odd Z nuclei (proton states).

levels, each of which yields the same spin and parity. It is convenient to think of states as defined by their wave functions and the resulting wave function being a combination of wave functions from states of equivalent spin and parity. This situation is referred to as configurational mixing and its inclusion in models of nuclear properties has been beneficial in describing electric quadrupole and magnetic dipole moments of nuclei.

The values of g_s that have been used for the neutron and proton are the values measured experimentally for free particles. It is interesting to note that these are both quite different than the value of 2.0, which is expected for point charges and that has been observed for the electron. In fact, the nonzero value of g_s for the uncharged neutron indicates that the internal structure of these particles plays an important role in determining g_s. It is not necessarily reasonable to assume that the values of g_s for nucleons in a nucleus should be the same as the free values. On the average, calculated moments show the best agreement with experimental results if gs values for the neutron and proton are taken to be about 60% of their free values.

The points described above improve the agreement between calculated and measured nuclear properties, but these (and some similar approaches) ultimately fall short of providing an ideal nuclear model. Another possible problem lies with the form of the nuclear potential that has been used in the shell model. This is taken to be spherically symmetric. It is known that nuclei with filled shells are reasonably spherical, but from figure 6.5 it is clear that nuclei with partially filled shells can show substantial nonspherical distortion. See further discussion on these points in the next section.

6.7 Other approaches to modeling nuclei

Both the liquid drop model and the shell model have met with some success in describing the properties of nuclei. The liquid drop model ignores the quantum mechanical properties of the individual nucleons and is therefore unable to make predictions of nuclear properties such as spin and parity. This model has, however, been successful in describing properties such as total binding energy. The shell model has considered the properties of the individual nucleons and, in particular, the behavior of the unpaired nucleon in odd A nuclei. This approach has been surprisingly successful in determining nuclear spins and parities and in describing the properties of many excited states. Although the shell model correctly predicts the change in sign of the quadrupole moment near filled shells, it is not appropriate for explaining the very large quadrupole moments for some nuclei and even under-estimates the small moments of nuclei with single-particle states. It has also been seen that magnetic dipole moment calculations provide order of magnitude agree-ment with experiment but fail to give accurate numerical values. It would thus seem that the two models, which take quite different approaches, each have their own strengths and weaknesses. The collective model has been an attempt to reconcile these two different approaches to better describe certain nuclear properties. The discussion here gives a brief overview of this model.

The basic idea of the collective model is to consider the filled shells of the nucleus as a central core that is described in terms of the liquid drop-like behavior of the nucleons. The quantum mechanical properties of the nucleus are described, as in the shell model, by the spins of the surface nucleons outside of the core. The motion of the surface nucleons introduces a nonsphericity to the central core and, in a practical sense, this type of behavior can be considered by a shell model with a nonspherically symmetric potential. This nonsphericity can account for the anomalous electric quadrupole moments that have been measured. It is important in such cases to realize that nonspherical nuclei can have energy associated with their rotational and/or vibrational degrees of freedom. In order to appreciate the importance of collective effects it is of interest to consider the excited states of some even–even nuclei. According to the shell model the ground states of even–even nuclei should all be 0^+. This is observed to be the case. We can consider the formation of excited states by the excitation of nucleons. The situation here is not so straightforward because, if a single nucleon is excited, then two levels will have unpaired nucleons that will contribute to the overall nuclear spin. From the shell model we have no guidelines for how to add these two contributions vectorially, although we can put limits on the resulting I value. An example helps to illustrate some general features of even–even nuclei.

The low-lying excited states of ^{38}Ar are illustrated in figure 6.9. The neutron ($N = 20$) and proton ($Z = 18$) configurations for ^{38}Ar are shown in figure 6.10. Since the neutron shell is filled, we expect that the low-lying excited states should be described by proton excitations. The simplest excitation would be an excitation of one of the $1d_{3/2}$ protons into the $1f_{7/2}$ state. This would give an integer spin between $7/2 – 3/2 = 2$ and $7/2 + 3/2 = 5$ depending on how the j vectors align. The parity of this state would be the product of the parities of $1d_{3/2}$ and $1f_{7/2}$ states; $(+1)(-1) = (-1)$. Figure 6.9 shows that this excitation probably corresponds to the 3^- state at 3.77 MeV. However, how can we explain the 2^+ state at 2.17 MeV? Another possible excitation would be a $2s_{1/2}$ proton into the $1d_{3/2}$ level resulting in $(2s_{1/2})^{-1}$ and $(1d_{3/2})^1$ states.

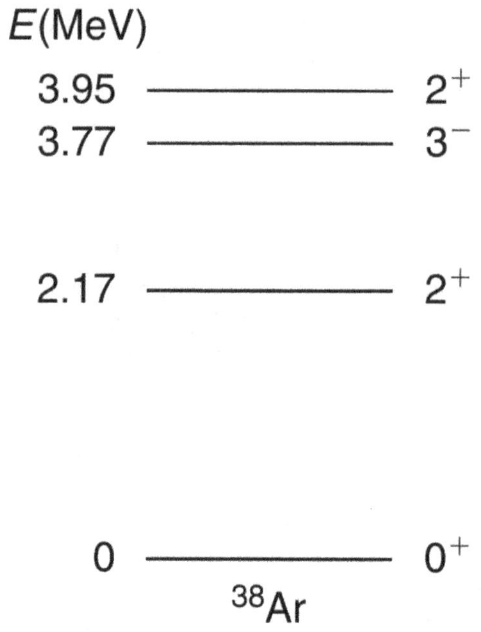

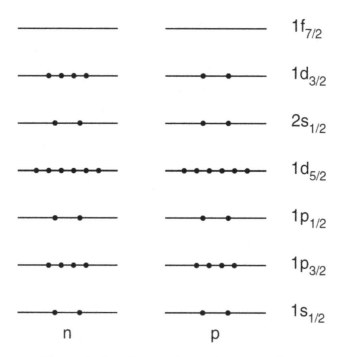

E(MeV)

| 3.95 | —————— | 2^+ |
| 3.77 | —————— | 3^- |

2.17 —————— 2^+

0 —————— 0^+

^{38}Ar

Figure 6.9. Low lying excited states of ^{38}Ar.

		$1f_{7/2}$
		$1d_{3/2}$
		$2s_{1/2}$
		$1d_{5/2}$
		$1p_{1/2}$
		$1p_{3/2}$
		$1s_{1/2}$

n   p

Figure 6.10. Ground state nucleon configuration for ^{38}Ar.

This would give possible spin values of 3/2–1/2 = 1 to 3/2 + 1/2 = 2 and a parity (+1)(+1) = (+1). This would explain the properties of the 2.17 MeV state. However, a detailed consideration of figure 5.12 shows that this state should be at an energy that is very similar to the 3⁻ state. Thus, this excitation nicely corresponds to the 2⁺ state at 3.95 MeV. It is, in fact, difficult to explain the 2⁺ state at 2.17 MeV. This is not an isolated case. In fact, the first excited state of virtually all even–even nuclei is 2⁺ states and these are difficult to explain on the basis of single-nucleon excitations. Notable exceptions to this feature occur for many doubly magic nuclei. The collective model is helpful in understanding the properties of even–even nuclei and makes predictions on the basis of the collective motion of a large number of nucleons rather than a single unpaired nucleon.

In even–even nuclei that are not doubly magic the unfilled shells introduce a nonspherical distortion. Classically, the energy associated with the rotation of an object with a moment of inertia I is given by,

$$E = \frac{I^2}{2I}$$

where I is the classical angular momentum. Quantum mechanically this is written in terms of the expectation values for I^2 as

$$E = \frac{I(I + 1)\hbar^2}{2I}. \tag{6.14}$$

Because both N and Z are even, the nucleus will have a symmetric wave function and the values of I will be constrained to be even. Values of $I = +2, +4, +6, +8, \ldots$ will give rise to excited rotational states with spins and parities of 2⁺, 4⁺, 6⁺, 8⁺,.... The lowest lying 2⁺ rotational state accounts for the lowest energy excited state in many deformed even–even nuclei. Another possibility for this 2⁺ state is discussed below. In light nuclei, the higher order rotational states are usually intermixed with states resulting from nucleon excitations. In heavy nuclei, purely rotational energy levels are sometimes seen as illustrated in figure 6.11 for ¹⁷⁴W. The relative spacing of the energy levels can be calculated from equation (6.14) and can be determined experimentally by normalizing measured energies to the energy of the 2⁺ state. Table 6.5 shows that the predictions of the collective model are good for small I but tend to overestimate the energies for larger I. While these results provide ratios of excited state energies, they do not give absolute energy values. This point will be discussed further below.

The energy of the first 2⁺ excited state of even–even nuclei shows a generally decreasing trend with increasing nuclear mass. This can be explained in terms of equation (6.14). With increasing mass, the corresponding increase in moment of inertia causes a reduction in the spacing of the rotational energy levels. The exceptions to this behavior occur for magic nuclei, especially those that are doubly magic, where the high degree of spherical symmetry reduces the moment of inertia, and the rotational modes are shifted to higher energy. The lowest lying states in these nuclei are typically the result of more complex behavior.

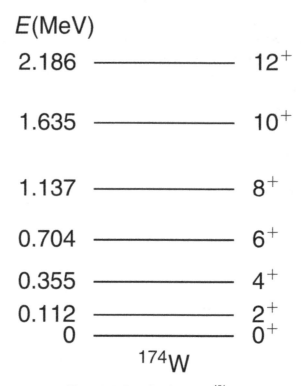

Figure 6.11. Rotational levels in ^{174}W.

Table 6.5. Measured and calculated excited state energies (relative to the 2^+ state) for the rotational modes of ^{174}W.

Level I^π	$E(I^\pi)/E(2^+)$	
	Measured	Calculated
2^+	1.00	1.00
4^+	3.17	3.33
6^+	6.29	7.00
8^+	10.2	12.0
10^+	14.6	18.3
12^+	19.5	26.0

Other possible excitation modes for nuclei are the vibrational modes. Since nuclear matter is relatively incompressible, radial oscillations of the nucleus are not typical. Rather, the simplest kind of nuclear vibrations would be shape oscillations analogous to the shape oscillations experienced by a liquid drop. The lowest order vibrations of this kind are quadrupole vibrations where the nucleus would oscillate between a prolate ellipsoid and an oblate ellipsoid. These oscillations may be described using a harmonic oscillator potential. The Schrödinger equation gives the splitting of the quantum mechanical energy levels for this potential in terms of the oscillation frequency, ω, as

$$\Delta E = \hbar\omega.$$

Such quantized nuclear vibrational states are referred to as phonons by analogy with quantized lattice vibrations in solid state physics. The $n = 1, 2, 3,...$ states have linearly increasing energy and are associated with 1, 2, 3,... phonons, respectively. The phonon has a spin and parity of $j^{\pi} = 2^+$ giving a spin and parity of the first excited vibrational state of 2^+. The spins of the two phonons associated with the second excited vibrational state can add vectorially to give a 0^+, 2^+, or 4^+ state. These represent a triplet at an energy of approximately twice the single-phonon state. Three phonons give a quintuplet of 0^+, 2^+, 3^+, 4^+, and 6^+ states at approximately three times the single-phonon energy. In the ideal harmonic oscillator potential, the energy states are degenerate in I but, in reality, perturbations that have not been discussed here lift the degeneracy and split each level into separate I states. An example of experimentally observed nuclear vibrational states is illustrated for ^{120}Te in figure 6.12. This shows the singlet 2^+ first excited state and the triplet second excited state. For higher energy levels the situation becomes more complex and pure vibrational modes are often not observed.

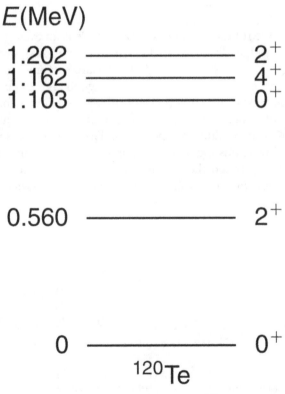

Figure 6.12. Vibrational levels in ^{120}Te.

Another approach to describing nuclear properties are the so-called α-cluster models. The α-particle, that is the nucleus of a ^{4}He atom (2 neutrons and 2 protons), is particularly stable. This was seen in figure 5.4 and is discussed in further detail in section 8.1. α-Cluster models view certain nuclei as comprised of α-particles bound together to form the nucleus. Such nuclei are even–even nuclei with $N = Z$ (sometimes referred to as 4N nuclei). These nuclei include, ^{8}Be, ^{12}C, ^{16}O, ^{20}Ne, ^{24}Mg, ^{26}Si, ^{32}S, ^{36}Ar and ^{40}Ca. This is an extension of the strong pairing tendency that is apparent in nuclear bonding, where pairs of neutrons bind with pairs of protons to form α-particle states within the nucleus. While the shell model itself has been very successful in describing the ground state properties of nuclei, the α-cluster models have been effective at describing the excited states of light nuclei.

Figure 6.13 shows the basic idea of an α-cluster model. The figure shows some interesting features of these models. Firstly, the lightest nucleus that can be described by the model, ^{8}Be, shows clear nonspherical symmetry as a result of the bonding of nucleons into α-particles within the nucleus, as seen in figure 6.13(b). Secondly, as seen in figure 6.13(c), the structure of ^{12}C can be viewed as comprised of three α-particles. However, as the figure shows, the three α-particles can be bound together with different geometries, leading to different overall nuclear shapes. This feature is sometimes referred to as shape coexistence, as discussed below (Jenkins (2016)).

Figure 6.13(b) illustrates the rotation of the ^{8}Be nucleus as viewed in the α-cluster model. Otsuka *et al* (2022) have provided first principles calculations based on the α-cluster model of the rotational states of ^{8}Be. As ^{8}Be is an even–even nuclide, it has a 0^+ ground state and as discussed previously, the rotational levels are 2^+, 4^+, ... Figure 6.14(a) shows the calculated and measured energy levels of ^{8}Be. The figure shows that the calculated values are in good agreement with those that have been measured experimentally. Figure 6.14(b) shows the results for the rotational levels of ^{10}Be. The ^{10}Be nucleus may be viewed as two α-particles (a ^{8}Be nucleus) with two additional neutrons. The strong interaction provided by the two additional neutrons increases the bonding between the two α-particles and reduces the distance between them compared to their distance in the ^{8}Be nucleus. This reduces the moment of inertia and shifts the rotational energy levels upward (see equation (6.14)).

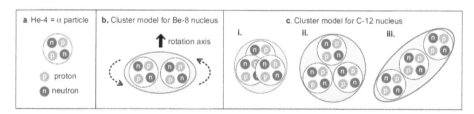

Figure 6.13. (a) α-Particle comprised of two neutrons and two protons, (b) ^{8}Be nucleus as a bound pair of 2 α-particles and (c) ^{12}C nucleus comprised of three α-particle. CC BY 4.0 Reprinted by permission from Otsuka *et al* (2022) CC BY 4.0.

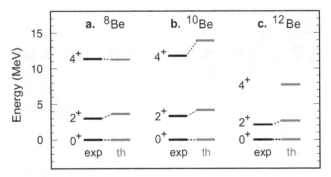

Figure 6.14. Experimental (exp) and theoretical (th) levels for even–even Be nuclei. Experimental values are from National Nuclear Data Center (2023). Reprinted by permission from Otsuka *et al* (2022) CC BY 4.0.

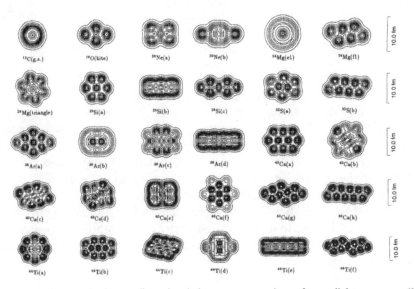

Figure 6.15. α-Particle density in two-dimensional cluster representations of some light even–even ($N = Z$) nuclei. Reprinted from Zhang and Rae (1993) Copyright (1993). With permission from Elsevier.

Following from figure 6.13(c), other even–even ($N = Z$) nuclei may be considered in terms of the α-clustering approach. Figure 6.15 shows the α-particle density for a two-dimensional α-cluster model for even–even ($N = Z$) nuclei from ^{12}C to ^{44}Ti (Zhang and Rae (1993)). This figure illustrates the concept of shape coexistence, where, for many nuclides different geometric configurations of α-particles are possible. This shape coexistence gives rise to states with different excitation energies.

Ikeda *et al* (1968) have developed the so-called Ikeda diagram as illustrated in figure 6.16. This diagram shows the various cluster structures that can exist in light nuclei. The hierarchy of states illustrated in the figure provides insight into the

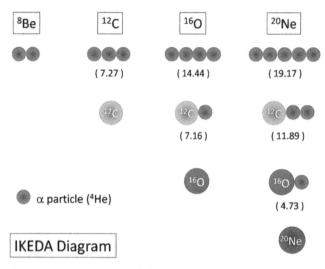

Figure 6.16. Ikeda diagram for even–even ($N = Z$) light nuclei modelled as α-clusters. The threshold energy for dissociating the ground state into the structures as illustrated are shown in parentheses (in MeV). From Ito and Ikeda (2014), copyright IOP Publishing. Reproduced with permission. All rights reserved.

interpretation of excited nuclear states as discussed below. The threshold energies for the formation of various structures, as given in the figure, are based on combined experimental and theoretical studies.

The dissociation of the nuclear ground state into various α-cluster structures, as illustrated in figure 6.16, can be viewed in the context of excited state formation. By analogy with the single-nucleon excitations illustrated in figure 6.3, excited states in the context of the α-cluster model represent the excitation of α-particles within the nucleus. In the case of the dissociation of a single α-particle in the nucleus, the state is often referred to as 4p-4h, indicating 4 particles (4p), that is two neutrons and two protons, in an excited state and four holes (4h) representing four vacant particle states. In the case of the dissociation of two α-particles, the state is correspondingly referred to as an 8p-8h state.

In order to investigate the formation of excited nuclear 4p-4h and 8p-8h, Middleton *et al* (1972) measured the α-particle spectrum from the reaction

$$^{12}\text{C} + {}^{32}\text{S} \rightarrow \alpha + {}^{40}\text{Ca}. \tag{6.15}$$

Figure 6.17 shows the resulting α-particle energy spectrum where the excitation energy given at the top of the figure represents the calculated energy available for excitation of the ^{40}Ca nucleus. The data show two rotational bands that correspond to 4p-4h and 8p-8h excitations. The energy levels in each band result from 0^+, 2^+ and 4^+ rotational states.

While observational evidence has suggested the validity of models involving α-particle clustering of neutrons and protons in nuclei, definitive experimental evidence for the existence of α-clusters is still lacking.

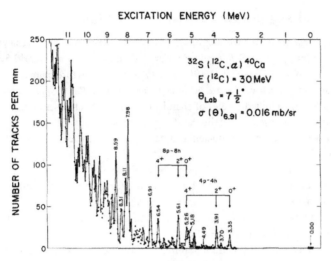

Figure 6.17. α-Particle energy spectrum from the reaction in equation (6.15) for an incident ^{12}C energy of 30 MeV. Reprinted from Middleton *et al* (1972), copyright (1972). With permission from Elsevier.

Problems

6.1. Consider the first eight excited states of a ^{13}C nucleus. Describe possible nucleon states for as many of these states as possible.

6.2.
    (a) For the data in figure 6.11 calculate the moment of inertia for a ^{174}W nucleus in each of the energy levels shown.
    (b) Assuming that a ^{174}W nucleus is approximately spherical, calculate the classical moment of inertia of a sphere of appropriate diameter and mass. Compare with part (a).

6.3. Using the data in figure 6.12, calculate the phonon frequency for the vibrational modes of a ^{120}Te nucleus.

6.4.
    (a) For the following odd–odd nuclei use the shell model to determine the expected nuclear parity and the range of possible values of the nuclear spin: ^{14}N, ^{20}F, ^{24}Na, and ^{26}Al.
    (b) Compare the results of part (a) with the actual measured spins and parities.

6.5. Use the shell model to find the ground state spins and parities of ^{91}Y, ^{91}Zr and ^{91}Nb.

6.6. Use the shell model to describe the spin and parity of the ground state ($3/2^-$) and first three excited states ($5/2^-$, $1/2^-$, and $3/2^-$, respectively, with increasing energy) of ^{59}Ni.

6.7. It is observed that Sb nuclei with odd A have $I^\pi = 5/2^+$ for $A \leqslant 121$ and $I^\pi = 7/2^+$ for $A > 121$. Explain.

References and suggestions for further reading

Enge H A 1966 *Introduction to Nuclear Physics* (Reading, MA: Addison-Wesley)

Henley E M and García 2007 *Subatomic Physics* 3rd edn (Singapore: World Scientific)

Ikeda K, Takigawa N and Horiuchi H 1968 The systematic structure-change into the molecule-like structures in the self-conjugate 4n nuclei *Prog. Theor. Phys. Suppl.* **E68** 464–75

Ito M and Ikeda K 2014 Unified studies of chemical bonding structures and resonant scattering in light neutron-excess systems, ^{10,12}Be *Rept. Prog. Phys.* **77** 096301

Jenkins D 2016 Alpha clustering in nuclei: another form of shape coexistence? *J. Phys. G: Nucl. Part. Phys.* **43** 024003

Krane K S 1988 *Introductory Nuclear Physics* (New York: Wiley)

Middleton R, Garrett J D and Fortune H T 1972 Search for multiparticle-multihole states of ^{40}Ca with the ^{32}S(^{12}C,α) reaction *Phys. Lett.* **39** 339–42

National Nuclear Data Center (NNDC) 2023 *Adopted Levels, Gammas for ^{8}Be* (https://nndc.bnl.gov/nudat3/getdataset.jsp?nucleus=8Be&unc=ND)

Otsuka T, Abe T, Yoshida T, Tsunoda Y, Shimizu N, Itagaki N, Utsuno Y, Vary J, Maris P and Ueno H 2022 α-Clustering in atomic nuclei from first principles with statistical learning and the Hoyle state character *Nat. Commun.* **13** 2234

Williams W S C 1991 *Nuclear and Particle Physics* (Oxford: Oxford University Press)

Zhang J and Rae W D M 1993 Systematics of 2-dimensional α-cluster configurations in 4N nuclei from ^{12}C to ^{44}Ti *Nucl. Phys.* A **564** 252–70

Part III

Nuclear decays and reactions

IOP Publishing

An Introduction to the Physics of Nuclei and Particles
(Second Edition)

Richard A Dunlap

Chapter 7

General properties of decay processes

7.1 Decay rates and lifetimes

In the past few chapters, we have seen that certain nuclei are unstable and decay to more stable configurations of neutrons and protons. In chapters 8–10 detailed accounts of the three most common nuclear decay processes will be given. Decay processes as related to fundamental particles will also be encountered in part IV of this book. In the present chapter some general properties of decays will be considered.

In a collection of identical unstable nuclei, the number of decays per unit time will be proportional to the number of nuclei of that species that are present as a function of time, $N(t)$; that is,

$$-dN(t) = \lambda N(t)dt \tag{7.1}$$

where the proportionality constant λ is the decay constant or decay rate. In general, the nuclear species that decays is called the parent and the nuclear species that is produced from the decay is referred to as the daughter. Equation (7.1) is easily integrated to yield,

$$N(t) = N(0)e^{-\lambda t} \tag{7.2}$$

where $N(0)$ is the number of parent nuclei present at $t = 0$. The half-life of the decay process, $\tau_{1/2}$, is the time required for the initial number of nuclei to decay to one half. That is, substituting $N(t) = N(0)/2$ gives

$$\tau_{1/2} = \frac{\ln 2}{\lambda}. \tag{7.3}$$

We can also define the mean lifetime, τ, which gives the mean time a nucleus survives in its initial state after creation. This is the integral of the decay time weighted by the decay rate,

doi:10.1088/978-0-7503-6094-4ch7

$$\tau = \frac{\int_0^\infty \left(-\frac{dN(t)}{dt}\right) t\,dt}{\int_0^\infty \left(-\frac{dN(t)}{dt}\right) dt}.$$

Substituting equations (7.1) and (7.2) into the above and integrating gives

$$\tau = \frac{1}{\lambda} = \frac{\tau_{1/2}}{\ln 2}. \tag{7.4}$$

This expression shows the relationship between lifetime and half-life as these terms are used in this book.

Several different experimental techniques have been used for the measurement of nuclear lifetimes, depending on the time scales involved. For moderate to long lifetimes, most techniques involve the measurement of the decay rate. This is defined as the number of decays per unit time and is defined from equations (7.1) and (7.2) as

$$\left|\frac{dN(t)}{dt}\right| = \lambda N(0)e^{-\lambda t}. \tag{7.5}$$

For nuclei with very long lifetimes, λ is small (see equation (7.4)) and the exponential in equation (7.5) can be approximated as unity so that,

$$\tau = \frac{1}{\lambda} = \frac{N(0)}{|dN(t)/dt|}.$$

Thus, a knowledge of the number of nuclei present, $N(0)$, and a measure of the decay rate provides the lifetime. This method is suitable for very long-lived nuclides, i.e. millions of years, and can be used for any lifetimes that are long compared to the time scale on which a measurement of $N(0)$ and $dN(t)/dt$ can be made, perhaps a few hours. A practical example is discussed in section 7.3.

Moderate lifetimes can usually be obtained directly from equation (7.5) by successive measurements of the decay rate over a time period comparable to the lifetime. Taking the natural logarithm of both sides of equation (7.5) gives

$$\ln\left|\frac{dN(t)}{dt}\right| = \ln\left|\frac{dN(t)}{dt}\right|_{t=0} - \lambda t \tag{7.6}$$

where we have written

$$\lambda N(0) = \left|\frac{dN(t)}{dt}\right|_{t=0}.$$

Figure 7.1 shows some experimental data for the decay of ^{64}Cu. The slope of the semilog plot yields the lifetime from equation (7.6). The lifetime of this nuclide is ideal for the application of this technique. Shorter lifetimes can certainly be measured, particularly with the use of appropriate computer-controlled instrumentation to make rapid successive measurements of $dN(t)/dt$. Lifetimes shorter than about 10^{-3} s require the use of techniques that, in many cases, depend on the nature

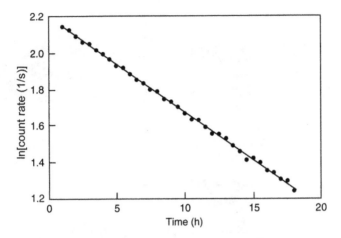

Figure 7.1. Results of a counting experiment to determine the lifetime of ^{64}Cu. The slope of the line gives the lifetime from equation (7.6) as 18.6 h.

of the decay processes and the method by which the parent nuclide is produced. These techniques can be roughly divided into two categories, direct and indirect measurements.

Direct measurements measure the time interval between the formation of the parent state and its decay to the daughter state. One example is the so-called coincidence technique. In many cases the parent state may be populated by the decay from another state and a particle (often a γ-ray) may be emitted during this process. The subsequent decay of the parent to the daughter may emit another particle and the time interval between these two events can be measured electronically.

Various indirect methods have been utilized for measuring lifetimes as short as 10^{-20} s and many are described in the nuclear instrumentation texts given in the bibliography. In general, these methods measure a quantity that can by appropriate theoretical considerations be related to the lifetime. These methods are not discussed in detail in this book, but much of the relevant background is presented in sections 7.2 and 11.5.

In many cases a single decay process between a parent state and a daughter state does not occur. Two more complex situations, multimodal decays and sequential decays, can also be considered. An example of a multimodal decay is illustrated in figure 7.2. Here ^{164}Ho is an odd–odd nucleus similar to that shown in figure 4.5 and may decay by either of two modes; β^- decay to ^{164}Er or electron capture to ^{164}Dy (see chapter 9 for further information about these processes). The former occurs for 53% of the ^{164}Ho nuclei and the latter for the remaining 47%. These percentages are referred to as the branching ratios (or when written as 0.47 and 0.53, as the branching fractions) of the decay and each of these processes is referred to as a decay mode. This decay process can be characterized by two decay constants, λ_1 and λ_2, referred to as partial decay constants, which correspond to the probability of decay by β^- and electron capture, respectively. Analogous to equation (7.1) we can write,

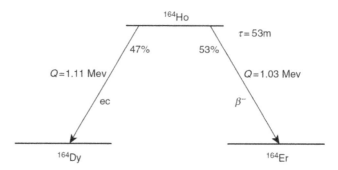

Figure 7.2. Multimodal decay of ^{164}Ho.

$$\frac{dN(t)}{dt} = -[\lambda_1 N(t) + \lambda_2 N(t)]$$

and

$$N(t) = N(0)e^{-(\lambda_1 + \lambda_2)t}$$

where a total decay constant, λ, is defined as the sum,

$$\lambda = \lambda_1 + \lambda_2. \tag{7.7}$$

From equation (7.4) this gives the total lifetime as

$$\tau = \frac{1}{\lambda_1 + \lambda_2}.$$

The branching fractions are given by the ratios of decay constants as

$$f_1 = \frac{\lambda_1}{\lambda} \text{ and } f_2 = \frac{\lambda_2}{\lambda}.$$

These derivations can be generalized for multimodal decays with more than two modes by replacing equation (7.7) with a sum of the partial decay constants for all modes.

Sequential decays occur when the daughter of one decay process is unstable and subsequently decays. We have seen such a situation in figure 4.3. A typical example of a sequential decay is illustrated in figure 7.3. This shows the α-decay of ^{218}Rn to ^{214}Po, which then α-decays to stable ^{210}Pb. Further details of the α-decay process will be discussed in chapter 8. The lifetimes for these two decays as given in the figure are related to corresponding decay constants, λ_1 and λ_2. The relevant differential equations for this process can be written as follows: the number of parent nuclei, N_1, decreases as the parent decays:

$$dN_1(t) = -\lambda_1 N_1(t)dt. \tag{7.8}$$

The change in the number of daughter nuclei can be written as the sum of the new nuclei formed by the decay of the parent and those that are lost due to the decay of the daughter,

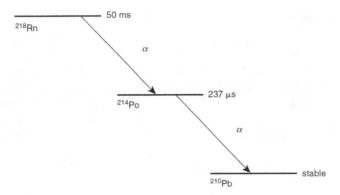

Figure 7.3. The sequential decay ^{218}Rn $\to$ ^{214}Po $\to$ ^{210}Pb.

$$dN_2(t) = \lambda_1 N_1(t)dt - \lambda_2 N_2(t)dt. \tag{7.9}$$

The solution to equation (7.8) can be found by direct integration:

$$N_1(t) = N_1(0)e^{-\lambda_1 t}. \tag{7.10}$$

Using this result, the solution to the first order differential equation (7.9) with the boundary condition that $N_2(0) = 0$ is found to be

$$N_2(t) = N_1(0)\frac{\lambda_1}{\lambda_2 - \lambda_1}(e^{-\lambda_1 t} - e^{-\lambda_2 t}). \tag{7.11}$$

The decay rates can be obtained from equations (7.10) and (7.11) by differentiation. The development given above can be extended (with considerable increase in mathematical complexity) to systems involving a greater number of sequential decays.

7.2 Quantum mechanical considerations

In chapter 5 we discussed the application of the Schrödinger equation to the description of nuclear states in terms of the spatial component of the nuclear wave function, $\Psi(r, t)$. If we consider the application of these ideas to the description of nuclear decay processes, it is obvious that the time development of the wave function is important. In general, the time dependent wave function can be expressed in terms of the stationary states, $\psi(r)$ as:

$$\Psi(r, t) = \psi(r)e^{-iE_0 t/\hbar} \tag{7.12}$$

where E_0 is the energy of the state. If the state is unstable then we must also consider the time dependence predicted by the radioactive decay law. Since the probability of finding the nucleus is given by the modulus square of the wave function, equation (7.2) requires that this probability decay with time as

$$|\Psi(r, t)|^2 = |\Psi(r, 0)|^2 e^{-\lambda t}.$$

This expression suggests that equation (7.12) can be modified to give

$$\Psi(r, t) = \psi(r)e^{-iE_0t/\hbar - \lambda t/2}. \tag{7.13}$$

Equation (7.13) implies that a time dependent state (one that is not absolutely stable) is not represented by a single value of the energy, E_0, but may be described by a distribution of energies $P(E)$, with a mean value E_0. To examine this energy dependent distribution, the time dependent form of the wave function in equation (7.13) is Fourier transformed to give

$$P(E) = (\text{constant}) \cdot \int_{-\infty}^{+\infty} e^{-iE_0t/\hbar - \lambda t/2} e^{iEt/\hbar} dt$$

where the constant is required to properly normalize $P(E)$. If the decay of the initial state begins at $t = 0$, then the lower limit of the integration can be set to zero. This integration yields,

$$P(E) = \frac{(\text{constant})}{i(E_0 - E) + \frac{\hbar\lambda}{2}}.$$

Normalization of $P(E)$ in the above expressions gives,

$$P(E) = \left(\frac{i}{2\pi}\right) \frac{1}{(E_0 - E) + \frac{i\hbar\lambda}{2}}.$$

The probability of finding the nucleus with a particular value of the energy, E, is given by the modulus of $P(E)$ squared,

$$|P(E)|^2 = \left(\frac{1}{4\pi^2}\right) \frac{1}{(E_0 - E)^2 + \frac{\hbar^2\lambda^2}{4}}. \tag{7.14}$$

This function is a Lorentzian, as illustrated in figure 7.4 and a simple inspection of equation (7.14) shows that this curve has a full width at half maximum, Γ, of

$$\Gamma = \hbar\lambda. \tag{7.15}$$

This equation gives the width of the energy distribution of an unstable state, $\Delta E \sim \Gamma$, and can be written in the more common form of the Heisenberg uncertainty principle as

$$\Delta E \Delta t = \hbar \tag{7.16}$$

where the uncertainty in time, Δt, is the lifetime, τ, of the unstable state. These concepts will become important in later chapters. However, for the present we should emphasize that the widths of nuclear state energy distributions are really very small. If we consider lifetimes greater than 10^{-15} s then the corresponding widths given by equation (7.15) are less than 10^{-6} MeV. As decay energies are commonly in the range of 10^{-2} to 10^{-1} MeV or greater, we can see that for such lifetimes the states are still very well defined, and it is suitable for most purposes to refer to the transitions as monoenergetic.

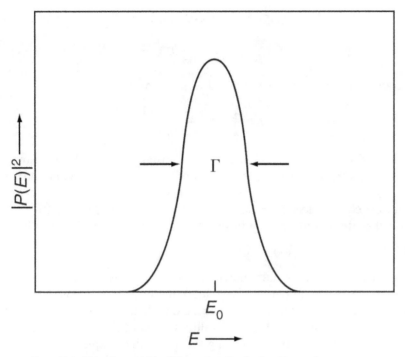

Figure 7.4. Heisenberg width of the energy distribution from a decay process.

7.3 Radioactive dating

An important application that makes use of the concepts described above is radioactive dating. This is a very useful technique for determining the age of organic systems or mineralogical samples. In the former case, ages of 10^4–10^5 years can be determined using ^{14}C dating methods (lifetime = 8270 years). In the latter case, various nuclides can be used for measuring ages up to a few billion years. Here we consider a simple example. A sample is produced at time $t = 0$, which contains a number of parent nuclei that decay to daughter nuclei with a known mean lifetime $\tau = 1/\lambda$. We make the following assumptions:

1. The sample contains no daughter nuclei at $t = 0$;
2. The daughter nuclei are not produced by any processes other than the decay of the parent;
3. The daughter nuclei are stable.

Thus, we can write the number of parent nuclei, $N_1(t)$ and daughter nuclei $N_2(t)$ at time t (the present) as

$$N_1(t) + N_2(t) = N_1(0). \tag{7.17}$$

The radioactive decay law gives the time dependence of the number of parent nuclei as

$$N_1(t) = N_1(0)e^{-\lambda t}. \tag{7.18}$$

These equations can easily be combined to give t as

$$t = \frac{1}{\lambda} \ln\left[1 + \frac{N_2(t)}{N_1(t)}\right].$$

Since the ratio $N_2(t)/N_1(t)$ and the decay constant, λ, can be measured in the laboratory the time between the formation of the sample and the present, t, can be calculated.

Unfortunately, in most cases some quantity of the daughter nuclei may also be present in the sample at the time of formation. Thus, equation (7.17) must be written as

$$N_1(t) + N_2(t) = N_1(0) + N_2(0). \tag{7.19}$$

This introduces an additional unknown into the equations above and does not allow for a direct determination of t from laboratory measurements. In many cases however, an additional, stable isotope of the daughter nuclide is also present in the sample at the time of formation. If additional quantities of this nuclide are not produced by any decay processes during the life of the sample, then this provides a means for determining the initial quantity of daughter nuclei in the sample. We refer to the number of nuclei of this stable isotope as N_s, and since this remains constant in time we can write,

$$N_s(t) = N_s(0)$$

and equation (7.19) can be written as

$$\frac{N_1(t) + N_2(t)}{N_s(t)} = \frac{N_1(0) + N_2(0)}{N_s(0)}.$$

This can be combined with equation (7.18) for $N_1(0)$ to give,

$$\frac{N_2(t)}{N_s(t)} = \frac{N_1(t)}{N_s(t)}[e^{-\lambda t} - 1] + \frac{N_2(0)}{N_s(0)}. \tag{7.20}$$

$N_2(t)/N_s(t)$ and $N_1(t)/N_s(t)$ can be measured in the laboratory. Assuming that samples with a common origin should have the same value of $N_2(0)/N_s(0)$ (which is generally believed to be a good assumption), then equation (7.20) allows for the calculation of the age, t.

As an example of the above, we consider the β^- decay ^{87}Rb $\rightarrow$ ^{87}Sr, which has a lifetime of 6.9×10^{10} years. Another stable isotope of Sr, ^{86}Sr, also exists. So, in the above equations $N_1(t)$ is the number of ^{87}Rb nuclei, $N_2(t)$ is the number of ^{87}Sr nuclei, and $N_s(t)$ is the number of ^{86}Sr nuclei as measured at the present time. Thus, plotting ^{87}Sr/^{86}Sr as a function of ^{87}Rb/^{86}Sr for various samples with a common origin will yield a straight line with a slope of $(e^{\lambda t}-1)$ from which the age, t, can be determined.

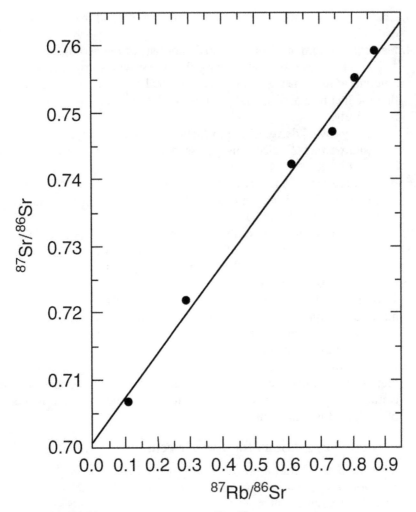

Figure 7.5. The ratio $^{87}Sr/^{86}Sr$ plotted as a function of $^{87}Rb/^{86}Sr$ for some meteorite samples. Data are from Wetherill (1975).

Figure 7.5 shows typical $^{87}Sr/^{86}Sr$ and $^{87}Rb/^{86}Sr$ ratios for some meteorite samples that are believed to be of common origin. These values show a straight line for $^{87}Sr/^{86}Sr$ as a function of $^{87}Rb/^{86}Sr$ where the slope yields an age for the meteorites of 4.4×10^9 years. This value is consistent with ideas concerning the origin of meteorites and other estimates of the age of the solar system.

Problems

7.1. Find a suitable reference and learn about ^{14}C dating methods. Write a two-to-three-page description of this method.

7.2. The human body contains about 20% carbon. Calculate the activity (in Curies) from ^{14}C for an average person.

7.3. Calculate the number of electrons emitted per second from a 1 gram sample of ^{137}Cs.

7.4. Natural uranium, as found on earth, consists of two isotopes in the ratio of ^{235}U/^{238}U $= 7.3 \times 10^{-3}$. Assuming that these two isotopes existed in equal amounts at the time the earth was formed, calculate the age of the earth. Note that the lifetimes of ^{235}U and ^{238}U are 1.03×10^9 years and 6.49×10^9 years, respectively.

7.5. ^{40}K decays by β^- decay with a branching ratio of 89%. One gram of natural potassium emits 27 electrons per second from this decay. Calculate the lifetime of ^{40}K.

7.6. Three radioactive sources each have an activity of 1 mCi at $t = 0$ and lifetimes of 1.0 m, 1.0 h, and 1.0 d. Calculate the decay rate for each source at $t = 1$ s, 1 m, 1 h, and 1 d.

7.7. Naturally occurring vanadium contains 0.25% ^{50}V. This nuclide decays by β^- decay with an estimated lifetime of 3×10^{16} y. Calculate the number of electrons emitted per second from a 1 g sample of natural vanadium as a result of the presence of this isotope.

7.8. Two radioactive nuclides, A and B, are present at $t = 0$ with the number of nuclei of A equal to twice the number of nuclei of B. The lifetimes of the two nuclides are τ for A and 3τ for B. Calculate the time (in units of τ) at which the number of nuclei of A and B are equal.

7.9. Consider the sequential decay where nuclide A decays to nuclide B with a lifetime τ_1, and nuclide B decays to nuclide C with a lifetime τ_2. If a sample contains only nuclide A at $t = 0$, calculate the time at which the decay rate of nuclide B is a maximum.

References and suggestions for further reading

Krane K S 1988 *Introductory Nuclear Physics* (New York: Wiley)

Segrè E 1977 *Nuclei and Particles—An Introduction to Nuclear and Subnuclear Physics* 2nd edn (Reading, MA: Benjamin/Cummings)

Wetherill G W 1975 Radiometric chronology of the early solar system *Ann. Rev. Nucl. Sci.* **25** 283–328

An Introduction to the Physics of Nuclei and Particles
(Second Edition)

Richard A Dunlap

Chapter 8

Alpha decay

8.1 Energetics of alpha decay

Alpha decay (α-decay) is the spontaneous emission of an α-particle, that is a ^{4}He nucleus or a bound system of two neutrons and two protons. The process can be described as

$$_Z^A X^N \rightarrow _{Z-2}^{A-4} Y^{N-2} + \alpha \tag{8.1}$$

and is known to occur in a few very light nuclides and many heavy nuclides. The energy release, Q, during α-emission can be calculated for the process in equation (8.1) as

$$Q = \left[m_N\left(_Z^A X^N\right) - m_N\left(_{Z-2}^{A-4} Y^{N-2}\right) - m_\alpha \right] c^2 \tag{8.2}$$

where the subscript N on the masses refers to nuclear masses. Since, in most cases, it is convenient to deal with measured atomic masses, equation (8.2) can be written as

$$Q = \left[m\left(_Z^A X^N\right) - m\left(_{Z-2}^{A-4} Y^{N-2}\right) - m(^4\text{He}) \right] c^2 \tag{8.3}$$

where the electron masses are properly accounted for, and the electronic binding energy has been ignored. It is assumed that the transitions are between nuclear ground states. Here it is important to realize that $m(^4\text{He})$ is the mass of a neutral ^{4}He atom and not the α-particle mass. If Q is negative, then the process is endothermic and cannot occur spontaneously. If Q is greater than zero, then the process is exothermic and can occur spontaneously (at least from an energetic standpoint). In this case, the excess energy is given up to the α-particle and the daughter nucleus in the form of kinetic energy. Because the number and identity of the nucleons do not

doi:10.1088/978-0-7503-6094-4ch8

change during α-decay, the nuclear binding energies can be directly substituted for the atomic masses in equation (8.3) (with an appropriate change of sign) as

$$Q = -\left[B\left({}^{A}_{Z}X^{N}\right) - B\left({}^{A-4}_{Z-2}Y^{N-2}\right) - B({}^{4}\text{He}) \right]$$

where the binding energy of ^{4}He is known to be 28.3 MeV. Alpha emission is energetically favorable in many instances while emission of other light nuclei from a heavy nucleus is rather unlikely. The reason for this is that the binding energy of the ^{4}He is anomalously large (since it is a doubly magic nucleus). Table 8.1 shows the Q for emission of various light nuclei from a ^{235}U nucleus.

The value of Q for an α-decay process as given in equation (8.3) can be calculated on the basis of the measured atomic masses. These results are shown as a function of A in figure 8.1. The following general features are seen in these data:

Table 8.1. Energy, Q, associated with the emission of various particles from a ^{235}U nucleus.

Emitted particle	Q (MeV)
n	−5.30
p	−6.70
^{2}H	−9.71
^{3}H	−9.97
^{3}He	−9.46
^{4}He	+4.68
^{6}Li	−3.85
^{7}Li	−2.88
^{7}Be	−3.79

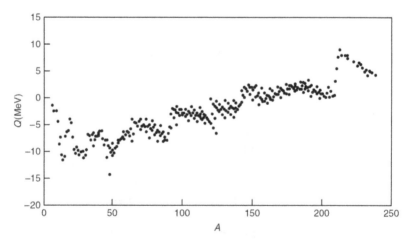

Figure 8.1. α-Decay energies as a function of A as determined from measured atomic masses. Data are shown for β-stable nuclides.

1. There is a crossing from negative Q to positive Q at around $A = 150$.
2. Local minima in the value of Q occur for some values of A.

These observations can be interpreted in the following terms. For A less than about 150, the process is endothermic, and α-decay does not occur; for A greater than about 150 the process becomes exothermic and can, in principle, occur with a Q value that generally increases with increasing A. The minima near particular values of A are the result of shell effects and correspond directly to the features in the binding energy as seen in figure 5.3.

In general, experimental information about α-decay includes a measurement of the kinetic energy, T_α, of the α-particles that are released during the decay process. It is important to understand the relationship between this quantity and the value of Q discussed above. The energy that is liberated during the decay goes into kinetic energy that is distributed between the α-particle and the daughter nucleus. A simple consideration of conservation of energy and momentum shows that,

$$T_\alpha = \frac{Q}{1 + \frac{m_\alpha}{m_D}}$$

where m_D is the mass of the daughter nucleus. For a typical α-decay process involving a heavy nucleus the daughter recoil energy accounts for about 2% of the total energy and the α-particle acquires about 98% of the energy.

The Geiger-Nuttall rule states that there is a dramatic decrease in the α-decay lifetime with increasing decay energy. An example is illustrated in figure 8.2. For even–even nuclei with a constant value of Z (for example, the various isotopes of Th) there is a smooth relationship between Q and τ as shown in the figure. The figure also shows that a similar relationship exists for a different value of Z, although data for

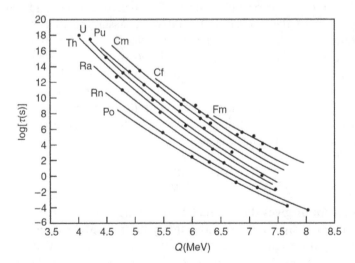

Figure 8.2. Geiger–Nuttall relationship between the α-decay lifetime and the decay energy for some even Z nuclei. Each line represents data for a different value of Z as indicated by the element name.

different Z do not fall on the same line. even–odd, odd–even, and odd–odd nuclei show the same general features, although for reasons described below there is some scatter introduced in the data if they are plotted on the same graph as the even–even nuclei. An interesting aspect of this figure is the range of values that are shown. For ^{232}Th with $Q = 4.08$ MeV the lifetime is 6×10^{17} seconds and for ^{218}Th with $Q = 9.85$ MeV the lifetime is 1.4×10^{-7} s. That is, a factor of about 2 in energy corresponds to a factor of 10^{24} in lifetime. This behavior explains the existence of many stable nuclei with A greater than 150 that have positive values of Q. For a value of Q less than about 4 MeV the lifetime becomes sufficiently long that the decay is, in practice, not observed. A description of the behavior illustrated by the Geiger–Nuttall relationship is one of the principal goals of the theory of α-decay presented below.

8.2 Theory of alpha decay

The basic theory of α-decay was developed in 1928 by George Gamow (1904–1968) and considers the probability that two neutrons and two protons will become bound together within a nucleus, thereby creating an α-particle and that this particle will then escape from the nucleus. Some ideas concerning the formation of α-particles within a nucleus have been presented in section 6.7. The lifetime for α-decay will then be simply given in terms of the time scale for α-particle formation within the nucleus, τ_0, and the probability that the α-particle having been formed will escape from the nucleus, P, as

$$\tau = \frac{\tau_0}{P}. \tag{8.4}$$

We will begin with a consideration of the escape probability, while the time τ_0 will be discussed briefly at the end of this section.

The behavior of the α-particle within the nucleus can be understood from a careful inspection of the nuclear potential well. Assuming that an α-particle has formed within the nucleus we must consider the behavior of this particle in the potential well of the daughter nucleus. This potential as illustrated in figure 8.3 consists of the more-or-less spherical square well dominated by the strong interaction when the α-particle is inside the nucleus; that is, for $r < a$ where $a = R_D + R_\alpha$ (R_D is the radius of the daughter nucleus and R_α is the radius of the α-particle). Outside of the nucleus the potential is described by the Coulomb interaction between the α-particle and the daughter nucleus:

$$V(r) = \frac{2Z_D e^2}{4\pi\varepsilon_0 r} \tag{8.5}$$

where the numerator gives the product of the charge of the α-particle $+2e$ and the daughter nucleus $+Z_D e$. The potential in equation (8.5) represents the Coulomb barrier. If the Q for α-emission is positive, then the α-particle will form inside the nucleus with an energy Q above zero potential as shown in the figure. Classically speaking, the α-particle can escape only if Q is greater than the maximum height of the Coulomb barrier, E_{max}. For a typical heavy nucleus, E_{max} is about 30 MeV.

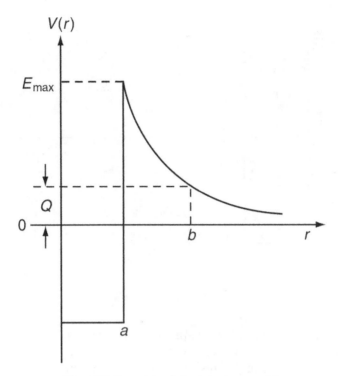

Figure 8.3. Potential well for the α-decay model.

A comparison with the α-decay energies shown in figure 8.1 demonstrates that this is never the case. The problem of α-decay then becomes a quantum mechanical problem of Coulomb barrier penetration by the α-particle.

Inside the nucleus, the potential can be approximated as a constant and the solutions for R, the radial part of the Schrödinger equation, follow along the lines described in chapter 5. Outside the nucleus, the radial part of the Schrödinger equation can be written as

$$-\frac{\hbar^2}{2m}\left[\frac{d^2R}{dr^2} + \frac{2}{r}\frac{dR}{dr}\right] + \left[\frac{2Z_\mathrm{D}e^2}{4\pi\varepsilon_0 r} + \frac{l(l+1)\hbar^2}{2mr^2}\right]R = QR \qquad (8.6)$$

where the reduced mass has been defined in terms of the α-particle mass and the daughter nucleus mass, as

$$m = \frac{m_\alpha m_\mathrm{D}}{m_\alpha + m_\mathrm{D}}.$$

We can define two regions outside the nucleus: $a < r < b$ and $r > b$ where b is defined as the radius at which $V(r) = Q$ or

$$b = \frac{2Z_\mathrm{D}e^2}{4\pi\varepsilon_0 Q}.$$

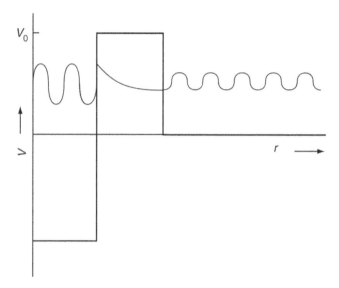

Figure 8.4. Tunneling of an α-particle wave function through a square barrier.

The radius a can be determined from equation (3.14). Solving equation (8.6) in these two regions is not simple. However, the inspection of a very simplified potential as shown in figure 8.4 is helpful. This shows the penetration of a square barrier of height V_0 by the α-particle wave function. We also begin with a consideration of the simple $l = 0$ case. Under these conditions the solutions to equation (8.6) are

$$R = \frac{1}{r} \exp\left[\pm\sqrt{\frac{2mr^2}{\hbar^2}(V_0 - Q)} \right] \tag{8.7}$$

giving an exponentially damped solution within the barrier and an oscillatory solution outside the barrier. The barrier can be approximated by a series of square barriers of diminishing height as described by equation (8.5). This is the so-called WKB (Wentzel–Kramers–Brillouin) approximation and is discussed in detail in most introductory quantum mechanics texts. In the limiting case this series is written as an integral and equation (8.7) becomes

$$R(r) = \frac{1}{r} \exp\left(\pm \int F(r')dr' \right) \tag{8.8}$$

where we have defined the function

$$F(r') = \sqrt{\frac{2m}{\hbar^2}\left(\frac{2Z_D e^2}{4\pi\varepsilon_0 r'} \right) - Q}.$$

Equation (8.8) can now be used to obtain the solution for $R(r)$ in the regions outside the nucleus. In principle we should obtain solutions for $r < a$, $a < r < b$, and $r > b$ and match the appropriate boundary conditions. However, for the purpose of calculating the tunneling probability it is sufficient to consider the solution inside the barrier. For $a < r < b$ we write the solution as the linear superposition of the positive and negative forms of equation (8.8) as

$$R(r) = \frac{A}{r} \exp\left(+\int_r^b F(r')dr'\right) + \frac{B}{r} \exp\left(-\int_r^b F(r')dr'\right). \tag{8.9}$$

The solution of equation (8.9) is dominated by the term with the coefficient A, and this allows us to ignore the second term on the right in this equation.

The tunneling probability, P, that is the probability that an α-particle that has formed near the surface of the nucleus (that is, at radius a) will escape from the nucleus (that is, it will get to radius b), is given as the ratio of the probability that the particle will exist at b to the probability that it will exist at a. These probabilities are given as the square of the radial part of the wave function multiplied by the surface area. That is,

$$P = \frac{4\pi b^2 \, |R(b)|^2}{4\pi a^2 \, |R(a)|^2} = e^{-G}$$

where the quantity of G is defined from the previous equation to be

$$G = 2 \int_a^b F(r')dr' = 2\sqrt{\frac{2mQ}{\hbar^2}} \int_a^b \left(\frac{b}{r'} - 1\right)^{1/2} dr'. \tag{8.10}$$

Integrating this expression yields

$$G = \frac{4Z_D e^2}{4\pi\varepsilon_0} \sqrt{\frac{2m}{\hbar^2 Q}} \left[\cos^{-1}\sqrt{\frac{a}{b}} - \sqrt{\frac{a}{b}\left(1 - \frac{a}{b}\right)} \right]. \tag{8.11}$$

The above derivation allows for the calculation of the barrier penetration probability in terms of known quantities.

It is now necessary to consider the time scale, τ_0, in equation (8.4). There are two factors that contribute to τ_0. The first depends on the details of the processes that cause the formation of the α-particle within the nucleus. Our knowledge of this is limited. The other factor may be viewed more or less classically in terms of an α-particle that is bouncing off the walls of the nuclear square well. The time scale for this process is related to the α-particle's velocity (which is related to the Q for the decay) and the radius of the nucleus. However, in order to consider the validity of the above description of α-decay we must make some assumptions concerning τ_0. We have no reason to expect that τ_0 will be substantially different in nuclei with similar mass and similar nucleon configurations. We will therefore consider the case where τ_0 is a constant and write,

$$\tau = \frac{\tau_0}{e^{-G}}. \tag{8.12}$$

It is most appropriate to consider the α-decay of even–even nuclei. The results of applying the tunneling model to the α-decay of a number of even–even nuclei are

Table 8.2. Measured and calculated α-decay lifetimes for some heavy nuclei.

Parent	Daughter	Q(MeV)	τ_{meas}(s)	τ_{calc} (s)
^{238}U	^{234}Th	4.27	2.0×10^{17}	3.0×10^{17}
^{234}U	^{230}Th	4.86	1.1×10^{13}	1.0×10^{13}
^{230}Th	^{226}Ra	4.77	3.5×10^{12}	3.5×10^{12}
^{226}Ra	^{222}Rn	4.87	7.3×10^{10}	6.6×10^{10}
^{222}Rn	^{218}Po	5.59	4.8×10^{5}	3.8×10^{5}
^{218}Po	^{214}Pb	6.11	2.6×10^{2}	1.4×10^{2}
^{214}Po	^{210}Pb	7.84	2.3×10^{-4}	1.0×10^{-4}
^{210}Po	^{206}Pb	5.41	1.7×10^{7}	5.2×10^{5}

given in table 8.2. The calculated lifetimes have been normalized to the lifetime of ^{230}Th and this gives the value of τ_0 from equation (8.12) as 6.3×10^{-23} s. With the exception of a slight discrepancy in the lifetime for ^{210}Po, the model results are amazingly consistent with experimentally measured values. The results presented in this section in the form of equations (8.11) and (8.12) provide a quantitative basis for the Geiger–Nutall rule. In cases where even–even nuclei decay by α-emission, the decay is almost always preferentially to the ground state of the daughter. This is reasonable, as the decay constant is such a sensitive function of the decay energy. If a nucleus were to decay to an excited state of the daughter, then the decay energy would be decreased proportionately, yielding a smaller branching ratio than a transition to the ground state. This is discussed further below.

The question of α-decay in even–odd, odd–even, and odd–odd nuclei is quite interesting. In general, lifetimes for these nuclei are as much as three orders of magnitude longer than those of even–even nuclei with similar mass and Q. The reason for this has to do with pairing effects and the formation of the α-particle within the nucleus. It appears that paired neutrons and protons, as exist in even–even nuclei, fairly readily form a bound α-particle state within the nucleus (see section 6.7). However, unpaired nucleons do not readily participate in the formation of an α-particle. As a result, the α-particle is most likely formed from lower lying paired nucleons leaving the unpaired nucleon in a higher energy state. This results in a daughter nucleus that is preferentially left in an excited state, and this reduces the decay energy and increases the lifetime.

8.3 Angular momentum considerations

In the discussion above it has been assumed that $l = 0$. Certainly, conservation laws impose some restrictions here, and α-decay processes are allowed only if total angular momentum and parity are conserved. This means that the angular momenta and parities of the daughter nucleus and the α-particle must combine to yield the angular momentum and parity of the parent nucleus. Since the orientation of the angular momenta of the various nuclei involved in the decay are not specified, we

Table 8.3. Decay constants for different values of α-particle angular momentum, λ_l, relative to the decay constant for $l = 0$, λ_0.

l	λ_l/λ_0
0	1.0
1	0.7
2	0.37
3	0.137
4	0.037
5	0.0071
6	0.0011

can merely put limits on the orbital angular momentum of the α-particle, l, in terms of the values of I for the parent (P) and daughter (D) nuclei. That is, conservation of angular momentum requires,

$$|I_D - I_P| \leqslant l \leqslant I_D + I_P.$$

As an example, we consider the α-decay ^{237}Np(5/2$^+$) $\rightarrow$ ^{233}Pa(3/2$^-$) + α. Conservation of angular momentum allows the α-particle to have $l = 1, 2, 3$, or 4 although conservation of parity requires the α-particle to have odd parity, constraining l to be 1 or 3.

For nonzero l, an inspection of equation (8.6) shows immediately that $l \neq 0$ has the effect of increasing the potential and constitutes a contribution to the overall potential barrier that the α-particle must tunnel through. This is the angular momentum barrier, and it has the effect of reducing, in some cases substantially, the value of the α-decay constant. An example of the decrease in the decay constant for α-decays producing α-particles with different l is given in table 8.3.

The rotational levels of heavy nuclei are an ideal example of this kind of behavior. Figure 8.5 shows the branching ratios for the α-decay of ^{244}Cm to the ground state and the first few excited states of ^{240}Pu. This figure shows that the α-decay branching ratio decreases consistently and substantially with increasing daughter excited state energy. This decrease is due to the combined effects of an increase in the angular momentum barrier and a decrease in the α-particle energy. In even–even nuclei the transition to the ground state is preferred, as the 0$^+$ to 0$^+$ transition eliminates the angular momentum term in the Schrödinger equation and the transition energy is maximized. In odd A nuclei, the situation is sometimes different, as shown in figure 8.6 for the decay of ^{243}Am. Here the principal decay mode is to the 0.075 MeV state of the ^{239}Np daughter, since the elimination of the angular momentum barrier for the (5/2$^-$) to (5/2$^-$) transition more than compensates for the 0.075 MeV decrease in transition energy. Note that the ground state transition is not preferred because the change in parity does not allow the 0$^+$ state for the α-particle.

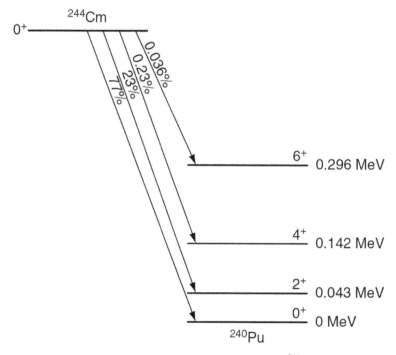

Figure 8.5. Branching ratios for the α-decay of ^{244}Cm.

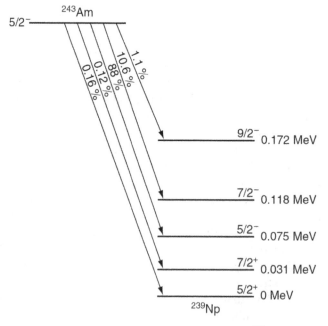

Figure 8.6. Branching ratios for the α-decay of ^{243}Am.

Problems

Discuss the relationship between Q and lifetime for each of these cases.

8.1. Tabulate all allowed α-particle spin and parity states for all transitions shown in figures 8.5 and 8.6.

8.2. Using known atomic masses calculate the value of Q for the α-decay of the following nuclides. Assume transitions between ground states.
   (a) ^{208}Po
   (b) ^{222}Ra
   (c) ^{240}Pu
   (d) ^{252}Fm.

8.3. From the *Table of Isotopes* (or another suitable reference) locate three examples of each of the following:
   (a) the α-decay of an even–even nucleus;
   (b) the α-decay of an even–odd nucleus;
   (c) the α-decay of an odd–even nucleus;
   (d) the α-decay of an odd–odd nucleus.

8.4. The lifetime for the α-decay of ^{226}Ra is 7.3×10^{10} s. Use this information to calculate the radius of a ^{222}Rn nucleus.

8.5. Calculate the kinetic energy of the α-particle released during the α-decay of ^{225}Ac.

8.6.
   (a) Consider the α-decay of a highly nonspherical nucleus (that is, one that has a large electric quadrupole moment). Calculate the relative α-decay rates along the semimajor axis and semiminor axis of a nucleus with a semimajor to semiminor radius ratio of 1.5.
   (b) Describe the α-decay radiation pattern that is expected from such a nucleus.

8.7. The α-decay of even–even parent nuclides preferentially populates the ground state of the daughter (rather than an excited state of the daughter). This is not necessarily the case for even–odd or odd–even parents. Explain.

8.8.
   (a) Calculate the Q for the α-decay of ^{241}Cm.
   (b) The branching ratios for the α-decay of ^{241}Cm to the excited states of ^{237}Pu are 13% to the 0.202 MeV state ($5/2^+$), 17% to the 0.156 MeV state ($3/2^+$), and 70% to the 0.145 MeV state ($1/2^+$). There is no decay to the ^{237}Pu ground state. Explain.

8.9. Using the semiempirical mass formula, derive an expression for the Q of an α-decay process. Use this expression to calculate the Q for the α-decay of ^{241}Am and compare with the value based on measured atomic masses.

8.10.
   (a) The ground state of ^8Be decays by splitting into two α-particles. Calculate the Q for this process.

(b) Excited states of ^{12}C are known to decay by splitting into three α-particles. Calculate the threshold energy for this process.

(c) Consider the possibility of the decay of excited states of ^{16}O into four α-particles.

Suggestions for further reading

Hodgson P E, Gadioli E and Gadioli Erba E 1997 *Introductory Nuclear Physics* (Oxford: Oxford)

Krane K S 1988 *Introductory Nuclear Physics* (New York: Wiley)

Williams W S C 1991 *Nuclear and Particle Physics* (Oxford: Oxford University Press)

IOP Publishing

An Introduction to the Physics of Nuclei and Particles
(Second Edition)

Richard A Dunlap

Chapter 9

Beta decay

9.1 Energetics of beta decay

Beta decay (β-decay) as described in chapter 4 represents the conversion of a neutron to a proton or the conversion of a proton to a neutron. The former situation is referred to as negative β-decay (β^- decay) and from a nuclear standpoint is described as

$$^A_Z X^N \rightarrow \, ^A_{Z+1}Y^{N-1} + e^- + \nu_e. \tag{9.1}$$

The conversion of a proton to a neutron represents positive β-decay (β^+ decay) and can be written as a nuclear process as

$$^A_Z X^N \rightarrow \, ^A_{Z-1}Y^{N+1} + e^+ + \bar{\nu}_e. \tag{9.2}$$

In both cases the appearance of the neutrino (or antineutrino) is necessary in order to conserve lepton number. Its existence is also needed to properly explain the distribution of electron (or positron) energies as discussed below. Another process which, in many ways, is analogous to β^+ decay is electron capture. This corresponds to the fundamental process,

$$p + e^- \rightarrow n + \nu_e. \tag{9.3}$$

In this case, a proton in the nucleus interacts with an atomic electron, typically an inner shell s-electron because the wave function will have the greatest probability of overlapping with the nuclear wave function. This represents the nuclear process,

$$^A_Z X^N + e^- \rightarrow \, ^A_{Z-1}Y^{N+1} + \nu_e \tag{9.4}$$

where it is seen that the identity of the parent and daughter nuclei are the same as for β^+ decay.

doi:10.1088/978-0-7503-6094-4ch9　　　　9-1

Here we consider the energetics of these three processes. For β^- decay we write the energy released, Q, in terms of the nuclear masses (subscript N):

$$Q = \left[m_N(_Z^A X^N) - m_N\left(_{Z+1}^A Y^{N-1}\right) - m_e \right] c^2. \tag{9.5}$$

It is assumed that the neutrino mass is negligible and that the transition is between nuclear ground states. The validity of this former assumption is discussed further in chapter 19. If we ignore the electronic binding energy, then equation (9.5) can be written in terms of atomic masses using the relation,

$$m(_Z^A X^N) = m_N(_Z^A X^N) + Z m_e.$$

This gives,

$$Q = \left[m(_Z^A X^N) - m(_{Z+1}^A Y^{N-1}) \right] c^2. \tag{9.6}$$

The omission of the electronic binding energy is justified since it is only the difference in binding energies between the right-hand and left-hand sides of equation (9.1) that is important, and this will be negligibly small.

For β^+ decay, Q can be written in terms of nuclear masses from equation (9.2) as

$$Q = [m_N(_Z^A X^N) - m_N(_{Z-1}^A Y^{N+1}) - m_e]c^2$$

noting that the positron and the electron have the same mass. Converting this expression to atomic masses gives

$$Q = [m(_Z^A X^N) - m(_{Z-1}^A Y^{N+1}) - 2m_e]c^2. \tag{9.7}$$

The energetics of electron capture is given by equation (9.4) as

$$Q = [m_N(_Z^A X^N) + m_e - m_N(_{Z-1}^A Y^{N+1})]c^2 - b_e \tag{9.8}$$

where b_e is the binding energy of the electron that is captured. In addition to the electronic binding energy terms that more or less cancel out in equations such as equation (9.6), equation (9.8) contains a term from the binding energy of the electron that is captured. As this is typically an inner shell (K-shell) electron, then this energy can be significant (many tens of keV) and is often included in the energy calculation for electron capture. In terms of atomic masses, the above becomes,

$$Q = [m(_Z^A X^N) - m(_{Z-1}^A Y^{N+1})]c^2 - b_e. \tag{9.9}$$

When Q for a β-decay is positive, the process is exothermic and can proceed spontaneously. When Q is negative, the process is endothermic and is not energetically favorable. A simple inspection of equations (9.7) and (9.9) shows that electron capture is energetically more favorable than β^+ decay by an additional $2m_e c^2$. An investigation of the nuclear data tables shows that for some β-unstable nuclei electron capture occurs (because Q in equation (9.9) is positive) while β^+ decay does not occur (because Q in equation (9.7) is negative).

In a process such as that shown in equation (9.1), the energy that is given up during the decay process appears as the kinetic energy of the electron and antineutrino and the recoil energy of the daughter nucleus. Because of the large mass associated with the daughter nucleus compared to the electron, its recoil energy is substantially smaller than the precision with which the electron's energy can be measured. Thus, in subsequent discussions in this chapter the nuclear recoil energy will be ignored. Therefore, for the process in equation (9.1), we consider the energy as being distributed between the electron and the antineutrino. ^{64}Cu is an interesting example, as it decays by both β^- decay to ^{64}Zn and by β^+ decay or electron capture to ^{64}Ni. The energy spectrum of electrons emitted by the β^- decay of ^{64}Cu is illustrated in figure 9.1. The plot shows a distribution of electron energies up to slightly less than 0.6 MeV. This maximum energy is referred to as the endpoint energy and, for ground state to ground state transitions, it is the value of Q as obtained from equation (9.6). A simple calculation based on tabulated atomic masses gives the endpoint energy of 0.579 MeV, in agreement with the figure. Decays that produce an electron with the endpoint energy also produce an antineutrino with negligibly small energy. On the other hand, decays that emit an electron with small energy emit an antineutrino with close to the endpoint energy. The measurement shown here indicates that it is much more probable for the electron to have a small amount of energy and the antineutrino to get the majority, than the other way around.

The measured energy spectrum of the positrons emitted from the β^+ decay of ^{64}Cu is shown in figure 9.2. A simple calculation using measured atomic masses in

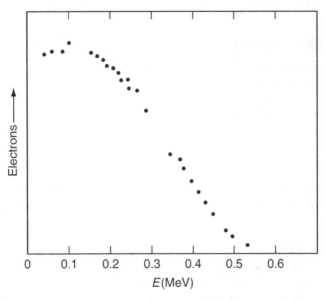

Figure 9.1. Energy spectrum of electrons from the β^- decay of ^{64}Cu. Data have been calculated from results given in Langer *et al* (1949).

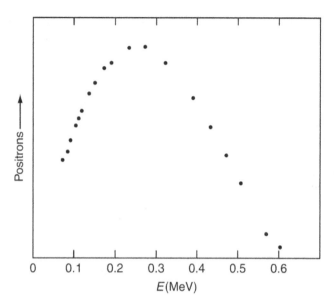

Figure 9.2. Energy spectrum of positrons from the β^+ decay of ^{64}Cu. Data have been calculated from results given in Langer *et al* (1949).

equation (9.7) gives a positron endpoint energy of 0.653 MeV, consistent with the data shown in the figure.

The theoretical model of β-decay as described below has been developed to explain the energy spectrum as shown in figures 9.1 and 9.2 and to explain the observed lifetimes for β-decay processes.

9.2 Fermi theory of beta decay

The basic theory of β-decay was developed in 1933 by Enrico Fermi (1901–54) and is described in this section. The decay constant for a decay process is merely the transition rate from the initial (parent) to the final (daughter) state. Perturbation theory gives the transition rate between states (Fermi's Golden Rule) as

$$\lambda = \frac{2\pi}{\hbar} |H_{if}|^2 \frac{dn}{dE} \tag{9.10}$$

where dn/dE is the density of final states and H_{if} is a matrix element given by the integral of the interaction operator, H,

$$H_{if} = G \int \psi_f^* H \psi_i d^3\vec{r}. \tag{9.11}$$

Here ψ_i and ψ_f are the wave functions of the initial and final states, respectively, and G is a constant representing the strength of the interaction. We will consider β^- decay as described in equation (9.1). The wave function of the initial state is the wave function of the parent nuclear state (P). The wave function of the final state is given by the wave functions of the daughter nuclear state (D), the electron, and the antineutrino, so that equation (9.11) can be written as

$$H_{if} = G \int \psi_D{}^* \psi_e{}^* \psi_{\bar{\nu}_e}{}^* H \psi_p d^3\vec{r}.$$

A similar matrix element could also be constructed for β^+ decay. The electron and antineutrino wave functions are taken to be plane waves (because these are free particles) and, as a matter of convenience, are normalized over the nuclear volume, V, as

$$\psi_e(\vec{r}) = \frac{1}{\sqrt{V}} e^{i\vec{p}\cdot\vec{r}/\hbar} \tag{9.12}$$

and

$$\psi_{\bar{\nu}_e}(\vec{r}) = \frac{1}{\sqrt{V}} e^{i\vec{q}\cdot\vec{r}/\hbar}. \tag{9.13}$$

Here p and q are the momentum of the electron and the antineutrino, respectively. The exponential in equation (9.12) can be expanded as

$$e^{i\vec{p}\cdot\vec{r}/\hbar} = 1 + \frac{i\vec{p}\cdot\vec{r}}{\hbar} + \ldots \tag{9.14}$$

and similarly for the antineutrino. For electrons with typical β^- decay energies, p is of the order of 1 MeV c^{-1} and for r of the order of the nuclear radius we find $pr/\hbar$ $=0.03$. Thus, taking only the leading terms in the expansions for the wave functions in equations (9.12) and (9.13) would seem to be a reasonable approximation and this gives,

$$H_{if} = \frac{GM_{if}}{V} \tag{9.15}$$

where we have defined

$$M_{if} = \int \psi_D^* H \psi_p d^3\vec{r}.$$

It is now necessary to consider the question of the density of final states. We can consider a simple analysis based on a free-particle Schrödinger equation,

$$-\frac{\hbar^2}{2m} \nabla^2 \psi = E\psi.$$

For the simple case of a three-dimensional infinite well in Cartesian coordinates this equation has momentum states that are quantized as

$$p_i = \frac{n_i \pi \hbar}{L} \tag{9.16}$$

where i is a coordinate, x, y, or z, and L is the length of the edge of the well. This means that the distance between allowed states in momentum space is p/L and the density of states is, therefore, L/p along each Cartesian direction. The density in three dimensions will be given by the cube of this quantity. The total number of allowed states with momentum less than a value $p = (p_x^2 + p_y^2 + p_z^2)^{1/2}$ is given by this density of states and the volume of a sphere in momentum space as

$$n_e = \left(\frac{1}{8}\right)\left(\frac{4}{3}\pi p^3\right)\left(\frac{L}{\pi\hbar}\right)^3.$$
(9.17)

The additional factor of 1/8 comes from the fact that only one quadrant in p space needs to be counted, as changing the sign of n_i in equation (9.16) does not yield new independent states. For the electron, the density of states per unit momentum is found by differentiating equation (9.17) to give,

$$dn_e = \frac{4\pi V p^2 \, dp}{(2\pi\hbar)^3}.$$
(9.18)

Correspondingly, the density of states for the antineutrino is,

$$dn_{\bar{\nu}_e} = \frac{4\pi V q^2 dq}{(2\pi\hbar)^3}.$$
(9.19)

In order to determine the transition rate as given by equation (9.10), it is necessary to consider the case where the final state is characterized by the correct electron and antineutrino energies and momenta. Specifically, we can write the antineutrino density of states corresponding to a particular final state energy as $\frac{dn_{\bar{\nu}_e}}{dE_f}$. The partial transition rate into the correct electron momentum state is, therefore,

$$d\lambda = \frac{2\pi}{\hbar} |H_{if}|^2 \frac{dn_{\bar{\nu}_e}}{dE_f} dn_e.$$
(9.20)

In principle, the total transition rate can be obtained by integrating the above, but this is not necessary in order to obtain a result that can be compared with the experimental energy spectrum. Substituting equations (9.15), (9.18), and (9.19) into (9.20) gives

$$d\lambda = \frac{G^2}{2\pi^3\hbar^7} |M_{if}|^2 p^2 q^2 \frac{dq}{dE_f} dp.$$
(9.21)

It is important to observe that the nuclear volume that appeared in the above calculations has cancelled out. The $\frac{dq}{dE_f}$ term on the right-hand side above can be dealt with by writing the total final state energy as

$$E_f = Q + m_e c^2$$
(9.22)

where the Q of the decay is written as the sum of the kinetic energies of the electron and the antineutrino,

$$Q = T_e + qc.$$
(9.23)

Using equation (9.23) in equation (9.22) and keeping T_e constant yields

$$\frac{dq}{dE_f} = \frac{1}{c} \qquad (9.24)$$

and

$$q = \frac{Q - T_e}{c}. \qquad (9.25)$$

Substituting equations (9.24) and (9.25) into equation (9.21) yields

$$\frac{d\lambda}{dp} = \frac{G^2}{2\pi^3\hbar^7c^3} |M_{if}|^2 p^2 (Q - T_e)^2. \qquad (9.26)$$

Experimentally, as shown in figure 9.1, the spectral intensity is measured as a function of energy, in this case the kinetic energy of the electron. Equation (9.26) can be rewritten in terms of energy. In general, the Q associated with β-decay electron is comparable to the electron rest mass energy and it is necessary to include relativistic effects in this calculation. Thus, the electron kinetic energy is expressed as

$$T_e = \sqrt{p^2c^2 + m_e^2c^4} - m_ec^2$$

and the measured spectral intensity is given as the energy dependent decay constant,

$$I(T_e) = \frac{d\lambda}{dT_e} = \frac{G^2}{2\pi^3\hbar^7c^6} |M_{if}|^2 (T_e^2 + 2T_em_ec^2)^{1/2}(Q - T_e)^2(T_e + m_ec^2). \qquad (9.27)$$

In order to make some numerical predictions using this theory it is necessary to have some information concerning the dependence of M_{if} on the electron kinetic energy (or equivalently on p and q). The simplest assumption that we can make is that M_{if} is independent of T_e and it turns out that this is applicable to a large number of β-decay processes. The electron energy spectrum as calculated from equation (9.26) using this approximation is shown in figure 9.3. Thus far we have included no factors that distinguish the electron from the positron, so we expect the same result as shown in the figure for the positron energy spectrum from β+ decay. Figure 9.3 shows somewhat better agreement with the positron spectrum in figure 9.2 than with the electron spectrum in figure 9.1. However, careful inspection shows that there is some discrepancy at low energy in both cases. There is a simple explanation for this behavior. Once produced by the decay of a proton the positively charged positron is repelled by the Coulomb interaction with the daughter nucleus. Recall that the positron, being a lepton, is not influenced by the strong nuclear force. This repulsive interaction literally pushes the positron out of the nucleus and diminishes the low energy portion of the spectrum. The electron, on the other hand, after being created by the decay of a neutron, is held back by the attractive Coulomb force with the

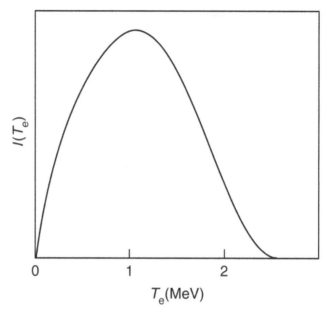

Figure 9.3. Calculated electron energy spectrum for β-decay.

daughter nucleus. This, therefore, augments the low energy portion of the spectrum. These Coulomb effects can be taken into account in a more quantitative way by introducing an additional factor, $F(Z, T_e)$, in equation (9.27) as

$$I(T_e) = \frac{G^2}{2\pi^3 \hbar^7 c^6} |M_{if}|^2 (T_e^2 + 2T_e m_e c^2)^{1/2} (Q - T_e)^2 (T_e + m_e c^2) F(Z_D, T_e). \quad (9.28)$$

$F(Z_D, T_e)$ is called the Coulomb factor or Fermi factor (or function) and is a function of the electron (or positron) kinetic energy and the charge on the daughter nucleus. A nonrelativistic calculation yields

$$F(Z_D, T_e) = \frac{2\pi\eta}{1 - \exp(-2\pi\eta)}$$

where

$$\eta = \pm \frac{Z_D e^2}{4\pi\varepsilon_0 \hbar v}.$$

The electron/positron velocity is given by v and the positive sign is used in this expression for electrons (β$^-$ decay) and the negative sign for positrons (β$^+$ decay). This expression is appropriate at low energies where the differences between figures 9.1 and 9.2 and figure 9.3 are most apparent. Results of a more general calculation that includes relativistic effects can be found in National Bureau of Standards (1952). Inclusion of this factor clearly shows that the low energy portion of the electron spectrum for β^- decay will be augmented while this portion of the positron spectrum for β^+ decay will be diminished.

9.3 Fermi–Kurie plots

The validity of this theory may be tested using a Fermi–Kurie plot. This is sometimes also called a Kurie plot or Fermi plot. The spectral intensity as a function of electron momentum is given from equation (9.26) combined with the Coulomb factor as

$$I(p) = \frac{G^2}{2\pi^3\hbar^7c^6} |M_{if}|^2 p^2 (Q - T_e)^2 F(Z_D, T_e). \tag{9.29}$$

Plotting $[I(p)/p^2F]^{1/2}$ as a function of T_e will yield a straight line if the assumptions of the theory are correct. The intercept on the horizontal axis will immediately give the value of Q. In cases where the decay is to an excited state of the daughter nucleus then it is essential to properly include this in a consideration of the electron energies. An example of a Fermi–Kurie plot for the β^- decay of ^{64}Cu is shown in figure 9.4. The linear relationship shown in this plot is evidence that the interpretation of the energy dependence of the right-hand side of equation (9.28) was correct. This energy dependence was based on the elimination of higher order terms in the expansion of the electron and antineutrino wave functions in equation (9.14). Beta transitions that are properly described by the linear relationship shown in the figure are said to be *allowed transitions*. This terminology is somewhat ambiguous as transitions that do not follow this linear behavior can still occur. This is discussed further in section 9.4.

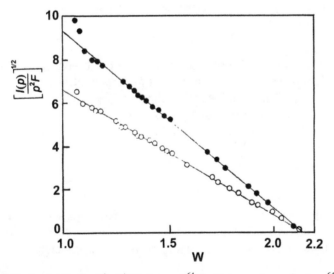

Figure 9.4. Fermi–Kurie plot for the $0^+\to0^+$ β^- decay of ^{64}Cu. Data are shown for two ^{64}Cu sources with different activities. Source thickness effects are seen by the nonlinearity at low energy. Note: the horizontal axis gives the total energy in units of the rest mass energy, $W = (T_e + m_ec^2)/(m_ec^2)$. Reprinted with permission from Langer *et al* (1949), copyright (1949) by the American Physical Society.

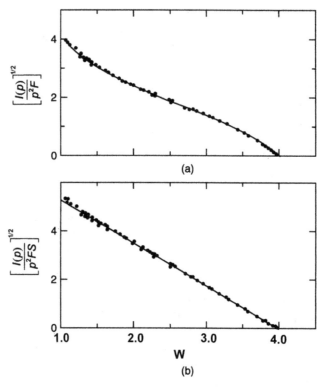

Figure 9.5. (a) Fermi–Kurie plot for the β^- decay of ^{91}Y and (b) corrected Fermi–Kurie plot for the β^- decay of ^{91}Y using the shape factor for a first forbidden decay. Note: the horizontal axis gives the total energy in units of the rest mass energy, $W = (T_e + m_e c^2)/(m_e c^2)$. Reprinted with permission from Langer and Price (1949). Copyright 1949 by the American Physical Society.

An example of a decay that is not described by the Fermi–Kurie plot is illustrated in figure 9.5(a). This decay is referred to as *first forbidden* (although it is obviously not forbidden entirely). The nonlinearities that are clearly seen are the result of additional energy dependence coming into equation (9.28). This nonlinearity can be accounted for by the inclusion of an additional term, $S(p, q)$, sometimes called a shape factor, in equation (9.29). In the case of the first forbidden decay shown in figure 9.5(a), the shape factor can be expressed as

$$S(p, q) = p^2 + q^2$$

and comes from the next highest order term in expansions such as equation (9.14). Including this in equation (9.29) and plotting $[I(p)/p^2 FS]^{1/2}$ as a function of energy as illustrated in figure 9.5(b) shows the expected linear dependence.

9.4 Allowed and forbidden transitions

The question of allowed and forbidden transitions can be described in more detail by considering angular momentum conservation. The elimination of higher order terms from equation (9.14) is valid in the limit as $r \rightarrow 0$. In this limit the electron (and antineutrino) can be viewed as being created at $r = 0$ and can therefore carry no orbital angular momentum. The total angular momentum of the electron–antineutrino pair is, therefore, the vector sum of their individual spins. Because the intrinsic spin is $s = 1/2$ then the total spin can be either $S = 0$ or $S = 1$. The spin 0 and spin 1 cases are referred to as Fermi decays and Gamow–Teller decays and correspond to singlet and triplet states for the electron antineutrino, respectively. Conservation of total angular momentum requires that the change in I between the parent, I_P, and daughter, I_D, be related to the spin of the electron–antineutrino pair, J, as

$$|I_D - I_P| \leqslant J \leqslant I_D + I_P. \tag{9.30}$$

For allowed decays the above shows that ΔI defined as $I_D - I_P$ can be 0 or ± 1. Fermi and Gamow–Teller transitions can occur for all cases that can satisfy equation (9.30) as indicated in table 9.1. However, it is important to note that only Fermi transitions are allowed for the case $I_D = I_P = 0$. Conservation of parity requires no change in parity between the parent and daughter nuclei since the parity of the electron–antineutrino pair (with $L = 0$) must be even.

Forbidden decays are not really forbidden but are merely less probable than allowed decays and therefore correspond to longer decay lifetimes. If the matrix element in equation (9.10) that corresponds to the allowed transition vanishes, then the only transition that can occur is a forbidden decay. This requires the inclusion of higher order terms in the expansion of equation (9.14). An example of the first forbidden decay where the allowed matrix element vanishes but the next order term in the expansion does not vanish was seen in the previous section. If the first two matrix elements vanish then the next order term must be included, and this is referred to as a second forbidden decay. This terminology continues to higher order forbidden decays. In all cases the lowest order nonvanishing term dominates and can be accounted for in the calculation of the energy spectrum by inclusion of the appropriate shape factor. Shape factors for the first few forbidden decays as obtained from the expansion of the exponential in equations (9.12) and (9.13) are given in table 9.2.

The most obvious cases where a first forbidden transition can occur is a situation where there is a change of parity between the parent and daughter nuclei. In this case the electron–antineutrino must have odd parity and since $\pi = (-1)^L$, L cannot be

Table 9.1. Properties of allowed and forbidden β-decays.

Decay	L	ΔI	Nuclear parity change
Allowed	0	0, ±1	No
1st Forbidden	1	0, ±1, ±2	Yes
2nd Forbidden	2	±1, ±2, ±3	No
3rd Forbidden	3	±2, ±3, ±4	Yes
4th Forbidden	4	±3, ±4, ±5	No

Table 9.2. Shape factors for forbidden β-decays.

Decay	$S(p,q)$
1st Forbidden	$(m_e c)^{-2}[p^2 + q^2]$
2nd Forbidden	$(m_e c)^{-4}[p^4 + q^4 + (10/3)p^2 q^2]$
3rd Forbidden	$(m_e c)^{-6}[p^6 + q^6 + 7p^2 q^2(p^2 + q^2)]$

zero. This requires that r cannot be approximated as zero in equation (9.14) and higher order terms must be included. Possible values for ΔI consistent with equation (9.30) for first forbidden decays are given in table 9.1.

In cases where $\Delta I = \pm 2$ and there is no parity change or $\Delta I \geqslant 3$ then an allowed or first forbidden decay cannot satisfy the necessary conservation laws. In this case a second or higher order forbidden decay may occur. Some of these are summarized in table 9.1. In all cases forbidden decays of a given order are less probable than those of the previous order, typically, by a factor of about 10^3 for decays with similar Q.

The distinction between allowed and forbidden decays can also be evaluated on the basis of equation (9.29). This is the decay rate per unit momentum so the total decay rate can be determined by integrating the right-hand side of this expression over momentum. This gives,

$$\lambda = \int_0^{p_{max}} I(p)dp = \frac{m_e^5 c^4 G^2}{2\pi^3 \hbar^7} |M_{if}|^2 f(Z_D, Q)$$

where $f(Z_D, Q)$ is defined as

$$f(Z_D, Q) = \int_0^{p_{max}} F(Z_D, T_e) \frac{p^2}{m_e^2 c^2} \frac{(Q - T_e)^2}{m_e^2 c^4} \frac{dp}{m_e c}.$$

The additional factors of m_e and c have been included in order to make the integrand dimensionless. The integration is up to a maximum value of momentum given (relativistically) in terms of Q as

$$p_{max} = \frac{1}{c}[(Q + m_e c^2)^2 - m_e^2 c^4]^{1/2}.$$

Numerically determined values of the function $f(Z_D, Q)$ have been summarized by Feenberg and Trigg (1950). From the relationship between the decay rate and the half-life, equation (7.3), the above gives

$$f(Z_D, Q)\tau_{1/2} = \frac{2\pi^3 \hbar^7 \ln 2}{m_e^5 c^4 G^2} \frac{1}{|M_{if}|^2}.$$

This expression shows that the value of $f(Z_D, Q)\tau_{1/2}$ depends only on the matrix element M_{if}. It is generally most convenient to deal with log $[f(Z_D, Q)\tau_{1/2}]$ and we would expect that allowed transitions would have similar values for this quantity because of similar matrix elements for the transitions. An analysis of

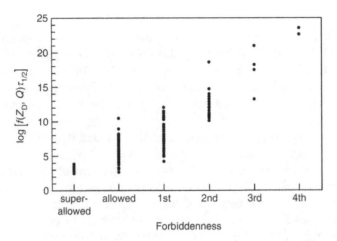

Figure 9.6. The quantity $\log[f(Z_D,Q)\tau_{1/2}]$ (with $\tau_{1/2}$ in seconds) plotted as a function of the degree of forbiddenness for some known β^- decays.

Table 9.3. Typical ranges of $\log[f(Z_D,Q)\tau_{1/2}]$ values for different β-decays. The half-life is in seconds.

Decay	$\log[f(Z_D,Q)\tau_{1/2}]$
Superallowed	3–4
Allowed	4–8
1st Forbidden	6–9
2nd Forbidden	10–14
3rd Forbidden	15–21
4th Forbidden	21–24

$\log[f(Z_D, Q)\tau_{1/2}]$ for different decays provides some useful information about the degree to which the transition is forbidden. Figure 9.6 shows $\log[f(Z_D, Q)\tau_{1/2}]$ for some known transitions. It would be convenient to find well-defined and different $\log[f(Z_D, Q)\tau_{1/2}]$ values for decays of different forbiddenness. Unfortunately, the distinction is not always clear as seen in figure 9.6. In terms of increasing $\log[f(Z_D, Q)\tau_{1/2}]$ values, the transitions with the smallest values are sometimes referred to as *superallowed*. These are followed by allowed transitions and transitions of increasing degree of forbiddenness as shown in table 9.3.

9.5 Parity violation in beta decay

Beta decay results from the weak interaction and, as indicated in chapter 2, is a process that does not necessarily conserve parity. If a process conserves parity, then the parity transform of any real experiment will yield the same results as the experiment itself. One way of observing these features is to investigate the properties of the mirror image of an experiment. The possibility of observing these effects in processes that are governed by weak interactions was first proposed in 1956 by Tsung-Dao (T D) Lee (b. 1926) and Chen-Ning (C N) Yang (b. 1922). The first experimental evidence demonstrating nonconservation of parity in a weak process, namely the β-decay of ^{60}Co, was reported in 1958 by Chien-Shiung (C S) Wu (1912–97) and co-workers. As described below the details of the experiment depend on the magnetic properties of cobalt.

Cobalt atoms possess a large atomic magnetic moment that can be readily aligned in an applied magnetic field. This atomic moment will couple to the nuclear magnetic moment and cause at least a partial alignment of the nuclear moments. This alignment is increased as the temperature is decreased due to a reduction in the thermal motion of the moments and, at sufficiently low temperatures (typically 0.01 K), a substantial fraction of the nuclear moments can be aligned. A measurement of the angular distribution of electrons from the β-decay of magnetically aligned ^{60}Co can be related to the preferential orientation of the emitted electrons and the direction of the nuclear magnetic moment. The case where there is a preferential emission antiparallel to the direction of the nuclear magnetic moment in the real experiment is illustrated in figure 9.7. It turns out that this is what is observed experimentally. The experiment as viewed in the mirror is illustrated on the right-hand side of figure 9.7 and it is seen in this case that the direction of the nuclear moment is opposite to that in the real experiment. Thus, the real experiment and the experiment as seen in the mirror do not yield the same results. This means that the assumed anisotropy of the electron emission violates parity conservation. From an experimental standpoint, the mirror image of the real experiment is created by

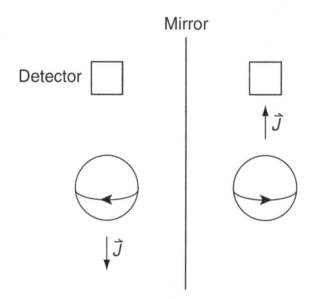

Figure 9.7. Comparison of the emission of electrons from the β⁻ decay of ^{60}Co as seen in a real experiment (left) and in a mirror (right).

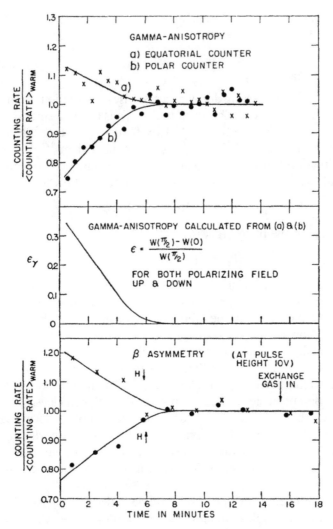

Figure 9.8. Results of parity violation experiments for the β^- decay of ^{60}Co as observed by γ-ray (top) and electron (bottom) count rates. Measurements were made for two magnetic field directions, indicated by $\times$ and ●. Data were obtained as a function of time as the sample warmed where the horizontal axis represents increasing temperature from left to right. Reprinted with permission from Wu *et al* (1958). Copyright 1958 by the American Physical Society.

changing the direction of the applied magnetic field by 180°. Results of such an experiment are shown in figure 9.8. The experiment was conducted by cooling the sample in an external field in order to align the nuclear moments. The sample was then allowed to warm up and the resulting thermal fluctuations randomized the ordering of the nuclear spins. The experiment was then repeated with the magnetic field applied in the opposite direction. The differences in the electron fluxes between the two field orientations is a clear demonstration of parity nonconservation.

The nonconservation of parity in weak processes will be discussed further in chapter 16 as it applies to the decay of mesons.

9.6 Double beta decay

The possibility of an interesting decay process arises in situations such as that shown in figure 4.5. Traditionally it is believed that ^{128}Te is β-stable since it is energetically unfavorable for it to decay either to ^{128}I by β^- decay or to ^{128}Sb by β^+ decay or electron capture. However, ^{128}Xe has a smaller mass than ^{128}Te and it would be energetically favorable for two ^{128}Te neutrons to become two protons and emit two electrons (and two antineutrinos). This double β-decay process is,

$$^{128}\text{Te} \rightarrow \,^{128}\text{Xe} + 2e^- + 2\bar{\nu}_e \tag{9.31}$$

and corresponds to the basic process,

$$2n \rightarrow 2p + 2e^- + 2\bar{\nu}_e. \tag{9.32}$$

Double β-decay as shown in equations (9.31) and (9.32) emits two electron antineutrinos and is, therefore, referred to as 2ν double β-decay ($2\nu\beta\beta$) or sometimes ordinary double β-decay. The possibility of neutrinoless double β-decay ($0\nu\beta\beta$) is discussed in section 14.2.

The theoretically predicted lifetime for double β-decay is sufficiently long that its observation will, at best, be very difficult. There are a number of possible candidates for observing double β-decay. As in the case of ^{128}Te, double β-decay is always from an even–even parent nucleus to an even–even daughter nucleus. The ground state transitions are, therefore, all $0^+ \rightarrow 0^+$. Possible double β-decay candidates are summarized in table 9.4. The decay rate is expected to increase with increasing decay energy and also to increase with decreasing barrier height, that is the energy (mass) barrier formed by the odd-odd nucleus (^{128}I in figure 4.5) between the parent and daughter nuclei.

The first experimental evidence for double β-decay was based on geological observations of rock samples containing a possible double β-decay parent nuclide. In such samples it is expected that an excess of the daughter nuclide (relative to the other isotopes of that element) would be found. The details of the analysis of this kind of experiment follow along the lines of our discussion of radioactive dating in section 7.3, although here we assume that we know the age of the rock and solve for the lifetime of the decay process. Some results of double β-decay lifetimes based on geological observations are given in table 9.5.

A number of experiments designed to directly observe the energy spectrum of electrons produced by a double β-decay process have been conducted and many are still in progress. The very long lifetimes mean very low decay rates and subsequently experiments are difficult because of interference from background radiation. Many experiments are conducted underground in mines to shield them from cosmic ray interference. ^{82}Se is particularly interesting as the decay energy is fairly large and the barrier energy is very small. The energy level diagram for this decay is shown in figure 9.9. Results of an experiment on ^{82}Se are illustrated in figure 9.10. Here the

Table 9.4. Possible double β-decays.

Parent	Parent natural abundance (%)	Daughter	Q (MeV)	Barrier energy (MeV)
^{46}Ca	0.003	^{46}Ti	0.985	1.382
^{70}Zn	0.62	^{70}Ge	1.01	0.653
^{76}Ge	7.67	^{76}Se	2.041	0.92
^{80}Se	49.82	^{80}Kr	0.138	1.87
^{82}Se	9.19	^{82}Kr	3.00	0.09
^{86}Kr	17.37	^{86}Sr	1.24	0.054
^{94}Zr	2.8	^{94}Mo	1.23	0.921
^{100}Mo	9.62	^{100}Ru	3.03	0.335
^{104}Ru	18.5	^{104}Pd	1.32	1.15
^{110}Pd	12.5	^{110}Cd	2.0	0.87
^{114}Cd	28.86	^{114}Sn	0.547	1.44
^{116}Cd	7.58	^{116}Sn	2.81	0.52
^{122}Sn	4.71	^{122}Te	0.349	1.622
^{124}Sn	5.98	^{124}Te	2.26	0.653
^{128}Te	31.79	^{128}Xe	0.876	1.26
^{130}Te	34.49	^{130}Xe	2.54	0.41
^{134}Xe	10.44	^{134}Ba	0.73	1.33
^{136}Xe	8.87	^{136}Ba	2.72	0.112
^{142}Ce	11.07	^{142}Nd	1.379	0.777
^{148}Nd	5.7	^{148}Sm	1.936	0.514
^{150}Nd	5.6	^{150}Sm	3.39	0.036
^{154}Sm	22.6	^{154}Gd	1.26	0.72
^{160}Gd	21.75	^{160}Dy	1.78	0.029
^{238}U	99.28	^{238}Pu	1.173	0.117

Table 9.5. Double β-decay lifetimes determined by geological experiments and calculated decay rates.

Decay	Lifetime (years)	Decays per year per gram
^{130}Te $\rightarrow$ ^{130}Xe	3.2×10^{21}	0.503
^{128}Te $\rightarrow$ ^{128}Xe	5.0×10^{24}	0.000 29
^{82}Se $\rightarrow$ ^{82}Kr	3.7×10^{20}	1.8

measured electron spectrum shows good agreement with the calculated energy spectrum for double β-decay. The total decay rate as observed in this experiment is consistent with the estimate based on geological observations.

Another situation can occur when single β-decay is energetically favorable but is unlikely because the spins and parities of the parent and daughter make the

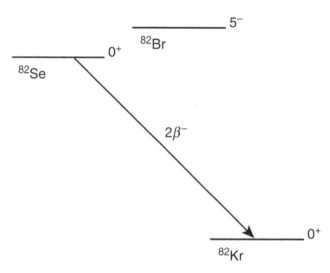

Figure 9.9. Possible double β-decay of ^{82}Se.

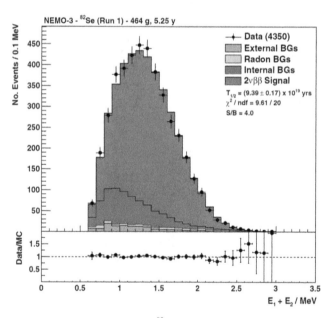

Figure 9.10. Summed energy for two electrons from ^{82}Se double β-decay and comparison with a Monte Carlo simulation. From Arnold *et al* (2018) CC BY 4.0.

transition highly forbidden. In such a case an allowed double β-decay might occur. Several such situations may be identified but the most notable is the possible double β-decay of ^{48}Ca to ^{48}Ti. The energy-level diagram is illustrated in figure 9.11. The ground state β-decay of ^{48}Ca to ^{48}Sc has a Q value of 0.28 MeV and the transitions to the excited states are correspondingly less. On the basis of spins and parities it can

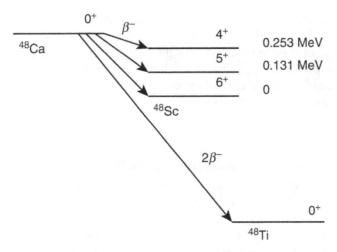

Figure 9.11. Possible double β-decay of ^{48}Ca.

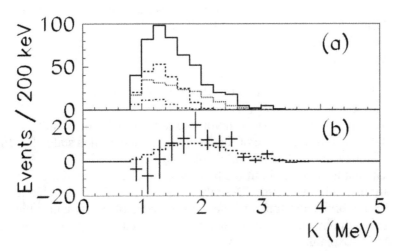

Figure 9.12. (a) Measured energy spectrum from the double β-decay of ^{48}Ca. Total spectrum (solid line) and various background corrections (broken lines). (b) Corrected double β-decay spectrum (data points) and calculated double β-decay spectrum (broken line). Reprinted with permission from Balysh *et al* (1996), copyright (1996) by the American Physical Society.

be seen that the ground state transition is a sixth forbidden decay and the excited state transitions are fourth forbidden decays. A possible alternative to these highly forbidden, low energy, single β-decays might be the higher energy, allowed double β-decay,

$$^{48}\text{Ca} \rightarrow {}^{48}\text{Ti} + 2e^- + 2\bar{\nu}_e.$$

Here the transition is allowed, as it is $0^+ \rightarrow 0^+$, and has a Q of 4.27 MeV. Some recent experimental results of the double β-decay electron energy spectrum from ^{48}Ca are shown in figure 9.12 along with a Monte Carlo simulation. In general, there is reasonable agreement shown and the measured lifetime, 6.2×10^{19} years, is consistent with theoretical predictions.

The number of double β-decays that have been detected experimentally is quite small because of the long lifetimes involved. However, experimental evidence as shown in figures 9.10 and 9.12 taken in conjunction with geological data offers clear evidence for double β-decay.

Problems

9.1. Describe all possible β-decay modes for ^{64}Cu.

9.2. Describe all possible decay schemes for tritium. Which of these is/are energetically favorable? Which is/are observed to occur?

9.3.
   (a) Using known masses determine the Q for the β-decay of ^{191}Os.
   (b) Electrons from the β-decay of ^{191}Os are observed experimentally. What is their maximum energy?

9.4. Determine the degree to which the following ground state to ground state β-decays are forbidden: ^{10}Be, ^{21}F, ^{37}S, and ^{60}Co.

9.5. For odd A nuclei it is normally expected that only one β-stable nuclide will exist for each value of A. The ground states of ^{113}Cd and ^{113}In are both considered stable.
   (a) Is it energetically favorable for one of these nuclides to decay to the other?
   (b) If one of these decays is favorable, why does it not occur?
   (c) What is the relevance of the fact that the first excited state of ^{113}In has an energy of 0.393 MeV and a spin and parity of $1/2^-$?

9.6. locate information about the β^- decay of ^{43}K from the *Table of Isotopes* or another suitable source. Describe as quantitatively as possible the reasons for the branching ratios to the various states of the daughter nuclide.

9.7. The endpoint energy for the electrons produced in the β^- decay of ^{8}Li is 13.1 MeV. Explain.

9.8. Show that ^{9}B may decay by proton emission, β^+ decay and electron capture but not by neutron emission or β^- decay.

9.9. locate information about the excited states of ^{46}Ti. Discuss the possibility of β^- decay of the ground state of ^{46}Sc.

9.10. Derive expressions for the Q and barrier height for double β-decay in terms of atomic masses.

References and suggestions for further reading

Arnold R *et al* 2018 Final results on ^{82}Se double beta decay to the ground state of ^{82}Kr from the NEMO-3 experiment *Eur. J.* C **78** 821

Balysh A *et al* 1996 Double beta decay of ^{48}Ca *Phys. Rev. Lett.* **77** 5186–9

Curran S C, Angus J and Cockroft A L 1949 Investigation of soft radiations—II. The beta spectrum of tritium *Phil. Mag.* **40** 53–60

Feenberg E and Trigg G 1950 The interpretation of comparative half-lives in the Fermi theory of beta-decay *Rev. Mod. Phys.* **22** 399–406

Krane K S 1988 *Introductory Nuclear Physics* (New York: Wiley)

Langer L M, Moffat R D and Price H C 1949 The beta-spectra of Cu^{64} *Phys. Rev.* **76** 1725–6

Langer L M and Price H C 1949 Shape of the beta-spectrum of the forbidden transition of Yttrium 91 *Phys. Rev.* **75** 1109

National Bureau of Standards 1952 *Tables for the Analysis of Beta Spectra* (Washington, DC: National Bureau of Standards Applied Mathematics Series) Report number NBS-AMS-13

Williams W S C 1991 *Nuclear and Particle Physics* (Oxford: Oxford University Press)

Wong S S M 2004 *Introductory Nuclear Physics* 2nd edn (New York: Wiley)

Wu C S, Ambler E, Hayward R W, Hoppes D D and Hudson R P 1958 Experimental test of parity conservation in beta decay *Phys. Rev.* **105** 1413–5

IOP Publishing

An Introduction to the Physics of Nuclei and Particles
(Second Edition)

Richard A Dunlap

Chapter 10

Gamma decay

10.1 Energetics of gamma decay

A nucleus in an excited state can decay to a lower energy (lower mass) state by gamma (γ) emission or internal conversion. The transition can be between an excited state and another lower energy excited state or between an excited state and the ground state. Such excited states can be readily formed in the daughter nucleus of an α- or β-decay process or by various nuclear reactions as discussed in detail in the next chapter. In the simplest case we can draw an analogy between nuclear transitions between single-particle states and atomic transitions between electronic energy levels. In the former case electromagnetic radiation is produced in the form of a γ-ray, in the latter case an x-ray or optical photon is emitted. Internal conversion is a process by which γ-decay energy liberates an atomic electron rather than a photon and is analogous to an Auger process for atomic transitions. For electronic transitions it is easy to see how transitions involving charged electrons can emit radiation and the analogy can be drawn with single-particle nuclear states involving a charged proton. However, single-particle nuclear transitions involving a neutron also produce radiation and in a classical sense this can be related to the fact that the neutron, although uncharged, carries a magnetic moment (which in turn is related to the neutron's internal structure). Many nuclear transitions are not between single-particle states but involve multiple nucleon excitations, rotational states, or vibrational states.

The energetics of γ-decay can be described in terms of the masses of the initial and final states, m_i and m_f, respectively, as

$$m_i c^2 = m_f c^2 + E_R + E_\gamma$$

where E_γ is the energy of the emitted γ-ray and E_R is the recoil energy of the nucleus. Conservation of linear momentum requires that,

doi:10.1088/978-0-7503-6094-4ch10

$$p_R = \frac{E_\gamma}{c}$$

where p_R is the momentum of the recoiling nucleus and E_γ/c is the momentum carried by the γ-ray. From this equation the nuclear recoil energy is expressed as

$$E_R = \frac{E_\gamma^2}{2m_f c^2}. \tag{10.1}$$

This derivation is nonrelativistic, and the recoil energy is sufficiently small to justify this assumption. Using a typical γ-decay energy of 1 MeV and a nuclear mass of $A = 100$ gives a recoil energy of about 5 eV. This is much smaller than the γ-ray energy and, to within the accuracy of most direct γ-ray energy spectrum measurements, this can be ignored. Thus, we can generally approximate

$$E_\gamma = (m_i - m_f)c^2.$$

However, it is important to note that this recoil energy actually decreases the γ-ray energy by the amount in equation (10.1) and that this is much larger than the typical Heisenberg line width of an excited nuclear state as given by equation (7.16). As a result, a consideration of the recoil energy can be extremely important for γ-resonance experiments such as Mössbauer effect spectroscopy.

10.2 Classical theory of radiative processes

A simple model of a γ-decay process can be derived on the basis of classical electrodynamics. A static electric field is produced by a distribution of charges and, as discussed in detail in chapter 6, the charge distribution can be described in terms of a multipole expansion. A time varying distribution of charges produces a time varying electric field, and this gives rise to the emission of radiation. When the time variation of the electric field is periodic (for example, sinusoidal) then a radiation field at the same frequency, ω, is produced. As for the static case, this radiation field can be described in terms of a multipole expansion. For example, consider the lowest order multipole term arising from an electric dipole. A simple model of an electric dipole consists of equal positive and negative charges separated by a distance a as shown in figure 10.1 and defines an electric dipole moment,

$$\vec{d} = q\vec{a} \tag{10.2}$$

along the z-axis. A radiation field at a frequency ω can be produced by allowing the dipole to oscillate along the z-axis so that,

$$\vec{d}(t) = q\vec{a} \sin \omega t.$$

A similar situation can be described by the multipole expansion of the magnetic moments of a system. These moments can be modeled on the basis of currents and in the simplest case the magnetic dipole moment may be thought of as arising from a

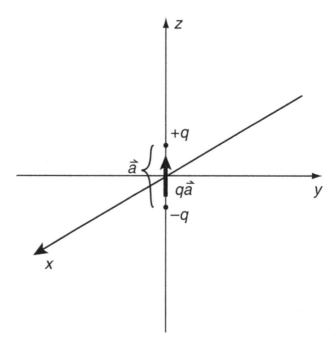

Figure 10.1. The electric dipole.

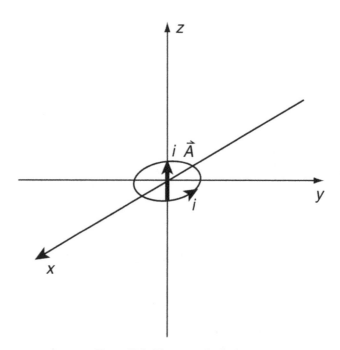

Figure 10.2. The magnetic dipole.

single loop of current. Figure 10.2 shows a magnetic dipole moment, $\vec{\mu}$, formed by a current, i, enclosing an area $\vec{A}$ where

$$\vec{\mu} = i\vec{A}. \tag{10.3}$$

A sinusoidally varying current produces a time varying magnetic dipole moment of the form,

$$\vec{\mu}(t) = i\vec{A} \sin \omega t$$

and a corresponding radiation field at frequency ω.

The basic properties of the radiation fields as described above can be determined by the application of classical electrodynamics. Of particular relevance to our future discussions is a calculation of the total radiated power. For the electric dipole this is found to be

$$P_e = \frac{1}{12\pi\varepsilon_0} \frac{\omega^4 d^2}{c^3}.$$

For the magnetic dipole this is

$$P_m = \frac{\mu_0}{12\pi} \frac{\omega^4 \mu^2}{c^3}.$$

Here the total radiated power represents the energy emitted in the form of radiation per unit time.

For a generalized charge and/or current distribution, higher order multipole moments exist as discussed in chapter 6. For example, an electric quadrupole moment can be constructed from four charges, as illustrated in figure 10.3. The order of the multipole moment, L, is defined as $L = 1$ for the dipole, $L = 2$ for the quadrupole, and so on. In general, the power radiated by an electric multipole moment of order L oscillating at a frequency ω can be calculated as

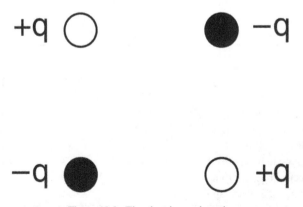

Figure 10.3. The electric quadrupole.

$$P_e(L) = \frac{2(L+1)c}{\varepsilon_0 L[(2L+1)!!]^2}\left(\frac{\omega}{c}\right)^{2L+2} Q_L^2 \tag{10.4}$$

where Q_L is the generalized electric multipole moment of order L and the double factorial is defined for odd or even arguments such as $7!! = 1 \times 3 \times 5 \times 7$ or $8!! = 2 \times 4 \times 6 \times 8$, respectively. Similarly for the magnetic case,

$$P_m(L) = \frac{2(L+1)\mu_0 c}{L[(2L+1)!!]^2}\left(\frac{\omega}{c}\right)^{2L+2} M_L^2 \tag{10.5}$$

where M_L is the generalized magnetic multipole moment of order L. The definition of Q_L and M_L for $L = 1$ differs from the definitions given in equations (10.2) and (10.3) only by some constant numerical factors.

The parity of the radiation field can be understood by investigating the effect of the transformation $r \to -r$ on the corresponding multipole moment. For example, for the electric dipole of figure 10.1, the application of this transformation results in $d \to -d$ and the radiation field is defined as having odd parity (-1). On the other hand, application of the transformation $r \to -r$ to the magnetic dipole shown in figure 10.2 causes no change in the resulting μ. This radiation field is defined as having even parity ($+1$). This procedure can be applied to the electric quadrupole shown in figure 10.3 leading to no change in the electric quadrupole moment and even parity. In general, the parity of an electric multipole radiation field of order L is given as

$$\pi_L = (-1)^L$$

and for a magnetic multipole field:

$$\pi_L = (-1)^{L+1}.$$

The radiation produced by an electric multipole is designated in terms of the value of L as EL. Magnetic multipole radiation is referred to as ML. Table 10.1 gives the designations and parities of the first few orders of multipole radiation.

Table 10.1. Properties of some multipole radiation types.

Type	Symbol	L	Parity
Electric dipole	E1	1	−1
Magnetic dipole	M1	1	+1
Electric quadrupole	E2	2	−1
Magnetic quadrupole	M2	2	+1
Electric octupole	E3	3	−1
Magnetic octupole	M3	3	+1
Electric hexadecapole	E4	4	−1
Magnetic hexadecapole	M4	4	+1

In order to relate these ideas to nuclear γ-decay, it is necessary to consider them in a quantum mechanical context as discussed in the next section.

10.3 Quantum mechanical description of gamma decay

Quantum mechanically the radiation given off by a varying electric or magnetic multipole is in the form of photons, each with energy $E_\gamma = \hbar\omega$. Thus, the decay constants are related to the power radiated, as given in equations (10.4) and (10.5), by the simple expressions,

$$\lambda_e(L) = \frac{P_e(L)}{\hbar\omega}$$

and

$$\lambda_m(L) = \frac{P_m(L)}{\hbar\omega}$$

for the electric and magnetic cases, respectively. Equations (10.4) and (10.5) can be used, more or less, for the quantum mechanical calculation of the decay constants except that the multipole moments must be replaced by appropriate multipole operators.

Looking first at the electric multipole case, we can write,

$$\lambda_e(L) = \frac{2(L+1)}{\varepsilon_0 \hbar L[(2L+1)!!]^2}\left(\frac{E_\gamma}{\hbar c}\right)^{2L+1}\left|Q_{if}(L)\right|^2 \tag{10.6}$$

where

$$Q_{if}(L) = \int \psi_f^* Q(L)\psi_i d^3\vec{r}. \tag{10.7}$$

Following along the ideas presented in the last chapter on decay probabilities, we have written the decay probability as a function of the matrix element $Q_{if}(L)$, which is defined in terms of an electric multipole operator, $Q(L)$. The electric multipole operator is found to be proportional to $r^L Y_{LM}^*(\theta, \phi)$ (see section 5.3). As in the discussion of β-decay, ψ_i and ψ_f are the wave functions of the initial and final states, respectively. In general, the calculation of the matrix element in (10.6) is not straightforward. However, some rough estimates (referred to as the Weisskopf estimates) can be made on the basis of assumptions that are suitable for many cases. These assumptions are:

1. The initial and final states are given by the single-particle wave functions $\psi_i = R_i(r)Y_{LM}(\theta,\phi)$ and $\psi_f = (4\pi)^{-1/2}R_f(r)$. The wave function indicates that the final state is assumed to be an s-state.
2. The radial part of the wave functions, $R_i(r)$ and $R_f(r)$, is a constant, $\sqrt{3}\,R_0^{-3/2}$, over the nuclear volume and zero outside the nucleus.

Equation (10.7) can then be written as the integral over the nuclear radius R_0:

Table 10.2. Energy and A dependence of multipole decay probabilities. The γ-ray energy is expressed in MeV.

L	$\lambda_e(L)$ (s^{-1})	$\lambda_m(L)$ (s^{-1})
1	$1.02 \times 10^{14}\, A^{2/3} E_\gamma^{\,3}$	$3.15 \times 10^{13}\, E_\gamma^{\,3}$
2	$7.28 \times 10^{7}\, A^{4/3} E_\gamma^{\,5}$	$2.24 \times 10^{7}\, A^{2/3} E_\gamma^{\,5}$
3	$33.9\, A^{2} E_\gamma^{\,7}$	$10.4\, A^{4/3} E_\gamma^{\,7}$
4	$1.07 \times 10^{-5}\, A^{8/3} E_\gamma^{\,9}$	$3.27 \times 10^{-6}\, A^{2} E_\gamma^{\,9}$

$$Q_{if}(L) = \frac{e}{\sqrt{4\pi}} \int_0^{R_0} R_f r^L R_i r^2 dr.$$

This is immediately integrated to yield,

$$Q_{if}(L) = \frac{e}{\sqrt{4\pi}} \left[\frac{3}{L+3} \right] R^L$$

where R_0 can be related to A as in equation (3.14), $R_0 = R_1 A^{1/3}$. Equation (10.6) can now be written as

$$\lambda_e(L) = \frac{2e^2(L+1)}{4\pi\varepsilon_0 \hbar L[(2L+1)!!]^2} \left[\frac{3}{L+3} \right]^2 \left(\frac{E_\gamma}{\hbar c} \right)^{2L+1} R_1^{2L} A^{2L/3}. \tag{10.8}$$

This expression can be evaluated numerically for different L as a function of E_γ and A and is summarized in table 10.2.

In the case of magnetic transitions, a similar approach can be taken where the electric multipole operator in equations (10.6) and (10.7) is replaced with the corresponding magnetic multipole operator,

$$M_{if}(L) = \int \psi_f^* M(L) \psi_i d^3 \vec{r}.$$

The magnetic multipole operator is found to be,

$$M(L) \propto r^L Y_{LM}^*(\theta, \phi) \nabla \cdot (\vec{r} \times \vec{j}\,)$$

where $\vec{j}$ is the current density inside the nucleus. It is also necessary to include a term to account for the intrinsic spin of the unpaired nucleon. After some approximations, the transition rate for the magnetic case, as obtained from the Weisskopf estimates, is,

$$\lambda_m(L) = \frac{20e^2\hbar(L+1)}{4\pi\varepsilon_0 c^2 m_p^2 L[(2L+1)!!]^2} \left[\frac{3}{L+3} \right]^2 \left(\frac{E_\gamma}{\hbar c} \right)^{2L+1} R_1^{2L-2} A^{(2L-2)/3}. \tag{10.9}$$

Numerically it is found from equations (10.8) and (10.9) that $\lambda_m(L) = 0.308 \cdot A^{-2/3} \lambda_e(L)$. These expressions provide the estimates of the decay constants, as given in table 10.2.

The predictions summarized in table 10.2 should not be taken too seriously as the assumptions that have been made do not allow for the calculation of precise numerical values. However, these results are useful in understanding the relative decay rates for E and M radiation as a function of L and can be summarized as follows:

1. For a given transition energy there is a substantial decrease in decay constant with increasing L.
2. Electric transitions have decay constants that are typically about two orders of magnitude higher than the corresponding magnetic transition constant.

The quantum mechanical selection rules, as discussed in the next section, must be considered in order to determine which of these transitions will actually occur.

10.4 Selection rules

Conservation of angular momentum requires that the total angular momentum of the photon, given by L in the above discussion, is related to the total angular momentum of the initial and final nuclear states as

$$|I_f - I_i| \leqslant L \leqslant I_f + I_i \tag{10.10}$$

as with $L = 1, 2, 3, \ldots$. This expression constrains the possible values of L for the transition on the basis of the properties of the initial and final states (although it puts no restrictions on whether the transition can be electric or magnetic). It is, however, also necessary to consider the question of parity. The parity of the photon for different types of radiation is given in table 10.1. Conservation of parity requires that a $\pi = -1$ photon may be emitted only when there is a change in the parity of the nucleus and a $\pi = +1$ photon may be emitted only when there is no change in the parity of the nucleus. This selection rule in conjunction with the conservation of angular momentum specifies what kinds of transitions can occur between certain nuclear states. It should be noted that these selection rules do not allow $0^+ \rightarrow 0^+$ transitions since equation (10.10) can only be satisfied for $L = 0$, which is not allowed because the photon has an intrinsic spin of 1. There are a small number of cases where there is a 0^+ ground state and a 0^+ first excited state. ^{40}Ca is one example; recall the discussion of rotational levels in even–even nuclei in section 6.7. In such cases the transition from the first excited state to the ground state must proceed by internal conversion as described below.

Often the above selection rules allow for more than one kind of decay. Some examples are given in table 10.3. The relative importance of the various allowed decay modes can be estimated on the basis of equations (10.8) and (10.9). These equations allow for a determination of partial decay constants (or partial lifetimes) and a calculation of the branching ratios for the various decay modes. From the equations it is clear that the lowest order multipolar radiation will dominate. In many cases, for example a low energy E1 + M2 transition (for example, $3/2^-$ to $1/2^+$) the lowest order mode dominates, and all other modes are negligible. In some cases,

Table 10.3. Examples of allowed γ-transitions.

I_i^π	I_f^π	Nuclear parity change	L	Allowed transitions
0^+	0^+	No	–	None
$1/2^+$	$1/2^-$	Yes	1	E1
1^+	0^+	No	1	M1
2^+	0^+	No	2	E2
$3/2^-$	$1/2^+$	Yes	1, 2	E1, M2
$5/2^+$	$1/2^-$	Yes	2, 3	M2, E3
2^+	1^+	No	1, 2, 3	M1, E2, M3
$3/2^-$	$5/2^+$	Yes	1, 2, 3, 4	E1, M2, E3, M4
$5/2^-$	$3/2^-$	No	1, 2, 3, 4	M1, E2, M3, E4

for example a fairly high energy M2 + E3 transition (for example, $5/2^+$ to $1/2^-$), the contribution from the second mode can be significant.

10.5 Internal conversion

For the decay from an excited state to a lower energy state internal conversion competes with γ-emission. For the $0^+ \to 0^+$ transition, that is an E0 transition, γ-decay is not allowed, and internal conversion is the only possibility for de-excitation. Internal conversion refers to the process by which the de-excitation of nuclear energy levels gives up its energy to an atomic electron. If the energy involved, E_γ, is greater than the binding energy of the electron, b_e (which it most commonly is), the electron is liberated with a kinetic energy,

$$T_e = E_\gamma - b_e. \tag{10.11}$$

This expression conserves energy and a consideration of the total angular momentum of the electron relative to the initial and final nuclear state is necessary to ensure that total angular momentum is conserved. It is electrons in the inner shells (specifically s-orbital electrons) that are involved in this process because it is these electrons that have wave functions that have largest values at the nucleus. Typically, only internal conversion electrons from K, L, and M shells are observed. The binding energy of these electrons is given as a function of Z in figure 10.4. Thus, the internal conversion electron spectrum consists of several peaks corresponding to different electron shells. An example is shown in figure 10.5. The relative intensity of the peaks is related to the probability of internal conversion with electrons from the different shells and the figure suggests that shells beyond M will contribute very little to the spectrum.

The total decay constant for γ-decay and internal conversion is given as the sum of the individual components,

$$\lambda = \lambda_\gamma + \lambda_e$$

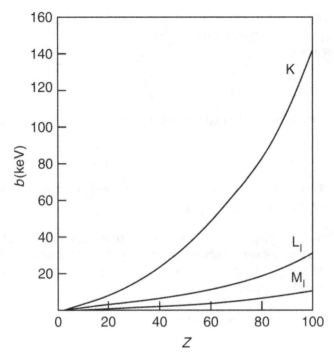

Figure 10.4. Electronic binding energies for different electron shells.

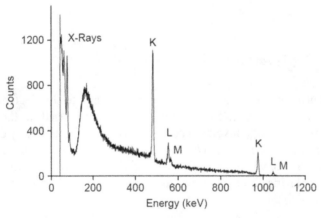

Figure 10.5. Internal conversion electron spectrum from ^{207}Bi showing two sets of peaks from two different energy transitions, corresponding to two different values of E_γ in equation (10.11). Reprinted from Wauters *et al* (2009). Copyright (2009). With permission from Elsevier.

where the γ and e subscripts refer to γ-decay and internal conversion, respectively. The internal conversion decay constant can also be expanded in terms of the contributions from the K, L, M,... shells as

$$\lambda_e = \lambda_K + \lambda_L + \lambda_M + \ldots.$$

The internal conversion coefficient is defined as the ratio of the total number of decays for a particular transition that proceed by internal conversion to those that proceed by γ-emission:

$$\alpha = \frac{\lambda_e}{\lambda_\gamma}.$$

The K-shell internal conversion coefficient can, therefore, be described as

$$\alpha_K = \frac{\lambda_K}{\lambda_\gamma}$$

and similarly for L, M, ... etc giving the total internal conversion coefficient,

$$\alpha = \alpha_K + \alpha_L + \alpha_M + ...$$

The calculation of the internal conversion decay constants is not simple but follows along the lines of the discussion above concerning the determination of appropriate matrix elements for γ-emission. A nonrelativistic approach simplifies matters and although it cannot be expected to give accurate numerical results at higher energies, it provides some insight into the relevance of the important parameters. Internal conversion coefficients for electric and magnetic transitions of order L can be found in this way to be,

$$\alpha(EL) = \frac{Z^3}{n^3}\left(\frac{L}{L+1}\right)\left(\frac{e^2}{4\pi\varepsilon_0\hbar c}\right)^4\left(\frac{2m_ec^2}{E_\gamma}\right)^{L+5/2}$$

and

$$\alpha(ML) = \frac{Z^3}{n^3}\left(\frac{e^2}{4\pi\varepsilon_0\hbar c}\right)^4\left(\frac{2m_ec^2}{E_\gamma}\right)^{L+3/2}$$

where n is the principal quantum number corresponding to the electron energy level; $n = 1, 2, 3,...$ for K, L, M,... etc. A general conclusion based on these expressions indicates that the internal conversion coefficient is greatest for heavy elements (large A and Z), small transition energy (E_γ), and inner shell electrons (small n). Numerous graphs of internal conversion coefficients for electric and magnetic transitions of different multipolarities and for different values of A can be found in the *Table of Isotopes*.

Problems

10.1. Discuss the decay of the ground state and the first excited state of ^{121}Sn.

10.2. The spins and parities of the ground state and first four excited states of ^{83}Kr are $9/2^+$, $7/2^+$, $1/2^-$, $5/2^-$ and $3/2^-$, respectively. For all γ-transitions ending at the ground state determine the allowed multipolarities.

10.3. The validity of the Weisskopf estimate for the M2 transition may be demonstrated by plotting the energy dependence of $\tau A^{2/3}$ for decays of different energies. From the *Table of Isotopes* or another suitable reference locate four M2 transitions and show that such a plot is consistent with the Weisskopf estimate. Note that the following points may be of relevance: (i) a logarithmic plot may best illustrate the features of the relationship; (ii) tables may give half-lives rather than lifetimes; (iii) in the case of multimodal decays, it is important to use the correct lifetime for the partial decay; (iv) it may be necessary to correct for the effects of internal conversion.

10.4. The first excited state of ^{58}Co has $I^{\pi} = 5^{+}$, $\tau = 12.0$ h and an energy of 0.025 MeV. For ^{58}Fe these values are 2^{+}, 10^{-11} s and 0.81 MeV, respectively. Does the Weisskopf estimate provide a good description of the decays of these two excited states?

10.5. Describe the type and multipolarity of possible γ-decays between the following spin and parity states: (a) $9/2^{-} \rightarrow 7/2^{+}$, (b) $1/2^{-} \rightarrow 7/2^{-}$, (c) $4^{+} \rightarrow 2^{+}$ and (d) $11/2^{-} \rightarrow 3/2^{+}$.

10.6.
  (a) Calculate the recoil energy for the transitions from the first excited state to the ground state for the following nuclides (energy given in parentheses): (i) ^{15}O (5.183 MeV), (ii) ^{19}O (0.0960 MeV), (iii) ^{57}Fe (0.0144 MeV), (iv) ^{70}Ge (1.0396 MeV), (v) ^{227}Th (0.0093 MeV), and (vi) ^{228}Th (0.0578 MeV).
  (b) Comment on the general trends that are expected for the importance of the recoil energy in terms of nuclear properties.

10.7. Calculate the Weisskopf value for the transition from the first excited state to the ground state for tungsten isotopes with $A = 180$, 182, 184, and 186. Compare these with experimentally measured values. What common features do these transitions have that could explain these results?

10.8. A nucleus has a $1/2^{-}$ ground state. The excited states are, more or less, evenly spaced in energy and have spins and parities of $5/2^{-}$, $3/2^{-}$, $7/2^{+}$ and $5/2^{+}$, respectively, with increasing energy. Draw an energy level diagram indicating all expected γ-transitions to the ground state, their multipolarities, and rough estimates of their expected decay rates.

References and suggestions for further reading

Dunlap R A 2019 *The Mössbauer Effect* (San Rafael, CA: Morgan & Claypool)

Heyde K 2004 *Basic Ideas and Concepts in Nuclear Physics—An Introductory Approach* (Bristol: IOP Publishing)

Wauters F, Kraev I S, Tandecki M, Traykov E, VanGorp S, Zákoucký D and Severijns N 2009 Performance of silicon PIN photodiodes at low temperatures and in high magnetic fields *Nucl. Instrum. Methods Phys. Res.* A **604** 563–7

IOP Publishing

An Introduction to the Physics of Nuclei and Particles
(Second Edition)

Richard A Dunlap

Chapter 11

Nuclear reactions

11.1 General classification of reactions and conservation laws

A nuclear reaction is a process that results from the interaction between a nucleus and a particle that is incident upon it. In general, a nuclear reaction can be represented by a particle, a, incident on a nucleus, A, which produces another particle, b, and a resulting nucleus, B. This can be written as

$$a + A \rightarrow B + b. \qquad (11.1)$$

A convenient shorthand notation that is in common use for the process in equation (11.1) is

$$A(a, b)B.$$

This convention will be used where appropriate throughout this chapter.

Nuclear processes can be roughly divided into two categories: *scattering* in which the incident particle and emitted particle are the same, and *reactions* in which the incident and emitted particles are different. Scattering processes can be either elastic or inelastic. Elastic scattering resulting from Coulomb interactions is referred to as Coulomb or Rutherford scattering and has already been discussed in chapter 3 in the context of experiments to determine the size of the nucleus. Such processes conserve kinetic energy. Inelastic scattering refers to the situation where kinetic energy is not conserved. In this case, the scattering particle will lose kinetic energy and the nucleus can be left in an excited state.

Nuclear reactions may be *direct reactions, compound nucleus reactions*, or *resonance reactions*. A *direct reaction* is one in which the incident particle interacts only with a limited number of valence nucleons in the target nucleus. This situation is most likely when the de Broglie wavelength of the incident particle is comparable to the size of an individual nucleon (about 1 fm) rather than the size of the nucleus

Table 11.1. Some examples of particles involved in nuclear reactions.

Symbol	Name	Nucleus of	Identity
n	Neutron	–	n
p	Proton	^{1}H	p
d	Deuteron	^{2}H	n + p
t	Triton	^{3}H	2n + p
^{3}He	–	^{3}He	n + 2p
α	α-particle	^{4}He	2n + 2p

(about 10 fm). This happens when the energy of the incident particle is relatively high. A *compound nucleus reaction* is one in which the incident particle becomes bound to the nucleus forming a compound nucleus before the reaction continues. Thus, the interactions between all the nucleons are important in determining the state of the compound nucleus. It is only the properties of the compound nucleus, and not how it was formed, that is of relevance in determining the way in which it decays. *Resonance reactions* are, in some ways, intermediate between these two extremes as the incident particle can become quasibound to the nucleus before the reaction proceeds. The distinction between these three situations is not always clear and in many cases a process can involve more than a single mode.

Several conservation laws must be considered when looking at the details of a reaction. Mass/energy and momentum must be conserved although kinetic energy, in many cases, is not. At sufficiently high energy new types of particles may be created (usually mesons). This can occur at energies above about 280 MeV, corresponding to the production of pions. These reactions are discussed further in section 16.2. At lower energies the number and identity of particles is normally conserved. Specifically, the number of neutrons and the number of protons will not change except in cases where the weak interaction is important. Charge must be conserved, but conservation of the number of protons will insure this in any case. Here we will consider relatively low energy reactions that involve the strong or electromagnetic interaction. High-energy reactions in which other kinds of particles can be created will be considered in part IV of this book. Low energy processes generally involve incident particles that are neutrons, protons, or bound systems consisting of a small number of nucleons. The most commonly encountered incident particles are summarized in table 11.1.

Elastic scattering has been considered previously. In the present chapter, inelastic scattering and some different low energy reactions are discussed in detail. Appendix C provides the details of the design of accelerators for producing the low- to medium-energy charged particles used for reactions as described here.

11.2 Inelastic scattering

Consider the simple inelastic scattering event A(a, a)A* as illustrated in figure 11.1 (the superscript * indicates that the nucleus is in an excited state). Here a particle

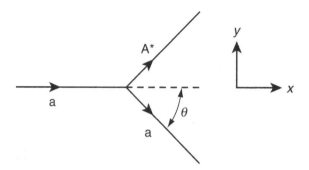

Figure 11.1. Geometry of an inelastic scattering event.

scatters from a nucleus inelastically, losing energy and leaving the nucleus in an excited state. We will deal with this problem nonrelativistically, as this is a good approximation for nucleons at the energies that we are considering. However, relativistic corrections can be included if necessary. The total energy in the laboratory frame before the collision, E, is given as the kinetic energy of the incident particle, E_i, and the rest mass energy of the system,

$$E = E_i + m_a c^2 + m_A c^2. \tag{11.2}$$

After the collision the total energy is

$$E = E_f + E_{A^*} + m_a c^2 + m_{A^*} c^2 \tag{11.3}$$

where E_f is the kinetic energy of the emitted particle and E_{A^*} is the kinetic energy of the nucleus in its excited state. The change in energy of the nucleus is represented by the change in its rest mass (that is, the energy of the excited state) and is found by equating equations (11.2) and (11.3) as

$$\Delta E = (m_A - m_{A^*}) c^2 = E_i - E_f - E_{A^*}. \tag{11.4}$$

This equation can be written in terms of the corresponding momenta as

$$\Delta E = \frac{p_i^2}{2m_a} - \frac{p_f^2}{2m_a} - \frac{p_{A^*}^2}{2m_{A^*}}. \tag{11.5}$$

From the figure we can see that conservation of the x-component of momentum gives

$$p_{A^*x} = p_i - p_f \cos\theta$$

and conservation of the y-component of momentum gives,

$$p_{A^*y} = p_f \sin\theta.$$

Substituting these expressions into equation (11.5) for p_{A^*} gives

$$\Delta E = \frac{p_i^2}{2m_a} - \frac{p_f^2}{2m_a} - \frac{1}{2m_{A^*}} \left[p_i^2 + p_f^2 - 2p_i p_f \cos\theta \right].$$

In terms of initial and final energies of the scattered particle this becomes

$$\Delta E = E_i\left(1 - \frac{m_a}{m_{A^*}}\right) - E_f\left(1 + \frac{m_a}{m_{A^*}}\right) + 2\frac{m_a}{m_{A^*}}\sqrt{E_iE_f}\cos\theta. \qquad (11.6)$$

Elastic scattering is seen to be a limiting case of this expression with $\Delta E = 0$ where, for a given scattering angle, θ, the difference between E_i and E_f accounts for the kinetic energy (recoil) of the nucleus. From equation (11.4) it is seen that in the case of inelastic scattering the actual mass of the nucleus after collision can only be calculated from equation (11.6) by an iterative process. However, to the accuracy of experimental measurements, the value of m_{A^*} on the right-hand side of equation (11.6) can be replaced by m_A, the known ground state mass. Since the atomic electrons are associated with the nucleus and there is no change in the identity of the nucleus, it is appropriate to use measured atomic masses in the analysis of the above equations to describe the dynamics of the interaction.

Experimentally the excited state energies of a nucleus can be investigated by allowing a beam of monoenergetic particles to be incident on the nucleus and measuring the energy spectrum of the particles scattered at a specific angle. As an example, we can consider the inelastic scattering of 10.02 MeV protons from a sample containing ^{10}B nuclei, ^{10}B (p, p)^{10}B*. For a fixed scattering angle, say 90°, protons that have a final energy E_f will give up an energy ΔE to the ^{10}B nucleus, as determined by equation (11.6). If ΔE is equal to the energy difference between the ^{10}B ground state and one of the ^{10}B excited states, then the energy given up by the proton can cause an excitation of the ^{10}B nucleus. Since the nucleus has a number of well-defined excited states, we expect that the protons scattered by a specific angle will have a series of well-defined energies, or resonances, related to the excited state energies by equation (11.6). The proton energy spectrum from such an experiment is shown in figure 11.2. The left-most resonance corresponds to elastic scattering where

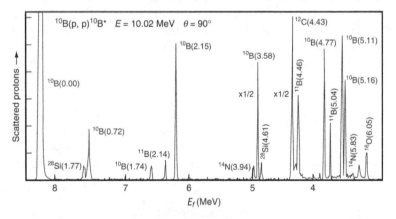

Figure 11.2. Number of 10.02 MeV protons scattered at 90° from a sample containing ^{10}B as a function of final proton energy. Decreasing proton energy, from left to right in the figure, corresponds to increasing ^{10}B excited state energy as shown in table 11.2. The ^{10}B peaks along with their corresponding excited state energies are indicated. Reprinted from Armitage and Meads (1962). Copyright (1962). With permission from Elsevier.

Table 11.2. Relationship of final proton energy to excited state energy for 10.02 MeV protons scattered from ^{10}B at an angle of 90°. Data are obtained from figure 11.2.

E_f (MeV)	ΔE (MeV)
8.19	0
7.53	0.72
6.61	1.74
6.23	2.15
4.93	3.58
3.85	4.77
3.54	5.11
3.50	5.16

the ^{10}B nucleus is left in the ground state. With decreasing energy each peak represents a larger ΔE and a higher energy excited state for the ^{10}B nucleus. An analysis on the basis of equation (11.6) gives the excitation energies in table 11.2 and the energy levels for ^{10}B, as shown in figure 11.3.

11.3 Nuclear reactions

Nuclear reactions can also occur where the identity of the incident and emitted particles is different. Several different situations are possible, and these may be described as follows:

1. An incident nucleon is absorbed by the nucleus leaving the nucleus in an excited state and the emitted particle is a γ-ray resulting from the de-excitation of the nucleus. A well-known example is the (n, γ) reaction.
2. An incident nucleon is absorbed by the nucleus and a different nucleon is emitted. An example is the (n, p) reaction.
3. The incident particle loses one (or more) of its nucleons to the nucleus; for example, the (d, p) reaction. This is referred to as a *stripping reaction* and is discussed further in section 11.4.
4. The incident particle gains one (or more) nucleon from the nucleus; for example, the (d, α) reaction. This is referred to as a *pick-up reaction.*
5. The reaction causes the component nucleons of the incident particle to become unbound. An example is the (d, np) reaction, where the terminology np refers to a neutron–proton pair that is not bound.

As an example of some of the above reactions let us consider a deuteron incident on an ^{16}O nucleus. The possible reactions involving the particles given in table 11.1 are as follows:

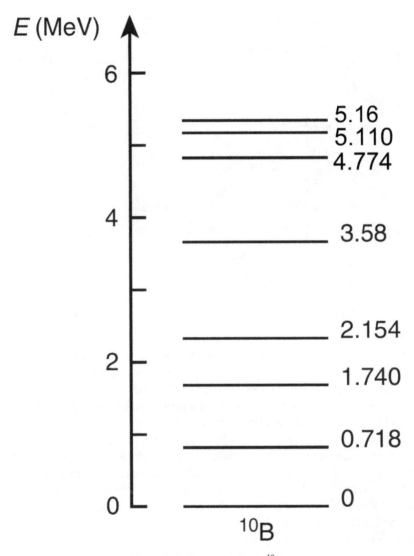

Figure 11.3. Energy levels for ^{10}B.

$$d + {}^{16}O \rightarrow {}^{16}O + d$$
$$d + {}^{16}O \rightarrow {}^{18}F$$
$$d + {}^{16}O \rightarrow {}^{17}O + p$$
$$d + {}^{16}O \rightarrow {}^{17}F + n$$
$$d + {}^{16}O \rightarrow {}^{14}N + \alpha \qquad (11.7)$$
$$d + {}^{16}O \rightarrow {}^{15}O + t$$
$$d + {}^{16}O \rightarrow {}^{15}N + {}^{3}He$$
$$d + {}^{16}O \rightarrow {}^{16}O + np.$$

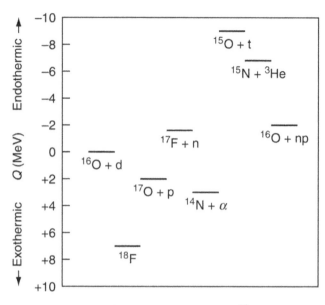

Figure 11.4. Schematic illustration of Q for various ^{16}O + d reactions.

The energetics of these processes can be readily established on the basis of measured atomic masses, although it is important to properly account for possible changes in the number of electrons associated with the nucleus before and after the reaction. A diagram indicating the Q for each of the reactions given in equation (11.7) is shown in figure 11.4. The vertical energy scale is referenced to ^{16}O + d. Reactions with positive Q have reaction products with a smaller mass than the initial components and reactions with negative Q have reaction products with a larger mass than the initial components. The former are exothermic reactions, and the latter are endothermic reactions. Unlike decay processes, endothermic reactions can occur because additional kinetic energy can be supplied by the incident particle.

It is the kinetic energy in the center of mass frame that is of importance in this problem. Since experimentally the incident particle energy is measured in the laboratory it is necessary to transform to the center of mass frame as,

$$E_{cm} = \frac{E_{lab}}{1 + \frac{m_a}{m_A}}.$$

Thus, the center of mass energy is always less than the energy measured in the laboratory frame.

Figure 11.4 shows that the (d, γ), (d, p), and (d, α) reactions are exothermic and the (d, n), (d, t), (d, ^3He), and (d, np) reactions are endothermic. These latter processes require a minimum center of mass kinetic energy of 1.62, 9.41, 6.63, and 2.22 MeV, respectively, in order to proceed. These center of mass energies correspond to laboratory energies of 1.82, 10.59, 7.46, and 2.50 MeV, respectively.

Energy above the amount required to cause a reaction to proceed may yield additional kinetic energy of the reaction by-products and/or may leave the resulting nucleus in an excited state. Any excited state at or below an energy of

$$E = E_{cm} + Q \tag{11.8}$$

is accessible. This requires the existence of an excited state of the resulting nucleus in this energy range. In general, we know that the density of excited states increases with increasing nuclear mass. It is also known that the density of excited states in a given nucleus generally increases with increasing energy. This is readily understood on the basis of the shell model. The lowest lying states are often single-nucleon states. Higher energy states generally involve multiple nucleon excitations. As the number of nucleons participating increases so does the number of combinations of state occupancies that will give similar but slightly different excited state energies. Thus, for heavy nuclei, and for cases where the available energy in equation (11.8) is fairly large (or both), the spacing (in energy) between the resonances is small. The practical implication of this will be seen in chapter 12.

The angular dependence of the energy spectrum of emitted particles in a nuclear reaction may be calculated (nonrelativistically) in a manner similar to the derivation in section 11.2. This calculation gives,

$$\Delta E = E_i\left(1 - \frac{m_a}{m_B}\right) - E_f\left(1 + \frac{m_b}{m_B}\right) + 2\frac{\sqrt{m_a m_b E_i E_f}}{m_B}\cos\theta + Q \tag{11.9}$$

where the nucleus B may be left in an excited state. A particularly interesting and useful example of the application of this expression deals with the deuteron stripping reactions (d, n) and (d, p). In these processes the incident deuteron loses one of its nucleons and this is absorbed by the nucleus. This is an effective way of looking at excited states of the resulting nucleus and is discussed further in the next section.

11.4 Deuteron stripping reactions

In the example of the (d, p) process in equation (11.7), deuteron stripping may be used to examine the excited states of ^{17}O. Since this reaction is exothermic, even low energy deuterons may excite any ^{17}O levels below about 1.9 MeV (although Coulomb effects, as discussed in section 11.6, are of some significance). Additional deuteron kinetic energy as measured in the center of mass frame may excite higher energy ^{17}O levels. As an example, the peaks in the energy spectrum of protons emitted by the ^{16}O(d, p)^{17}O reaction at an angle of 25° with respect to a beam of 10 MeV (laboratory frame) deuterons are summarized in table 11.3. Excited state energies as calculated using equation (11.9) are summarized in the table and are illustrated in figure 11.5.

It is important to note the location of the neutron separation energy for ^{17}O at 4.15 MeV in the figure. Excited ^{17}O states above 4.15 MeV are sometimes thought of as n-^{16}O resonances as the neutron is only quasibound to the ^{16}O nucleus. Since ^{17}O is β-stable, excited states below 4.15 MeV may decay only by γ-decay (or internal conversion, as appropriate). However, it is energetically favorable for excited states

Table 11.3. Final proton energy and excited state energy for the $^{16}O(d, p)^{17}O*$ reaction for a scattering angle of 25°.

E_f (MeV)	ΔE (MeV)
11.69	0.00
10.81	0.871
8.58	3.06
7.77	3.85
7.05	4.56
6.50	5.08
6.19	5.38
5.85	5.71
5.66	5.89
5.60	5.94

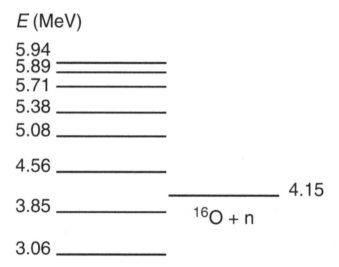

Figure 11.5. Energy level diagram for ^{17}O.

above 4.15 MeV to decay either by γ-decay or by neutron emission. The interaction between the neutron and the ^{16}O nucleus can be understood better by looking at the reactions involving neutrons incident upon nuclei, as discussed in the next section.

11.5 Neutron reactions

A neutron that is incident on an ^{16}O nucleus (for example) may be absorbed forming ^{17}O. The energy available in this process is given by equation (11.8) and in this example is the center of mass energy of the neutron plus 4.15 MeV. When the available energy is equal to the energy of an ^{17}O excited state then a resonant condition exists, and the neutron absorption cross section is greatly enhanced. The measured neutron absorption cross section for ^{16}O as a function of incident neutron energy is illustrated in figure 11.6. The resonant peaks correspond to the formation of ^{17}O excited states. Only those states above the ^{17}O neutron separation energy of 4.15 MeV are accessible as this is the Q available from a reaction with a very low energy neutron. The resonant peaks in figure 11.6 may be related to the excited state energy levels of ^{17}O, as shown in figure 11.5. For example, the lowest energy peak in figure 11.6 at has energy E_{cm} = 0.93 MeV and gives the ^{17}O excited state at 0.93 + 4.15 = 5.08 MeV. Once ^{17}O has been formed in an excited state above 4.15 MeV it may decay by one of two modes (which are referred to as channels); neutron emission back to ^{16}O or γ-decay to a lower energy level of ^{17}O. This former is referred to as the incident channel for the reaction. If the decay is by γ-decay (or a series of γ-decays) to an energy level below 4.15 MeV, then the ^{17}O nucleus can no longer decay by neutron emission and the neutron is said to have been captured. This process is the so-called (n, γ) reaction and has important applications for the study of excited state structure, the formation of radioactive nuclides, chemical analysis by neutron activation analysis and the operation of nuclear fission reactors.

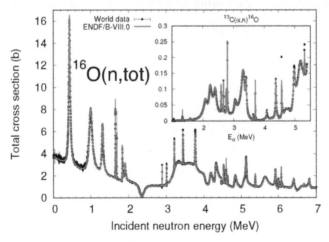

Figure 11.6. Total neutron cross section for ^{16}O as a function of incident neutron energy. Reprinted from Brown *et al* (2018). Copyright (2018). With permission from Elsevier.

It is interesting to note the values of the cross sections, as shown in figure 11.6. The radius of an ^{16}O nucleus is about 3 fm, giving a cross sectional area of about 0.3 barns (30 fm^2). The figure shows cross sections that are more than an order of magnitude larger. In fact, (n, γ) cross sections for some nuclei can be in excess of 10^5 barns. Thus, although the cross section is expressed in units of area it should not be viewed as the physical cross section. Rather, the cross section results from the quantum mechanics of the interaction between the neutrons and the nucleus and is a measure of the reaction probability.

In addition to providing information about the energies of the excited states, the energy dependence of the neutron cross section can provide information about the stability of these states. In chapter 7 it was shown that the energy width of an unstable state is related to its lifetime by the Heisenberg uncertainty principle. A generic example of neutron absorption is illustrated in figure 11.7. A neutron is absorbed by the nucleus ^AX forming an excited state of ^{A+1}X at an energy above the neutron separation threshold of ^{A+1}X. Following along the lines of equations (7.7) and (7.15), the total width, Γ, of the ^{A+1}X excited state is the sum of the partial widths for decay by γ-emission, Γ_γ, and decay by the incident channel, Γ_i:

$$\Gamma = \Gamma_\gamma + \Gamma_i$$

From the line shape given in equation (7.14), we expect an energy dependent cross section for neutrons incident upon ^AX with an energy in the vicinity of the ^{A+1}X excited state energy E_0,

$$\sigma_i(E) = \frac{C}{(E - E_0)^2 + \frac{\Gamma^2}{4}} \tag{11.10}$$

where the constant C is related to the density of states in the system. The density of states is found in the following way. The reaction is assumed to take place within the volume of the nucleus, V. Following along the lines of the discussion in section 9.2,

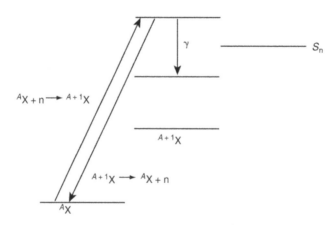

Figure 11.7. (n, γ) and (n, n) reaction channels for a neutron incident on ^AX.

the density of states per unit momentum for fermions within this volume is given by equation (9.18) as

$$\frac{dn}{dp} = \frac{4\pi p^2}{(2\pi\hbar)^3} V.$$

If the neutron, moving with velocity v within the volume V, has a cross section σ_i, then it sweeps out a volume $\sigma_i v$ per unit time. The decay rate per unit momentum is then given as the fraction of the volume V swept out per unit time multiplied by the density of states:

$$d\lambda = \frac{\sigma_i v}{V} dn = \sigma_i v \frac{4\pi p^2}{(2\pi\hbar)^3} dp.$$

This gives the total decay rate by integrating over momentum as

$$\lambda = \frac{4\pi}{(2\pi\hbar)^3} \int_{-\infty}^{+\infty} v\sigma_i p^2 \, dp.$$

Substituting (11.10) for the cross section and changing the integration over momentum to an integration over energy gives

$$\lambda = \frac{8\pi CmE}{\Gamma} \tag{11.11}$$

where m is the reduced mass of the system:

$$m = \frac{m_n m_X}{m_n + m_X}.$$

In a system with a large number of neutrons and nuclei the rate of formation of the excited state would be balanced by the rate of decay back through the incident channel. This is given as

$$\lambda = \frac{\Gamma}{\hbar}. \tag{11.12}$$

Equating equations (11.11) and (11.12) yields an expression for C and the cross section from equation (11.10) becomes,

$$\sigma_i(E) = \frac{1}{8\pi mE\hbar^2} \frac{\Gamma_i \Gamma}{(E - E_0)^2 + \frac{\Gamma^2}{4}}. \tag{11.13}$$

A more detailed analysis of this problem shows that Γ must be replaced by $g\Gamma$ where the statistical factor g is given in terms of the spins of the incident neutron, s_n, and target nucleus, s_X, and the total angular momentum of the excited state, I,

$$g = \frac{2I + 1}{(2s_n + 1)(2s_X + 1)}.$$

The above expression for the cross section is known as the Breit–Wigner formula and allows for the analysis of the resonance line shapes in data such as that shown in figure 11.6.

The (n, γ) process is important for low energy neutrons as energy levels very near the neutron separation energy are populated and γ-decay becomes the likely decay mode. The relative scale of neutron energies depends on the nature of the target nucleus. In the case of neutrons incident on ^{16}O the first available energy level in ^{17}O above the neutron separation energy is at 4.56 MeV and a center of mass neutron energy of 0.41 MeV is required to populate this level. On the average, the neutron energy necessary to populate the first available state above the neutron separation energy is inversely proportional to the density of states near the energy Q. For reasons described previously, we expect that the density of states will be greater when the value of Q places the resulting nucleus well above the ground state energy and when the value of A is large. An example of this is illustrated in figure 11.8. It is seen in this figure that the location of the first resonance peak and the spacing between the peaks for ^{238}U is of the order of tens of eV, compared with 100s of keV for ^{16}O, as seen in figure 11.6.

Neutrons for experiments can be produced by a variety of neutron-emitting sources. Unlike charged particles (for example, electrons, protons, deuterons) neutrons cannot be accelerated by means of an electric field. However, higher energy neutrons produced by a radioactive source, can be slowed to produce a source of lower energy neutrons. This is done by allowing the neutrons to lose kinetic energy through a variety of scattering mechanisms in an appropriate material, called a moderator. In general neutrons are classified according to their kinetic energy, as summarized in table 11.4. The ability to reduce the kinetic energy of a neutron from high energy (a few MeV) to a few eV or less is of crucial importance to the operation of a fission reactor and will be discussed in detail in the next chapter.

Neutron cross sections at high energies can be quite complex. At higher energies the increasing density of states decreases the average distance between resonance peaks. As the distance between the resonances becomes comparable to their halfwidth, the resonances overlap and at some point, the concept of distinct energy levels becomes ambiguous. This will be discussed further in the next chapter with specific reference to ^{235}U and ^{238}U.

An important point to consider when dealing with low energy neutron reactions is that of thermal effects. The thermal motion of the atoms in the target will influence

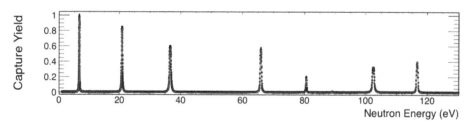

Figure 11.8. Low energy (n, γ) capture yield for ^{238}U. CC BY 4.0 Reprinted with permission from Mingrone *et al* (2017) CC BY 4.0.

Table 11.4. Categories of neutrons based on their kinetic energy.

Class	Typical E
Thermal	0.02 eV
Epithermal	1 eV
Slow	1 keV
Fast	>100 keV

the center of mass energy of the incident neutrons. If the motion of a neutron is described by a velocity $\vec{v}_n$ and the motion of an atom is described by a velocity vector $\vec{v}_A$, then the center of mass energy is given by,

$$E = \frac{1}{2}m(\vec{v}_n - \vec{v}_A)^2$$

where m is the reduced mass. This may be written in terms of the center of mass energy when thermal motion is neglected, E_{cm}, and the thermal energy of the atom, $E_A = k_B T$ as

$$E = E_{cm} + \frac{m}{m_A}E_A - 2\left[\frac{m}{m_A}E_{cm}E_A\right]^{1/2}\cos\theta \tag{11.14}$$

where θ is the angle between $\vec{v}_n$ and $\vec{v}_A$. In general m/m_A will be small and except in the case of very low energy neutrons, E_A will be small compared to E_{cm}. Thus, the second term on the right-hand side of equation (11.14) can normally be ignored. The $\cos\theta$ can take on values from -1 to $+1$ so that the actual energy may be reduced or increased relative to the value of E_{cm}. This results in a broadening of the resonance peak, the so-called Doppler broadening, where the width of the distribution of $\cos\theta$ values is approximately unity. Thus, we can write the increased width of a resonance at E_{cm} due to thermal effects as

$$\Delta E = 2\sqrt{\frac{m}{m_A}E_{cm}k_B T}.$$

An example of the thermal broadening of a low energy neutron resonance in ^{238}U is shown in figure 11.9.

11.6 Coulomb effects

In contrast to neutron reactions, reactions that involve charged particles are affected by Coulomb interactions. This is true both in the case when the incident particle is charged and when the emitted particle is charged. The interaction potential between the charged nucleus and the charged particle forms a Coulomb barrier, as for the case of α-decay described in chapter 8. Unless the energy of the particle is greater

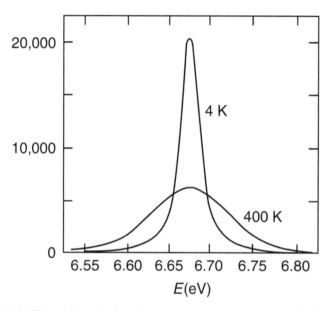

Figure 11.9. Effects of Doppler broadening on the (n, γ) resonance at 6.67 eV in ^{238}U.

than the barrier height, the calculation of reaction cross sections must be considered as a quantum mechanical tunneling problem. At low energies, charged particles that are incident on a nucleus predominantly undergo elastic scattering because of their inability to tunnel through the Coulomb barrier. With increasing particle energy, the tunneling probability, and hence the reaction cross section, increases. The calculation of the tunneling probability for charged particles incident upon a nucleus follows along the lines of the development in chapter 8 for α-decay. It is customary to parameterize the cross section due to Coulomb interactions at low energies as

$$\sigma(E) = \frac{S(E)}{E} e^{-G} \tag{11.15}$$

where the Breit–Wigner equation is written as $S(E)/E$ and is modified by the e^{-G} factor to account for the tunneling probability. Here G is proportional to $E^{-1/2}$ as given in equation (8.11) with Q replaced by the incident particle energy E. At low energies $S(E)$ is more or less constant or varies slowly (as long as no resonance reactions are present in the energy range) and equation (11.15) may be written as

$$\sigma(E) = \frac{S(0)}{E} e^{-a/\sqrt{E}} \tag{11.16}$$

where the parameter a is determined from equation (8.10) as a function of the nuclear and particle masses, charges, and dimensions. For a given process, the result is that there is a threshold energy below which the reaction cross section is virtually zero because the tunneling probability is virtually zero. The resonant peaks corresponding to the population of certain states appear at higher energies

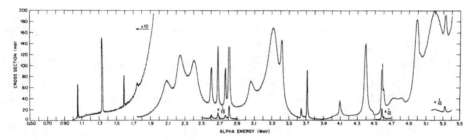

Figure 11.10. Energy dependence of the cross section for the reaction $^{13}C(\alpha, n)^{16}O$. Reprinted with permission from Bair and Haas (1973), copyright (1973) by the American Physical Society.

superimposed on a background given by equation (11.16). An example of this kind of behavior for a charged incident particle (α-particle) is shown in figure 11.10. Coulomb effects are seen in the low-energy portion of the spectrum where the cross section drops to zero below about 1.0 MeV.

Problems

11.1. Derive equation (11.9) in the text.

11.2.
   (a) An α-particle is incident on a ^{13}C nucleus leading to the reaction $\alpha + {}^{13}C \rightarrow {}^{17}O^*$. Show that this is immediately followed by the decay $^{17}O^* \rightarrow {}^{16}O + n$.
   (b) The cross section for the $^{13}C(\alpha, n)^{16}O$ reaction is shown in figure 11.10. Determine the energy of the ^{17}O excited state corresponding to the lowest energy peak in the figure.

11.3. For a tritium (^3H) nucleus incident on a ^{12}C nucleus construct a graph in the style of figure 11.4.

11.4.
   (a) Determine the missing particle/nucleus to complete the following reactions:
      (i) $^{29}Si(\alpha, n)X$
      (ii) $^{60}Ni(x, n)^{60}Cu$
      (iii) $^{111}Cd(n, x)^{112}Cd$
      (iv) $X(p, d)^{188}Os$
      (v) $^{156}Gd(d, x)^{157}Tb$
   (b) For each reaction calculate the value of Q.

11.5. Compare the (p, n) reaction to β^- decay. How are the Q values for these two processes related?

11.6. The (d, p) reaction is used to study the excited states of ^{29}Si.
   (a) What target nuclei are used in this experiment?
   (b) What is the Q for this reaction?

(c) For an incident deuteron energy of 10 MeV and a scattering angle of 30°, what is the final proton energy that corresponds to the ground and first three excited states of ^{29}Si?

11.7. Protons with an initial energy of 8 MeV are scattered elastically from a variety of different nuclei. For a scattering angle of 90°, what is the scattered proton energy when the target nucleus is (a) ^{7}Li, (b) ^{25}Mg, (c) ^{95}Mo, and (d) ^{208}Pb?

11.8. The 4.56 MeV state of ^{17}O is populated by the (n, γ) reaction with ^{16}O (see figure 11.5). Calculate the excess line width introduced by Doppler broadening at room temperature. Compare with the information for the (n, γ) reaction with ^{238}U shown in figure 11.9.

References and suggestions for further reading

Armitage B H and Meads R E 1962 Energy levels of B^{10} (I) the reactions B^{10}(p, p′)B^{10}, B^{10}(d, d′) B^{10}, and C^{12}(d, α)B^{10} *Nucl. Phys.* **33** 494–501

Bair J K and Haas F X 1973 Total neutron yield from the reactions ^{13}C(α, n)^{16}O and ^{17,18}O(α,n)^{20,21}Ne *Phys. Rev. C* **7** 1356–64

Brown D A *et al* 2018 ENDF/B-VIII.0: The 8th major release of the nuclear reaction data library with CIELO-project cross sections, new standards and thermal scattering data *Nucl. Data Sheets* **148** 1–142

Enge H A 1966 *Introduction to Nuclear Physics* (Reading, MA: Addison-Wesley)

Mingrone F *et al* 2017 Neutron capture cross section measurement of ^{238}U at the CERN n_TOF facility in the energy region from 1 eV to 700 keV *Phys. Rev. C* **95** 034604

Ripani M 2018 Energy from nuclear fission *EPJ Web Conf.* **189** 00013

IOP Publishing

An Introduction to the Physics of Nuclei and Particles
(Second Edition)

Richard A Dunlap

Chapter 12

Fission reactions

12.1 Basic properties of fission processes

Fission is the splitting of a relatively heavy nucleus into two lighter nuclei. This can be a spontaneous process, or it can be induced by the reaction of the nucleus with an incident particle (usually a neutron). Alpha decay is an extreme case of spontaneous fission where one of the nuclei is very light. Fission usually refers to the situation where the two fragments are of more similar mass. A simple fission process can be described as

$$^{A}X \rightarrow {}^{B}Y + {}^{A-B}Z. \tag{12.1}$$

An investigation of figure 4.1 shows that if a heavy nucleus breaks up into two lighter nuclei, then excess binding energy will be released, that is, the process will be exothermic. Typically, a nucleus with (say) $A = 236$ has a binding energy of about 7.6 MeV per nucleon or a total binding energy of 1.78 GeV. If this breaks up into two equal fragments then the two nuclei, with $A = 118$, have binding energies of 8.5 MeV per nucleon or a total of 1.99 GeV. This gives a net energy release of about 210 MeV per fission. The semiempirical mass formula can be used to estimate the net energy release as a function of the size of the fission fragments, as shown in figure 12.1. Here the net energy release as a function of B in equation (12.1) is plotted for $A = 236$. This indicates that the energy release is greatest for fission fragments of equal size. The semiempirical mass formula can also be used to determine the energy release for symmetric fission as a function of A in equation (12.1). This is illustrated in figure 12.2. Clearly this is consistent with the binding energy shown in figure 4.1, since, if the binding energy per nucleon for nuclei with $A/2$ nucleons becomes less than that for nuclei with A nucleons, then the process will not produce excess energy and is not energetically favorable.

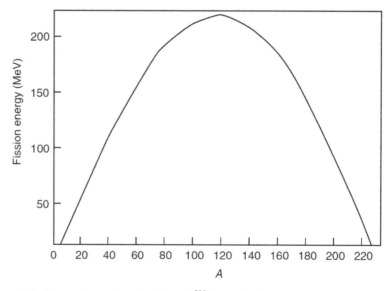

Figure 12.1. Energy release from the fission of ^{236}U as a function of A of one of the fragments.

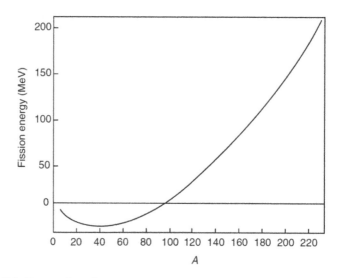

Figure 12.2. Energy release from symmetric fission as a function of a of the parent nucleus.

The physics of spontaneous fission is much like that of α-decay because the charged fragments must overcome a Coulomb barrier before they can separate. The calculation of the barrier height is not straightforward but can be accomplished by means of the semiempirical mass formula. The results of such a calculation are shown in figure 12.3. A more detailed calculation involving shell effects gives a similar curve with some additional structure near magic nuclei, as shown in the

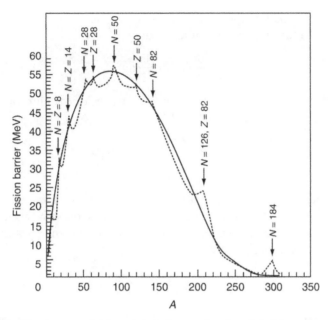

Figure 12.3. Coulomb barrier energy for symmetric fission as a function of A of the parent as calculated from the semiempirical mass formula (solid curve) The dotted curve illustrates the effect of the nuclear shell structure. Reprinted from Myers and Swiatecki (1966). Copyright (1966). With permission from Elsevier.

Figure 12.4. Schematic illustration of nuclear shapes during the fission process.

figure by the broken line. These results show that the barrier height drops to zero for A around 300, indicating that nuclei with a mass greater than this are unstable and will undergo spontaneous fission with a very short lifetime. This instability can be understood using the following model.

In the context of the liquid drop model, fission can be viewed as the fragmentation of a drop, as illustrated in figure 12.4. Any oscillations in the shape of the drop will tend to grow and to cause the drop to fragment if the deformation from spherical symmetry makes the drop more stable. We can calculate the energy associated with nuclear deformation (or stretching as shown in the figure) in terms of nuclear binding energies as

$$\Delta E = B(\varepsilon) - B(\varepsilon = 0) \tag{12.2}$$

where ε is a deformation parameter. If it is assumed that the deformations are elliptical, then ε is the eccentricity. Since nuclear matter is not compressible the volume of the nucleus does not change during deformation. We can, therefore, equate the volume of the spherical nucleus of radius R_0 before stretching to the volume of the stretched ellipsoidal nucleus as

$$\frac{4\pi}{3}R_0{}^3 = \frac{4\pi}{3}ab^2$$

where a and b are the semimajor and semiminor axes of the ellipsoid, respectively. The eccentricity is then related to a and b as

$$a = R_0(1 + \varepsilon) \tag{12.3}$$

$$b = R_0(1 + \varepsilon)^{1/2}.$$

Using the semiempirical mass formula we can now calculate the energy associated with the deformation. We anticipate that changes in the binding energy will be manifested in the surface and Coulomb terms of the semiempirical mass formula. The surface area of an ellipsoid can be expressed as

$$S = 4\pi R_0{}^2\left(1 + \frac{2}{5}\varepsilon^2 + \ldots\right)$$

where R_0 is related to ε by equation (12.3). The Coulomb energy of an ellipsoid relative to that of a sphere of the same volume can be expressed as

$$\frac{E_{\text{ellipse}}}{E_{\text{sphere}}} = 1 - \frac{1}{5}\varepsilon^2 + \ldots.$$

Taking terms to order ε^2, the semiempirical mass formula gives ΔE from equation (12.2) as

$$\Delta E = -a_S A^{2/3}\left(1 + \frac{2}{5}\varepsilon^2\right) - a_C Z(Z-1)A^{-1/3}\left(1 - \frac{1}{5}\varepsilon^2\right) - a_S A^{2/3} + a_C Z(Z-1)A^{-1/3}$$

or

$$\Delta E = \left[-\frac{2}{5}a_S A^{2/3} + \frac{1}{5}a_C Z(Z-1)A^{-1/3}\right]\varepsilon^2.$$

ΔE will be positive and the nucleus will be unstable to stretching if

$$\frac{Z(Z-1)}{A} > \frac{2a_S}{a_C}.$$

Using values of the semiempirical mass formula parameters from chapter 4 gives

$$\frac{Z(Z-1)}{A} > 50 \tag{12.4}$$

Heavy nuclei have $Z/A \approx 0.4$ and equation (12.4) gives $A \approx 300$ as the maximum mass of fission stable nuclei, consistent with the barrier energy as shown in figure 12.3.

12.2 Induced fission

In some nuclei, fission can be induced by bombardment with neutrons. Uranium provides an interesting example as it is commonly used for fuel in commercial fission reactors. Natural uranium consists of approximately 0.72% ^{235}U and 99.27% ^{238}U. The relative natural abundances are related to their α-decay lifetimes of approximately 1×10^9 y and 6.5×10^9 y, respectively. A low energy neutron can be captured by either of these nuclei yielding the following reactions:

$$n + {}^{235}U \rightarrow {}^{236}U + 6.46\ \text{MeV} \tag{12.5}$$

and

$$n + {}^{238}U \rightarrow {}^{239}U + 4.78\ \text{MeV}. \tag{12.6}$$

In heavy nuclei the density of states, especially at energies a few MeV above the ground state, is generally quite high and there will be many excited states available that can be occupied. An inspection of figure 12.3 shows that the barrier energy for $A = 236$ is about 6.2 MeV. Thus, the process in equation (12.5) leaves the ^{236}U nucleus with enough excess energy to induce fission without the need to tunnel through the Coulomb barrier. The process in equation (12.6) does not produce enough energy to induce fission and about 1.4 MeV additional is needed (which can be supplied by additional kinetic energy associated with the neutrons). Nuclides in which fission can be induced by low energy (that is, thermal) neutrons are referred to as *fissile* while those materials in which thermal neutrons will not induce fission because the energy is below the barrier are called *non-fissile*. For heavy nuclides, odd A nuclides are generally fissile while even–even nuclides are non-fissile. This is seen in the example in equations (12.5) and (12.6) and can be understood by a consideration of the pairing term in the semiempirical mass formula.

In a sample of fissile material, a spontaneous fission will produce excess neutrons, and these will induce further fissions. Under appropriate conditions a chain reaction can occur. Since, from figure 3.1, heavier β-stable nuclei have a greater ratio N/Z, the two fragments resulting from the fission of a heavy nucleus will require less than the total number of neutrons available. Thus equation (12.1) may be more correctly written as

$$^AX \rightarrow {}^BY + {}^{A-B-\nu}Z + \nu n \tag{12.7}$$

where, on the average, ν excess neutrons are given off in a fission process. Even considering the excess neutrons given off during the fission process, the fission fragments Y and Z are typically too rich in neutrons and will decay towards the β-stability line by β⁻ decay. As these nuclides approach the β-stability line, the lifetime becomes longer, and this is the source of the long-lived radioactive waste produced by fission reactors.

12.3 Fission processes in uranium

In most nuclear reactors it is fission of uranium nuclei that supplies the energy, and this is primarily a result of the behavior of ^{235}U. The induced fission process in ^{235}U is written as

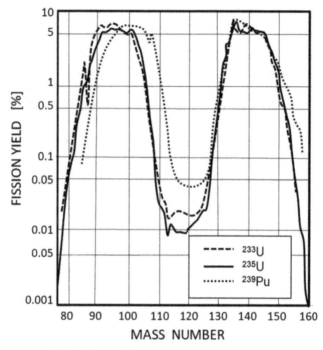

Figure 12.5. Fission yield for ^{233}U, ^{235}U and ^{239}Pu. CC BY 4.0 Reproduced from Woźnicka (2019) CC BY 4.0.

$$n + {}^{235}U \rightarrow {}^{B}Y + {}^{236-B-\nu}Z + \nu n$$

where, on the average ν is about 2.5. These excess neutrons are referred to as prompt neutrons and are given off on the time scale of the fission process, about 10^{-14} seconds. The distribution of masses for the fission fragments from ^{235}U (and some other fissile nuclides) is referred to as the fission yield and is illustrated in figure 12.5. This figure illustrates that the fission process does not result in equal-sized fragments. By definition this graph must be approximately symmetric since every fission that produces a fragment that is larger than $A/2$ must also produce a fragment that is smaller than $A/2$ by the same amount. Minor variations result because of differences in the number of excess neutrons that are emitted.

The fission fragments are normally left in excited states that γ-decay to their ground states. These γ-rays are referred to as prompt γ-rays. The energy that is immediately released by the fission process is distributed between the kinetic energy of the fission fragments, the kinetic energy of the prompt neutrons and the energy of the prompt γ-rays, as shown in table 12.1. Primarily, it is this energy that becomes available as heat that can be extracted from the reactor. The neutron rich fission fragments decay by a series of β^- and γ-decays toward β-stability. This results in a delayed release of energy due to electron and antineutrino kinetic energy (the latter of which is lost into space) and γ-ray energy, as given in the table. In some cases, a nuclide in the β-decay sequence is left in an excited state that is above the neutron

Table 12.1. Distribution of energy from the fission of ^{236}U.

Source of energy	E (MeV)
Fission fragment kinetic energy	167
Prompt neutron kinetic energy	5
Prompt γ-ray energy	6
Delayed β-decay energy	8
Delayed γ-ray energy	7
Delayed antineutrino energy	12
Total	205

separation threshold and can therefore decay by neutron emission. These are referred to as delayed neutrons and, on the average, there are about 0.02 neutrons per fission given off on a time scale of tens of seconds after the fission event. Although these represent a negligible amount of energy, they are very important in controlling the chain reaction, as discussed below.

12.4 Neutron cross sections for uranium

In order to understand the details of induced fission in uranium it is essential to investigate the neutron cross section as a function of energy. There are a number of possible processes by which the neutrons can interact with nuclei. These are discussed below.

Elastic scattering—It is important to realize that elastic scattering conserves kinetic energy. However, as the target nucleus gains energy when it recoils, the neutron loses some small amount of energy.

Inelastic scattering—In this process the neutron gives up some of its kinetic energy to the target nucleus leaving it in an excited state. This process has a threshold energy equal to the energy of the first excited state above the ground state of the nucleus. For ^{235}U and ^{238}U, the inelastic scattering threshold is 14 and 44 keV, respectively.

Radiative capture—In the (n, γ) reaction, a neutron is absorbed and the resulting nucleus decays by γ-emission to a state below the neutron separation energy, thereby capturing the neutron. The cross section for (n, γ) process is characterized by a series of resonances as discussed in the last chapter. The lowest energy resonance occurs at an energy equal to the difference between the neutron separation energy and the energy of the next available state.

Fission—In this process the neutron leaves the nucleus in an energy state above the fission barrier and fission proceeds spontaneously on a very short time scale. The energy available is the sum of the center of mass energy and the Q for the process. The energy dependent neutron induced fission cross sections for ^{235}U and ^{238}U are illustrated in figure 12.6.

The total neutron cross section is the sum of the four contributions described above. In order to understand the relative importance of these four processes at

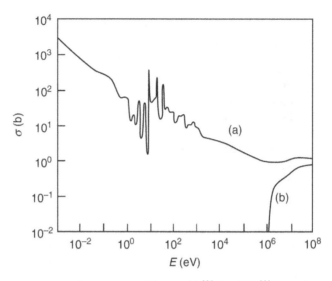

Figure 12.6. Fission cross sections for neutrons incident on (a) ^{235}U and (b) ^{238}U as a function of energy. The figure illustrates the general features but not the details of the cross section for ^{235}U in the region of (n, γ) resonances.

different energies we divide the energy scale into four regions; $E < 1$ eV, 1 eV$< E <$ 100 eV, 100 eV $< E < 1$ MeV, and $E > 1$ MeV, and discuss the behavior of ^{235}U and ^{238}U for each of these regions.

12.4.1 Cross sections for ^{235}U

$E < 1$ eV—The cross section is dominated by the fission cross section, and this accounts for about 85% of the total. The remaining cross section is predominantly due to (n, γ) processes as the result of a ^{236}U resonance lying just below $E = 0$.

    1 eV $< E < 100$ eV—This region is dominated by the (n, γ) resonances with the remaining cross section being fission and to a lesser extent, elastic scattering. It is important to realize that scattering processes reduce the energy of the neutrons while (n, γ) processes absorb them.

    100 eV $< E < 1$ MeV—(n, γ) processes are important here as well but the energy levels of the excited states overlap considerably, and the resonances no longer show discrete peaks. Above the 14 keV threshold, inelastic scattering processes are important. Fission processes become less important as the energy increases.

    $E > 1$ MeV—Processes are similar to those in the previous energy range but with a decreasing probability of (n, γ) processes.

    Cross Sections for ^{238}U

    $E < 1$ eV—Elastic scattering is the only possible process in this region.

    1 eV $< E < 100$ eV—Resonances due to (n, γ) processes and some elastic scattering; see figure 11.8.

    100 eV $< E < 1$ MeV—Unresolved (n, γ) resonances, some elastic scattering and inelastic scattering above the threshold energy of 44 keV.

$E > 1\,\mathrm{MeV}$—Predominantly inelastic scattering and a reduced probability for (n, γ) reactions. Fission above a threshold energy of about 1.4 MeV.

Neutrons that are given off during the fission process in $^{235}\mathrm{U}$ or $^{238}\mathrm{U}$ have a kinetic energy of about 2 MeV. Based on the information above we can understand the interaction of these neutrons with uranium nuclei. A simple example is given in the next section.

12.5 Critical mass for chain reactions

In a sample of uranium, a chain reaction is produced if the $(\nu - 1)$ excess neutrons produced by a fission event (see equation (12.7)) induce more than one additional fission. In a controlled chain reaction, the neutrons produced by one fission will induce exactly one more fission. The remaining neutrons will be lost either by (n, γ) processes or by exiting the sample. In order to understand this problem quantitatively we can look at an example of a piece of pure $^{235}\mathrm{U}$. If each fission neutron will, on the average, produce $q\nu$ neutrons then there will be a net gain of $(q\nu - 1)$ neutrons on the time scale of the fission, τ. The value of $q < 1$ and accounts for the neutrons that are lost. Thus, if $n(t)$ neutrons are present at time t then at time $t + dt$ there will be

$$n(t + dt) = n(t)\left[1 + (q\nu - 1)\frac{dt}{\tau}\right]$$

neutrons. This can be written as

$$\frac{dn}{dt} = \frac{(q\nu - 1)n(t)}{\tau}$$

and has the solution

$$n(t) = n(0)\exp\left[\frac{(q\nu - 1)t}{\tau}\right]. \qquad (12.8)$$

If $q\nu - 1$ is negative then $n(t)$ becomes small and the chain reaction dies out, if $q\nu - 1$ is positive then $n(t)$ becomes very large very fast, and the chain reaction is uncontrolled. It is only when $q\nu = 1$ that the number of neutrons remains constant, and the chain reaction is controlled. Since $\nu = 2.5$ this implies a value of q about 0.4 for a controlled chain reaction. In order to see how $q\nu$ can be controlled we need to determine how large τ is and how far the neutrons travel during that time. From figure 12.6 we see that a 2 MeV neutron in $^{235}\mathrm{U}$ has a fission cross section of about 1 barn. This may be compared with the total neutron cross section for $^{235}\mathrm{U}$ at this energy, which is about 7 barns (the nonfission part being primarily inelastic scattering). Thus, on the average, a neutron will undergo one fission per seven interactions. The distance that the neutron travels between interactions (that is, the mean free path) is

$$l = \frac{1}{\rho\sigma}$$

where ρ is the number density of nuclei in the material and σ is the total cross section. For uranium $\rho = 4.8 \times 10^{28}$ m^{-3} and, using $\sigma = 7$ barns, we find a mean free path of 0.03 m. A simple nonrelativistic calculation shows that a 2 MeV neutron will traverse this distance in 1.5×10^{-9} s. Because, on the average, seven interactions are necessary to induce one fission, then the total time between fissions is $7 \times 1.5 \times 10^{-9} = 10^{-8}$ s (since the actual fission process is much faster). The total distance traveled is found by considering the neutron's path as a random walk, because each inelastic scattering event will change the direction of the neutron's path in a random way. The total distance traveled from the origin is given as $\sqrt{7} \times 0.03$ m $= 0.079$ m. Since the (n, γ) reaction is not a significant contribution to neutron loss in this energy region, then a sphere of ^{235}U with a radius much less than 0.079 m will lose most of the fission neutrons before they induce further fission events, and the chain reaction will die out. A sphere of radius much more than 0.079 m will have q greater than unity and the reaction will be uncontrolled. The size of the ^{235}U sample at which the chain reaction is sustained is called the critical radius (and the mass is called the critical mass). A more detailed calculation for ^{235}U gives a critical radius of 0.087 m corresponding to a critical mass of about 52 kg.

12.6 Moderators and reactor control

In a natural mixture of ^{235}U and ^{238}U there is a small probability of fission being induced by 2 MeV neutrons. Since the concentration of ^{235}U is small (0.72%), the probability of ^{235}U fission is also small. The majority of neutrons will scatter inelastically with ^{238}U nuclei until their energy is below the fission threshold of 1.4 MeV. The greatest probability of inducing fission is at low energy in ^{235}U. Even though the fraction of ^{235}U is small the fission cross section is about three orders of magnitude greater than at high energy for either ^{235}U or ^{238}U. In a reactor using natural uranium or uranium only slightly enriched in ^{235}U there are two major design problems to overcome: how to reduce the energy of the high energy neutrons to the range below about 1 eV without losing neutrons to the (n, γ) process (primarily in the ^{238}U) and how to sustain the chain reaction in a controlled way.

The most common design of a nuclear reactor is the thermal reactor that uses natural or only slightly ^{235}U enriched uranium as a fuel. This reactor design utilizes a core consisting of number of uranium fuel rods, each of which is less than the critical mass (typically about 1 cm in diameter). The rods are bundled together to form fuel assemblies in arrays of typically 8×8 to 16×16 rods (depending on the reactor design). The fuel assemblies are surrounded by a moderator that is contained in a reactor vessel and are separated by control rods, as illustrated in figure 12.7. Fast neutrons within a fuel rod have a small probability of inducing fission within the rod but will most likely be emitted into the moderator while their energy is still in the MeV range. The moderator is a material that effectively decreases the energy to a fraction of an eV. These thermalized neutrons are then incident upon another fuel rod where they will most likely induce a fission process in ^{235}U, because of the large fission cross section in this isotope at low energies (see figure 12.6). The control rods are made of a material that will effectively stop the neutrons and are used to limit the

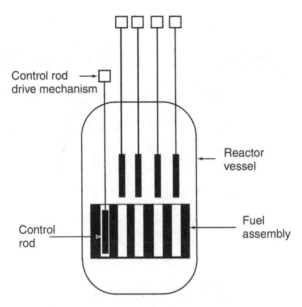

Figure 12.7. Basic design of the core of a fission reactor.

number of neutrons that travel between fuel rods allowing the value of q in equation (12.8) to be controlled. We briefly discuss the criteria for suitable materials for the moderator and control rods below.

Moderator—The purpose of the moderator is to allow the neutron energy to be reduced in the region outside the fuel rod, thus avoiding neutron loss from (n, γ) reactions in the uranium. The moderator is a material that interacts with neutrons over a wide range of energies by elastic and/or inelastic scattering, thus reducing the neutron energy without reducing the number of neutrons. Desirable criteria for the choice of moderator are as follows:

1. Inexpensive and easily obtained;
2. Small (n, γ) cross section over a wide range of energies;
3. Small nuclear mass, in order to maximize the energy transfer per scattering event;
4. Reasonably high density;
5. Chemically stable with a low level of toxicity.

Various materials satisfy some of these criteria although no material is ideal. Graphite satisfies all criteria except number 3 requiring the use of a larger quantity of moderator material. Water satisfies all criteria except number 2 and requires the use of ^{235}U enriched fuel. Heavy water (D_2O) has a lower (n, γ) cross section than (H_2O) but the processes that do occur (n + d → t + γ) produce radioactive tritium as an undesirable by-product. In general, the properties of the moderator determine the quantity of moderator needed as well as the degree to which the uranium has to be enriched in ^{235}U for the reactor to operate efficiently.

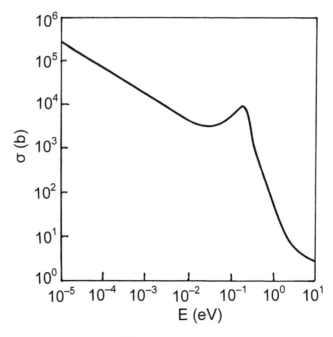

Figure 12.8. Total neutron cross section for ^{113}Cd as a function of energy. Data from National Nuclear Data Center (2011).

Control rods—The control rods should have a large neutron absorption (that is (n, γ)) cross section, especially at low energies. Cadmium metal is the most commonly used material and has a large low energy neutron cross section due to the presence of 12.3% ^{113}Cd in natural Cd. The total neutron cross section for ^{113}Cd is shown in figure 12.8. The large peak at a fraction of an eV is due to an (n, γ) process and the low energy 1/E dependence as given in equation (11.13) is seen in the figure.

12.7 Reactor stability

In principle it should seem that it would merely be a matter of regulating the position of the control rods between the fuel elements in order to control the number of neutrons that can induce fission and thereby adjust the value of $q\nu$ in equation (12.8). The problem here, however, is a matter of the time scale involved. As we have seen, the average lifetime of a fission neutron in ^{235}U is about 10^{-8} s. In a fission reactor the neutrons spend much of their time in the moderator and their lifetime can be several orders of magnitude longer than given by our simple calculation. However, this time scale (perhaps 10^{-4} s) is still too short to allow the position of control rods in the reactor to be adjusted mechanically. The key to reactor control lies with the delayed neutrons. Each fission process produces ν' delayed neutrons (about 0.02) in addition to the ν prompt neutrons. The critical condition as described above is now

$$(\nu + \nu')q = 1. \tag{12.9}$$

If the reactor is designed in such a way that $q\nu = 0.99$ then the control rods can be utilized on a time scale of tens of seconds to regulate the delayed neutrons to ensure that the condition in equation (12.9) is met.

An important factor in reactor stability is the influence of temperature on the parameter q. As a result of energy release due to fission the reactor components naturally increase in temperature, T. For safety reasons it is important that

$$\frac{dq}{dT} < 0.$$

If the opposite were true, then temperature increases would lead to an increased q and could result in an uncontrolled chain reaction. One of the major factors influencing dq/dT is the Doppler broadening of the (n, γ) resonances in the fuel rods. Although most of the neutrons that are lost are absorbed in the control rods, a small number pass through a fuel rod when their energy is in the range of the (n, γ) resonances in ^{238}U and are lost by radiative capture. As the temperature increases and these resonances broaden (see section 11.5) the probability that a neutron will have the correct energy to be absorbed increases. This additional probability of neutron absorption corresponds to a decrease in q as temperature increases.

12.8 Current fission reactor designs

Thermal neutron reactors fall into several categories, including military reactors that are used for the production of weapons grade fissile material and commercial power reactors for electricity production. Here we look at the application of nuclear reactors for commercial purposes. There are several designs of thermal neutron fission reactors, as described above, that are in common use worldwide as power reactors. These reactors can be categorized in terms of the moderator material that they utilize. The moderators that are in use, as described in section 12.6, are water (i.e., light water, H_2O). heavy water (D_2O) and graphite (i.e., carbon).

Light water moderators were largely developed in the United States and there are two basic designs that are in use, boiling water reactors and pressurized water reactors, that are distinguished on the basis of the design of their cooling system. In both designs the water which serves as the moderator, also serves as the cooling medium. In the simplest design, the boiling water reactor (BWR), the water that serves as the moderator is boiled in the reactor core by the heat produced by the fission reactions. The steam that is produced is used directly to turn a turbine, which drives a generator to produce electricity. The basic design of a boiling water reactor is shown in figure 12.9. After driving turbines, the steam is cooled and condensed in a heat exchanger and then returned to the reactor core. Heat that is dissipated through the heat exchanger, is transferred to either a significant body of water, such as a river or the ocean, or is dissipated in the atmosphere through cooling towers. The use of cooling towers has become the standard approach to cooling fission reactors in most parts of the world as it is, in general, less of an environmental concern than cooling using a body of water.

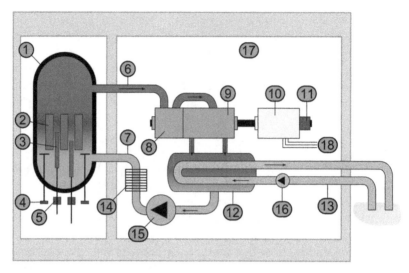

Figure 12.9. Diagram of boiling water reactor. (1) Reactor vessel, (2) fuel core element, (3) control rod element, (4) circulation pumps, (5) control rod motors, (6) steam, (7) inlet circulation water, (8) high pressure turbine, (9) low pressure turbine, (10) electric generator, (11) electrical generator exciter, (12) steam condenser, (13) cold water for condenser, (14) pre-warmer, (15) water circulation pump, (16) condenser cold water pump, (17) concrete chamber, (18) connection to electricity grid. This [boiling water reactor system diagram] image has been obtained by the author from the Wikimedia website where it was made available by Steffens (2011) under the GNU Free Documentation licence. It is included within this chapter on that basis. It is attributed to Robert Steffens.

The pressurized water reactor (PWR) is somewhat more complex as illustrated in figure 12.10. In this design, the water that is heated in the reactor core is kept under pressure and is prevented from boiling. This superheated water transfers its heat, by means of a heat exchanger, to water which is not under pressure, and which boils to produce steam to drive turbines. This design reduces the possibility of radioactive contamination from the core being released to the environment.

In both the boiling water reactor and the pressurized water reactor, the properties of H_2O as a moderator require that the uranium in these reactors be enriched to 2% to 3% ^{235}U. This requirement specifically deals with the higher (n, γ) cross section for water, which results from the properties of ^1H and which limits the amount of moderator which can be incorporated into the reactor design.

A reactor design similar to the pressurized water reactor has been developed in Canada and utilizes heavy water (D_2O) as the moderator and the cooling medium. This design is known as a CANDU reactor (CANadian Deuterium–Uranium reactor) and the basic design is illustrated in figure 12.11. As in the pressurized light water reactor, the heavy water is contained in a sealed system and is kept under pressure to prevent it from boiling. This superheated heavy water then transfers heat to light water through a heat exchanger. The steam output is connected to a turbine and generator to produce electricity and the water is cooled in a condenser and returned to the heat exchanger, as for the pressurized water reactor in figure 12.11.

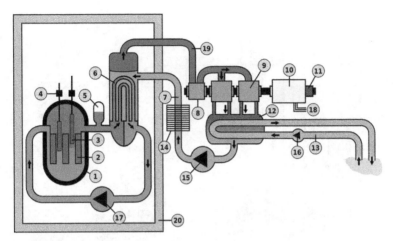

Figure 12.10. Diagram of a pressurized water reactor. (1) Reactor pressure vessel, (2) fuel assembly, (3) control rod, (4) control rod drive mechanism, (5) pressurizer, (6) steam generator, (7) feedwater, (8) high-pressure turbine, (9) low-pressure turbine, (10) generator, (11) exciter, (12) condenser, (13) cooling water, (14) feedwater heater, (15) feedwater pump, (16) cooling water pump, (17) reactor coolant pump, (18) generated electricity, (19) main steam line, (20) containment. This [schematic diagram of a pressurized water reactor] image has been obtained by the author from the Wikimedia website where it was made available by Lardot (2006) under a CC BY-SA 3.0 licence. It is included within this chapter on that basis. It is attributed to Nicolas Lardot.

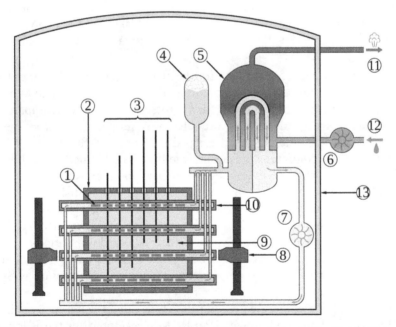

Figure 12.11. Diagram of a CANDU heavy water reactor. (1) Nuclear fuel rod, (2) calandria, (3) control rods, (4) pressurizer, (5) steam generator, (6) light water pump, (7) heavy water pump, (8) nuclear fuel loader, (9) heavy water, (10) pressure tubes, (11) steam, (12) condensate, (13) containment. This [schematic diagram of the pressurised heavy water cooled version of a CANDU (CANada Deuterium–Uranium) nuclear reactor] image has been obtained by the author from the Wikimedia website where it was made available by Inductiveload (2007) under a CC BY-SA 2.5 licence. It is included within this chapter on that basis. It is attributed to Inductiveload.

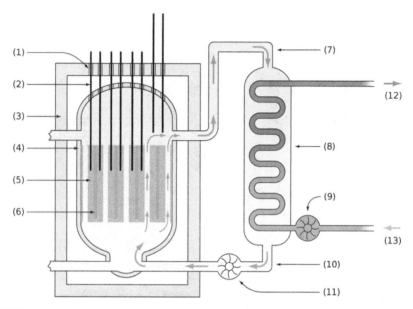

Figure 12.12. Diagram of a Magnox gas cooled graphite moderated fission reactor. (1) Charge tubes, (2) control rods, (3) radiation shielding, (4) pressure vessel, (5) graphite moderator, (6) fuel rods, (7) hot gas duct, (8) heat exchanger, (9) water circulator, (10) cool gas duct, (11) gas circulator, (12) steam, (13) water. This [schematic diagram of a Magnox nuclear reactor] image has been obtained by the author from the Wikimedia website where it was made available by Emoscopes (2005a) under a CC BY-SA 3.0 licence. It is included within this chapter on that basis. It is attributed to Emoscopes.

The use of a pressurized closed system for the heavy water not only serves to reduce potential environmental radioactive contamination but is necessary to avoid loss of the expensive heavy water. The lower (n, γ) cross section for ^{2}H (compared to ^{1}H) permits a reactor design that can more effectively moderate neutron energy and subsequently allows for the use of unenriched uranium as a fuel.

Thermal neutron fission reactors designed in the United Kingdom and the former Soviet Union utilize graphite as a moderator. Since graphite, is a solid, it cannot be used as both the moderator and the cooling medium. The British reactor design is gas cooled and uses carbon dioxide as the coolant. These reactors, shown in figure 12.12, are known as Magnox reactors (the name was derived from the magnesium–aluminum alloy used to clad the fuel rods) and were originally designed as dual-use reactors to produce both electricity and weapons grade ^{239}Pu. Because of the moderating properties of graphite, natural (unenriched) uranium is used as a fuel.

The Soviet graphite moderated reactor design uses light water as a coolant. This type of reactor is commonly known as an **RBMK** reactor, after the Russian name Reaktor Bolshoy Moshchnosti Kanalnyy (high-power channel-type reactor) and was also originally designed as a dual-use reactor. Like the gas cooled graphite reactor, the water-cooled version also uses natural uranium as a fuel. The design of an **RBMK** reactor is shown in figure 12.13.

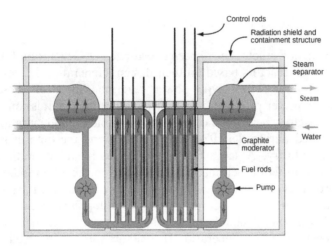

Figure 12.13. Diagram of an RBMK water cooled graphite moderated rector. This [diagram of RBMK nuclear reactor] image has been obtained by the author from the Wikimedia website where it was made available by Emoscopes (2008) under a CC BY-SA 3.0 licence. It is included within this chapter on that basis. It is attributed to Emoscopes.

Table 12.2. Number of different types of nuclear fission reactors that are operational (as of 2014).

Type	Number
Boiling water reactor (light water)	80
Pressurized water reactor (light water)	277
Heavy water reactor	49
Gas-cooled graphite reactor	15
Water cooled graphite reactor	15
Fast breeder reactor	2

At present there are abound 440 operational fission power reactors worldwide. Table 12.2 provides a breakdown of the number of each type of reactor, as described above, that is in use. Fast breeder reactors are discussed in the next section.

12.9 Advanced fission reactor designs

The utilization of nuclear fission power grew substantially in the 1970s but has not grown significantly in the past 35 years or so. Part of the lack of nuclear growth has been the result of shifting public opinion as a result of several serious reactor accidents. Principal among these were Three Mile Island in 1979, Chernobyl in 1986 and Fukushima in 2011. Going forward, future nuclear energy development must certainly consider reactor safety. However, another important concern is the utilization of uranium resources. These resources are limited and current reactor designs (at least most of them) utilize the energy content of less than 1% of the uranium.

Historically, fission reactors have been categorized as Generation I, Generation II and Generation III, reactors based on their design. Generation I reactors were very early experimental and prototype power reactors. Some of these were used commercially for power production but all have been decommissioned as of 2015. Generation II reactors were more commercially viable than Generation I and were largely constructed during the 1970s and 1980s during the time of growth of the nuclear power industry. A large fraction of these is still operational and they constitute the majority of functional power reactors in the world today. Generation III reactors are an evolutionary development from Generation II reactors that are designed to deal with a number of issues including, higher thermal efficiency, improved fuel technology, significantly enhanced safety and reduced capital and maintenance costs. The first Generation III reactors were constructed in the mid- to late-1990s, although, to date, only a dozen or so have become operational.

Generation IV reactors incorporate significant changes to previous designs and may be considered as more revolutionary than evolutionary. The main goals of Generation IV reactors are to improve the economics, safety, reliability and sustainability of nuclear energy. The Generation IV International Forum was initiated in 2000 by the United States Department of Energy's Office of Nuclear Energy and is an international co-operative endeavor aimed at investigating options for the design and construction of advanced nuclear fission reactors. The Forum has identified six basic design approaches that could lead to significantly improved nuclear power reactors. The reactor designs that are being considered are:

- Very-high-temperature gas-cooled reactor;
- Supercritical-water-cooled reactor;
- Molten-salt reactor;
- Gas-cooled fast reactor;
- Lead-cooled fast reactor;
- Sodium-cooled fast reactor.

The first design of reactor is a thermal neutron reactor which operate on basically the same physical principles as the reactors described above. The second two reactor designs can be either a thermal neutron reactor or a fast neutron reactor. The last three designs are fast neutron reactors. Reactors in many of the categories shown above are still in the stage of concept development and design. In some cases, experimental reactors have been constructed. In one category commercial power reactors are already in operation, as described below. While a detailed discussion of all of these possible designs is beyond the scope of the present book, it is of interest to consider two designs that have attracted considerable attention, in order to appreciate the innovative approach that Generation IV reactors can take. Here we look at the design of a very high temperature gas cooled reactor and a molten salt fast reactor. The latter description includes an overview of the basic operation of a fast neutron reactor. The use of thorium as a reactor fuel is also considered.

Pebble bed reactor scheme

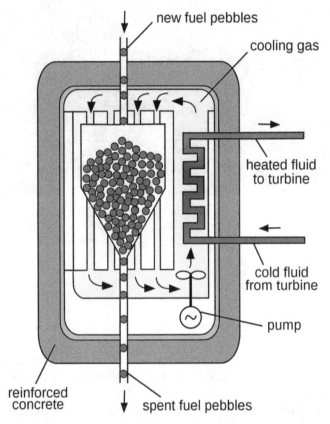

Figure 12.14. Diagram of a pebble bed reactor, one of the designs for a possible Generation IV fission reactor in the very-high-temperature gas-cooled reactor category. This [pebble bed reactor scheme] image has been obtained by the author from the Wikimedia website https://commons.wikimedia.org/wiki/File: Pebble_bed_reactor_scheme_(English).svg (Picoterawatt 2012), where it is stated to have been released into the public domain. It is included within this chapter on that basis.

The basic design of the pebble bed reactor, which is a type of very high temperature gas cooled reactor, is shown in figure 12.14. In this design, the fissile fuel is in the form of spheres of about 0.5–0.8 mm in diameter. Each sphere contains a small grain of fuel which is coated with several layers of carbon and finally a layer of silicon carbide. About 15 000 of these fuel spheres are combined together in a graphite matrix and formed into a spherical fuel element (called a pebble) about 6 cm in diameter (see figure 12.15). About 300 000 of these graphite pebbles are contained in the core of the reactor, as shown in figure 12.14. The graphite acts as the moderator, and heat is extracted by circulating gas (usually helium) through the core and then transferring the heat through a heat exchanger to water to operate a turbine. New fuel pebbles are added to the core from the top and spent fuel pebbles

Figure 12.15. Graphite fuel pebble for a pebble bed fission reactor. This [graphite pebble for pebble bed reactor] image has been obtained by the author from the Wikimedia website where it was made available by Weirdmeister (2017) under a CC BY-SA 4.0 licence. It is included within this chapter on that basis. It is attributed to Weirdmeister.

are removed from the bottom. The reactor is designed to operate at temperatures up to around 1000 °C. Reactor control is passive. There are no control rods, and temperature is maintained by negative feedback due to reduced efficiency resulting from increased Doppler broadening with increasing temperature, as illustrated in figure 11.9. The fuel grains are designed to withstand temperatures of up to about 1600 °C and the ceramic outer layer acts as a seal to prevent the escape of radioactive material. It is important to note that the high operating temperature of these reactors allows them to achieve much better thermodynamic efficiencies for electricity generation than existing lower temperature reactors.

Several experimental reactors have been built along the lines of the pebble bed design. There are some concerns about this approach to a fission reactor. Graphite is combustible and it is essential that oxygen is prevented from coming in contact with the fuel pebbles when they are at elevated temperature. Friction between fuel pebbles can cause the formation of graphite dust by abrasion. It is important to ensure that dust is properly contained, as there is a possibility that the abrasion could release radioactive dust as well. Finally, a major drawback of the design is the inability to accurately monitor the operation of the of the reactor in real time and the integrity of the fuel pebbles can only be assessed infrequently.

Fast breeder reactors are designed to make use of much of the energy content of the ^{238}U component of natural uranium. As natural uranium is 99.28% ^{238}U, the ability to utilize the energy from this isotope will very significantly increase the longevity of this energy source. The basic principle of the fast breeder reactor is that it breeds fissile fuel from non-fissile material, referred to, in this case, as fertile material. A simple example of this process involves the reaction of neutrons with ^{238}U. The (n, γ) process for neutrons incident on ^{238}U is

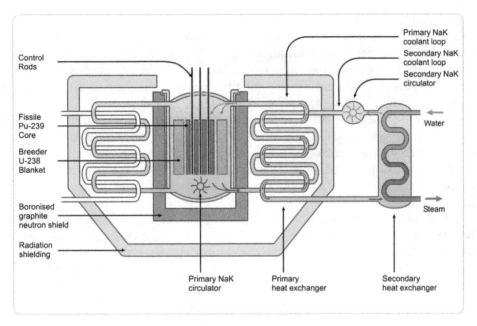

Figure 12.16. Diagram of a sodium-cooled fast breeder reactor. This [schematic diagram showing the operation of the DFR (Dounreay Fast Reactor)] image has been obtained by the author from the Wikimedia website where it was made available by Emoscopes (2005b) under a CC BY-SA 3.0 licence. It is included within this chapter on that basis. It is attributed to Emoscopes.

$$n + {}^{238}U \rightarrow {}^{239}U + \gamma.$$

The ${}^{239}U$ produced in this reaction decays by the β-decay,

$${}^{239}U \rightarrow {}^{239}Np + e^- + \bar{\nu}_e.$$

with a lifetime of about 40 min. This decay is followed by the β-decay,

$${}^{239}Np \rightarrow {}^{239}Pu + e^- + \bar{\nu}_e$$

with a lifetime of about four days. The ${}^{239}Pu$ formed in this decay process has a lifetime of about 35 000 years and, on the time scale of the operation of a nuclear reactor, can be considered stable. As expected for an actinide nucleus with an odd value of A, ${}^{239}Pu$ is fissile and is suitable for use in place of ${}^{235}U$ as a fuel in a fission reactor.

The design of a fast breeder reactor, in this case a sodium-cooled fast breeder reactor, is shown in figure 12.16. As seen in the diagram the reactor core is surrounded by a breeder blanket containing fertile ${}^{238}U$. The core contains fissile material such as ${}^{235}U$ or ${}^{239}Pu$ which produces neutrons by fission. A fraction of those neutrons induces additional fissions in the core in order to maintain a sustained chain reaction and produce energy, while excess neutrons travel to the breeder blanket where they breed new fissile fuel as described above.

Since it is essential that excess neutrons make it to the breeder blanket and are not lost in various reactions in the core, the fast breeder reactor is designed without the inclusion of a moderator. Following along the lines of the discussion above for the operation of thermal neutron reactors, the lack of a moderator implies that the fuel in fast breeder reactor must be substantially enriched in fissile nuclides. An enrichment of around 25% is typical for most breeder reactor designs.

The figure of merit for the breeder reactor is the breeding ratio, that is, the ratio of the number of new fissile nuclei created in the blanket to the number of fissile nuclei that undergo fission in the core. A breeding ratio greater than unity means that the reactor produces more fissile fuel than it expends. Since the reactor breeds ^{239}Pu, it would be most practical, in a sustainable fast breeder reactor program, to utilize fissile fuel in the core that is largely ^{239}Pu.

Of all of the approaches to Generation IV reactors, the sodium-cooled fast breeder reactors have the longest history of development. The first functional test reactor (ER-1) was built in the United States in 1950. Since then, more than twenty sodium-cooled fast breeder reactors have been built worldwide. Nearly all of these have been small, experimental facilities. However, three commercial scale fast breeder reactors have been constructed. The Superphénix (or SPX) fast breeder reactor on the Rhône river at Creys-Malville in France, was a 1242 MW (electric) fast breeder reactor that was operational and connected to the grid from 1986 to 1997. The Soviet/Russian fast breeder reactor program has been the most active and two notable fast breeder reactors are currently providing electricity to the grid (see table 12.2). The BN-600 reactor is a 600 MW (electric) fast breeder reactor located at the Beloyarsk Nuclear Power Station, in Zarechny, Sverdlovsk Oblast, Russia. It became operational in 1980. The BN-800 is an 880 MW (electric) fast breeder reactor, also located at the Beloyarsk Nuclear Power Station that became operational in 2016. The reactor is shown in figure 12.17.

While the implementation of fast breeder reactors to produce fissile ^{239}Pu from naturally occurring non-fissile ^{238}U is a potentially viable approach to vastly extend the longevity of nuclear fission power, the use of thorium as a source of power is also an important possibility. Naturally occurring thorium is 100% ^{232}Th which decays by α-decay with a lifetime of about 2×10^{10} y. Because of its long lifetime, thorium is more abundant on earth than uranium. As expected for an even–even actinide nucleus, ^{232}Th is non-fissile. However, the availability of ^{235}U and fissile material which can be bred from this nuclide, makes it possible to breed fissile reactor fuel from fertile ^{232}Th. The basic approach is to use excess neutrons from a fissile source (typically ^{235}U or ^{239}Pu) to react with thorium in a breeder blanket according to the (n, γ) process,

$$n + {}^{232}\text{Th} \rightarrow {}^{233}\text{Th} + \gamma.$$

The ^{233}Th produced in this reaction decays by the β-decay,

$$^{233}\text{Th} \rightarrow {}^{233}\text{Pa} + e^- + \bar{\nu}_e.$$

with a lifetime of about 32 min. This decay is followed by the decay

Figure 12.17. The BN-800 is sodium-cooled fast breeder reactor at the Beloyarsk Nuclear Power Station, in Zarechny, Sverdlovsk Oblast, Russia. This reactor became operational in 2016. This [reactor BN-800] image has been obtained by the author from the Wikimedia website https://commons.wikimedia.org/wiki/File:BN-800_reactor.jpg (Rosatom 2016), where it is stated to have been released into the public domain. It is included within this chapter on that basis.

$$^{233}\text{Pa} \rightarrow \,^{233}\text{U} + e^- + \bar{\nu}_e$$

with a lifetime of about 39 days. The ^{233}U formed in this decay process has a lifetime of about 230 000 years and is, therefore, suitable as a fissile fuel for a fission reactor. This ^{233}U can be utilized in a thermal reactor to produce energy or in the core of a fast breeder reactor to produce energy and additional ^{233}U fuel in a ^{232}Th blanket. The induced fission process in ^{233}U is

$$n + \,^{233}\text{U} \rightarrow \,^{234}\text{U} \rightarrow \text{fission}.$$

In order to implement a sustainable thorium fast breeder reactor program, the breeding ratio must be greater than unity so that more fissile material is produced than is consumed. The program will be sustainable, without additional input of ^{235}U or ^{239}Pu, when sufficient ^{233}U has been bred from natural ^{232}Th. Estimates for the time scale of this process are in the range of 30–40 years.

India has the largest thorium reserves in the world and also has the most ambitious program to develop thorium as a reactor fuel. The Fast Breeder Test Reactor (FBTR) located at Kalpakkam, Tamil Nadu, India is a 13 MW (electric) facility completed in 1985 that is still operational. A larger Prototype Fast Breeder Reactor (PFBR) is a 500 MW (electric) reactor that is nearing completion. Both reactors are part of the second stage of India's three-stage plan to develop thorium fueled fission reactors. The first stage was the utilization of thermal neutron reactors to breed ^{239}Pu from fertile ^{238}U. The second stage is the use of fast breeder reactors to breed fissile ^{233}U from fertile ^{232}Th. The final stage is to use the ^{233}U as a fuel to provide energy, with excess neutrons to be used to breed additional ^{233}U from a ^{232}Th blanket. Once sufficient ^{233}U is produced the last stage of this program is sustainable.

Many existing operational nuclear power reactors are Generation II reactors and a significant fraction of these are nearing the end of their expected life and are scheduled for decommissioning in the not-so-distant future. The continuation of nuclear fission as an energy source depends on the development and construction of Generation IV reactor, as these are necessary to ensure safe efficient reactor operation and to provide breeding options to increase the lifetime of uranium and thorium resources.

Problems

12.1. Figure 11.8 shows the neutron absorption cross section for ^{238}U as a function of the center of mass energy. Calculate the ^{239}U excited state energies corresponding to the peaks shown in the figure. Comment on the results of this calculation.

12.2.

(a) A neutron with initial kinetic energy T_i collides elastically with a stationary nucleus of mass M. On the average the neutron energy after the collision may be written as

$$T_f = \frac{(M^2 + m_n^2)}{(M + m_n)^2} T_i.$$

For a uranium fission neutron travelling in a water moderator, estimate the number of elastic collisions necessary to reduce energy from 2.0 MeV to 0.01 eV. Assume that the elastic scattering cross section for neutrons on ^{1}H is 0.33 b and is independent of energy below 2.0 MeV. Assume the cross section for ^{16}O is negligible.

(b) Estimate the time required for the reduction of neutron energy in part (a).

12.3. Use the semiempirical mass formula to derive an expression for the curve shown in figure 12.1.

12.4. Use the semiempirical mass formula to derive an expression for the curve shown in figure 12.2.

12.5. A fission reactor produces 3×10^9 W of electrical power. Assuming that the steam generator has an efficiency of 30%, calculate the mass of ^{235}U consumed by the reactor in one day.

12.6. ^{256}Fm decays almost exclusively by spontaneous fission with a lifetime of 3.8 h. Assuming that Fm has a molar specific heat of 25 J $(\text{mole} \cdot \text{K})^{-1}$ and assuming that all fission energy is converted into thermal energy within the Fm, estimate the increase in the internal temperature of a 1 μg sample of Fm due to fission during 1 s.

12.7. Consider a sample of ^{235}U with a mass of 1000 kg, that is, much greater than the critical mass so that $q \approx 1$. At $t = 0$ a single spontaneous fission neutron initiates a chain reaction. Calculate the time required for all ^{235}U nuclei in the sample to be consumed in fission reactions.

12.8. Determine which isotopes of Pu with lifetimes of greater than a day are fissile.

References and suggestions for further reading

Bodansky D 2004 *Nuclear Energy—Principles, Practices and Prospects* 2nd edn (New York: Springer)

Emoscope 2005a *Schematic Diagram of a Magnox Nuclear Reactor* (https://commons.wikimedia. org/wiki/File:Magnox_reactor_schematic_(int).svg)

Emoscopes 2005b *Schematic Diagram Showing the Operation of the DFR (Dounreay Fast Reactor)* (https://commons.wikimedia.org/wiki/File:DFR_reactor_schematic.png)

Emoscopes 2008 *Diagram of RBMK Nuclear Reactor* (https://commons.wikimedia.org/wiki/File: RBMK_reactor_schematic.svg)

Inductiveload 2007 *Schematic Diagram of the Pressurised Heavy Water Cooled Version of a CANDU (CANada Deuterium-Uranium) Nuclear Reactor* (https://commons.wikimedia.org/ wiki/File:CANDU_Reactor_Schematic.svg)

Lardot N 2006 *Schematic Diagram of a Pressurized Water Reactor* (https://commons.wikimedia. org/wiki/File:Schema_reacteur_eau_pressuris%C3%A9e.svg)

Murray R L and Holbert K E 2015 *Nuclear Energy—An Introduction to the Concepts, Systems, and Applications of Nuclear Processes* 7th edn (Waltham, MA: Butterworth-Heinemann)

Myers W D and Swiatecki W J 1966 Nuclear masses and deformations *Nucl. Phys.* **81** 1–60

National Nuclear Data Center 2011 (https://nndc.bnl.gov/useroutput/sigma/endf-6[78043].txt)

Picoterawatt 2012 *Pebble Bed Reactor Scheme* (https://commons.wikimedia.org/wiki/File: Pebble_bed_reactor_scheme_(English).svg)

Rosatom 2016 *Reactor BN-800* (https://commons.wikimedia.org/wiki/File:BN-800_reactor.jpg)

Steffens R 2011 *Boiling Water Reactor System Diagram* (https://commons.wikimedia.org/wiki/ File:Boiling_water_reactor_no_text.svg)

Weirdmeister 2017 *Graphite Pebble for Pebble Bed Reactor* (https://commons.wikimedia.org/wiki/ File:Pebble_Bed_Graphite_Ball.jpg)

Woźnicka U 2019 Review of neutron diagnostics based on fission reactions induced by fusion neutrons *J. Fusion Energy* **38** 376–85

IOP Publishing

An Introduction to the Physics of Nuclei and Particles
(Second Edition)

Richard A Dunlap

Chapter 13

Fusion reactions

13.1 Fusion processes

We saw in the last chapter how energy can be extracted by breaking up heavy nuclei into lighter nuclei and thereby making use of the increase in binding energy per nucleon. Along similar lines, we can see from figure 4.1 that combining two light nuclei to make a heavier nucleus (up to $A = 55$) also yields an increase in binding energy per nucleon and this process may be used as a source of energy. This process is referred to as fusion and is the process by which the Sun and other stars produce energy. Fusion is a desirable method of producing energy and has several significant advantages over fission. These include:

1. An inexpensive and plentiful supply of fuel;
2. Reactions that are inherently easier to control and are therefore much safer;
3. Substantially reduced environmental hazards from reactor by-products.

Unfortunately, at present, fusion power is not technologically feasible. This is because of the fundamental differences between induced fission and the fusion process. For a fissile material, fission is induced by a thermal neutron because the resulting nuclear state is above the Coulomb barrier. For fusion, the Coulomb barrier, which is in the range of a few MeV or a few tens of MeV, is always a consideration. It is certainly straightforward to accelerate nuclei to these energies in even very small particle accelerators and to collide them with other nuclei to produce fusion reactions. Unfortunately, in such a situation, the energy expenditure is substantially greater than the energy gain. Thus, although this is a useful way of learning about fusion reactions in the laboratory, fusion, in the practical sense, always involves energies below the Coulomb barrier and inevitably involves a tunneling process. A treatment of this phenomenon is analogous to that for α-decay where the barrier height can be expressed as

doi:10.1088/978-0-7503-6094-4ch13

$$V_C = \frac{e^2}{4\pi\varepsilon_0} \frac{Z_1 Z_2}{R_1 + R_2} \tag{13.1}$$

Here Z_i and R_i are the charges and radii of the two nuclei. In order to maximize the tunneling probability it is necessary to minimize the barrier height. From equation (13.1) it is, therefore, obvious that nuclei with small values of Z are the most interesting. These are involved in the processes that are primarily responsible for energy production in the Sun and also those that have attracted interest as possible sources of fusion power. In this section we consider some of the possible fusion reactions involving isotopes of hydrogen.

The simplest fusion process might appear to be the fusion of two protons, the so-called p–p process. However, two protons cannot form a bound state and p–p fusion is analogous to β^+ decay where one of the protons is converted to a neutron to give,

$$p + p \rightarrow {}^2H + e^+ + \nu_e \tag{13.2}$$

The energy release from this process is 0.42 MeV per fusion. Normally, an additional 1.02 MeV will become available due to the annihilation of the positron with an electron. Like the β-decay processes discussed previously, the presence of leptons in the reaction indicates that the weak interaction is responsible. As a result, the cross section for this reaction is very small. Fusion processes involving deuterons are of importance for our further discussions. The simplest of these is,

$$p + d \rightarrow {}^3He + \gamma \quad (Q = 5.49 \text{ MeV}). \tag{13.3}$$

We will see in section 13.3 that this is an important reaction in stars. The most obvious process involving the fusion of two deuterons is the formation of ${}^4He$,

$$d + d \rightarrow {}^4He + \gamma \quad (Q = 23.8 \text{ MeV}). \tag{13.4}$$

This process is unlikely, as the energy release is well above the neutron and proton separation energies of ${}^4He$. Consequently, there are two possible modes of d–d fusion,

$$d + d \rightarrow {}^3He + p \quad (Q = 4.0 \text{ MeV}) \tag{13.5}$$

and

$$d + d \rightarrow {}^3He + n \quad (Q = 3.3 \text{ MeV}). \tag{13.6}$$

A final process of importance is the fusion of a deuteron with a triton (d–t fusion):

$$d + t \rightarrow {}^4He + n \quad (Q = 17.6 \text{ MeV}). \tag{13.7}$$

This process releases a substantial amount of energy and is of particular importance, as we will see in section 13.4, for the operation of a controlled fusion reactor.

It is important to understand how the fusion energy is distributed among the fusion by-products. In the case of a process, such as given in equation (13.7), where two relatively massive particles appear after the fusion process, a simple consideration of energy and momentum conservation is useful. For a generic reaction X(a, b)Y the Q of the reaction is given in terms of the reactant masses by

$$Q = [m_X + m_a - m_Y - m_b]c^2.$$

The total kinetic energy of the fusion by-products is, therefore,

$$Q = \frac{1}{2}m_Y v_Y^2 + \frac{1}{2}m_b v_b^2$$

where the v are the velocities and we have assumed that the incident kinetic energy is small. This is a reasonable assumption for problems dealing with stellar processes and controlled fusion reactors. Conservation of momentum gives,

$$m_Y v_Y = m_b v_b.$$

The above equations can be solved to give the energy of the two particles as

$$E_Y = \frac{Q}{1 + \frac{m_Y}{m_b}}$$

and

$$E_b = \frac{Q}{1 + \frac{m_b}{m_Y}}.$$

From these expressions the ratio of kinetic energies is given by

$$\frac{E_b}{E_Y} = \frac{m_Y}{m_b}.$$

It is readily seen that the less massive particle acquires the larger fraction of the energy. This is particularly important, for example, in the case of d–t fusion where 80% of the energy is carried away by the neutron. Because of the neutron's low reaction cross section, the method by which this energy may be extracted requires careful consideration.

13.2 Fusion cross sections and reaction rates

The cross section for fusion follows from the cross section for charged particle reactions as given by equation (11.15),

$$\sigma \propto \frac{1}{v^2}e^{-G}. \tag{13.8}$$

Some numerical factors have been omitted here but the expression above contains all energy or velocity dependencies. The factor G in the exponent is determined by the probability of tunneling through the Coulomb barrier between the two nuclei. This follows directly from the discussion of α-decay and equation (8.11) as,

$$G = \frac{2Z_1 Z_2 e^2}{4\pi\varepsilon_0 \hbar}\sqrt{\frac{2m}{E}}\left[\cos^{-1}\sqrt{\frac{a}{b}} - \sqrt{\frac{a}{b}\left(1 - \frac{a}{b}\right)}\right] \tag{13.9}$$

where m is the reduced mass of the system. Along the lines of the discussion in chapter 8, the quantities a and b are defined as

$$a = R_1 + R_2$$

and

$$b = \frac{Z_1 Z_2 e^2}{4\pi\varepsilon_0 E}.$$

Since the kinetic energy of the nuclei involved is much less than the Coulomb barrier energy, the term in brackets in equation (13.9) can be approximated as $\pi/2$. Using the center of mass kinetic energy, $E = mv^2/2$, the cross section for fusion is given by equation (13.8) with

$$G = \frac{2Z_1 Z_2 e^2}{4\pi\varepsilon_0} \frac{\pi}{\hbar v}.$$

Some examples of energy dependent cross sections of important fusion reactions are shown in figure 13.1. The reaction rate for fusion can be determined by considering a beam of particles of species '1' with a flux Φ_1 incident on a collection of particles of

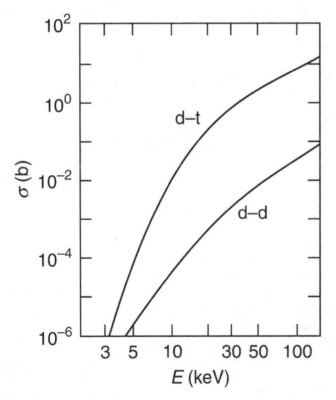

Figure 13.1. Energy dependence of the reaction cross section for d–d fusion (total of equations (13.5) and (13.6)) and d–t fusion (equation (13.7)).

species '2' with a number density n_2 and a relative reaction cross section σ. In this case, the reaction rate per unit volume will be given by

$$R = \sigma \Phi_1 n_2 \tag{13.10}$$

For a plasma consisting of two types of particles, as would be appropriate for the interior of a star or the fuel of a fusion reactor, the flux in equation (13.10) is written in terms of n_1, the number density of particles of species '1', and the relative particle velocity, v, as

$$\Phi_1 = n_1 v.$$

This gives a reaction rate of

$$R = \frac{n_1 n_2 \langle \sigma v \rangle}{1 + \delta_{12}} \tag{13.11}$$

where the delta function in the denominator is 0 for the case where the two species of particles are different and 1 for the case where they are the same. This avoids double counting for a plasma consisting of a single type of nuclei. Since both the cross section and the velocity are functions of the particle energy, it is appropriate to determine the quantity $\langle \sigma v \rangle$ in equation (13.11) in terms of the energy distribution of the particles in the plasma. This quantity is referred to as the reactivity. The Maxwell–Boltzmann distribution gives the probability of finding a particle with a velocity between v and $v + dv$ at a temperature T as

$$P(v)dv = \left(\frac{2}{\pi}\right)^{1/2} \left(\frac{m}{k_B T}\right)^{3/2} v^2 e^{-E/k_B T} dv.$$

This gives the average value of $\langle \sigma v \rangle$ as

$$\langle \sigma v \rangle = \int \sigma v P(v) dv$$

or

$$\langle \sigma v \rangle \propto \int e^{-G} e^{-E/k_B T} dE.$$

Thus, the reaction rate is determined by the product of the energy dependent cross section and the Maxwell–Boltzmann distribution. An example of combining these two factors is illustrated in figure 13.2. This indicates that there is a maximum in the reaction rate as a result of the decreasing cross section at low energies and the decreasing Maxwell–Boltzmann distribution of high energies. In most cases, the energies involved in stellar fusion and fusion reactors are relatively low and it is the portion of the reaction rate curve that increases with energy that is of importance. This is illustrated in figure 13.3 where the product $\langle \sigma v \rangle$ is given for d–d, d–t, and other fusion reactions over the range of energies that is most relevant. The peak reaction rate for d–t fusion occurs at around 30 keV, while the peak for d–d fusion occurs well into the MeV range and is not seen in the graph.

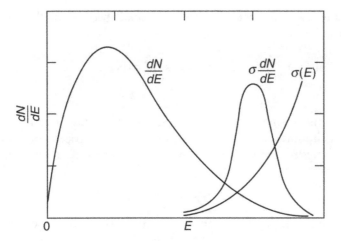

Figure 13.2. Effect of combining the Maxwell–Boltzmann energy distribution, dN/dE, and the fusion cross section, $\sigma(E)$, to give the overall reaction rate, $\sigma dN/dE$.

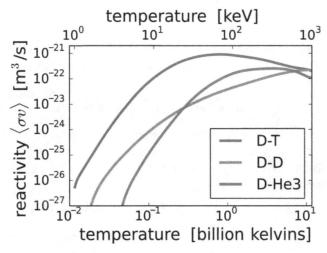

Figure 13.3. Reactivities for some fusion reactions as a function of plasma temperature. Note $1\ K = 8.3 \times 10^{-5}\ \mathrm{eV}$. This [plot of the fusion reactivity vs. temperature for three common reactions] image has been obtained by the author from the Wikimedia website https://commons.wikimedia.org/wiki/File:Fusion_rxnrate.svg where it was made available by (Dstrozzi 2009) under a CC BY 2.5 licence. It is included within this chapter on that basis. It is attributed to Dstrozzi.

13.3 Stellar fusion processes

The Sun, like most other stars, produces energy by fusing four hydrogen nuclei ($^1\mathrm{H}$) into one helium nucleus ($^4\mathrm{He}$). Fusion processes involving heavier nuclei are important in some stars but are relatively unimportant in the Sun for two reasons:

1. The concentration of heavier nuclei is relatively small (the Sun is about 92% hydrogen and about 8% helium, in terms of the number of atoms).

2. Reaction rates for heavier nuclei are small because of the larger Coulomb barriers involved.

The fusion of four ^{1}H nuclei represents the process

$$4p \rightarrow {}^4He + 2e^+ + 2\nu_e \tag{13.12}$$

where two of the fusing protons must be converted into neutrons by β^+ decay processes. It is important to note that this is a nuclear process and in order to deal with the energetics in terms of atomic masses it is necessary to add four electrons to each side to give,

$$4^1H \rightarrow {}^4He + 2e^- + 2e^+ + 2\nu_e.$$

The four hydrogen nuclei do not fuse simultaneously to form helium. Instead, the helium forms in a series of steps. The first step of this fusion process is the fusion of two protons as given by equation (13.2). In principle, two deuterons could then fuse according to equation (13.4) to form a ^{4}He nucleus. However, the low concentration of deuterons in the Sun as well as the large value of Q for this process, as discussed previously, make this process highly unlikely. A more likely process is p-d fusion as given by equation (13.3) to form ^{3}He. Because of the high concentration of protons, the most logical process involving ^{3}He would seem to be the formation of ^{4}Li. However, ^{4}Li (three protons and one neutron) does not form a bound state and immediately leads to the process,

$$p + {}^3He \rightarrow {}^4Li \rightarrow {}^3He + p.$$

The reaction of ^{3}He with a deuteron is unlikely as the deuterons that are formed relatively quickly fuse with protons to form more ^{3}He. Thus, a ^{3}He nucleus's most likely fate will be to eventually react with another ^{3}He nucleus to form ^{4}He according to the reaction,

$$^3He + {}^3He \rightarrow {}^4He + 2^1H + \gamma \quad (Q = 12.86 \text{ MeV}).$$

The overall process described above is the most common method of energy production in the Sun and is referred to as the proton–proton cycle. It has the net result of fusing four protons and, in conjunction with two β-decay processes, forms a ^{4}He nucleus. The total energy associated with this process is $Q = 26.7$ MeV. Most of this energy is eventually converted into solar radiation, while a small amount is carried away as kinetic energy by the neutrinos from the β-decay processes and is lost. The rate at which this process can proceed is limited by the weak interaction cross section for deuteron production as indicated in equation (13.2).

Other reactions that ultimately result in the fusing of four hydrogen into one helium are also possible in the Sun and these are described in figure 13.4. The branching ratios for the various reactions are indicated in the figure. Details of some of these reactions will be discussed in chapter 19.

In stars with higher internal temperatures, particle energies are higher and there is a greater probability of fusion processes involving heavier nuclei. One process,

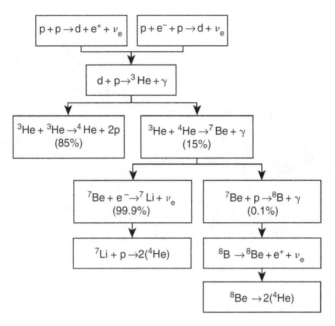

Figure 13.4. Contributions to energy production in the Sun.

known as the carbon–nitrogen-oxygen cycle (or CNO cycle), is equivalent to the proton–proton cycle since it ultimately fuses four hydrogen into one helium. It proceeds as follows:

$$^{12}C + p \rightarrow {}^{13}N + \gamma$$

$$^{13}N \rightarrow {}^{13}C + e^+ + \nu_e$$

$$^{13}C + p \rightarrow {}^{14}N + \gamma \qquad (13.13)$$

$$^{14}N + p \rightarrow {}^{15}O + \gamma$$

$$^{15}O \rightarrow {}^{15}N + e^+ + \nu_e$$

$$^{15}N + p \rightarrow {}^{12}C + {}^4He.$$

Again, two β-decay processes are required to convert two of the protons to neutrons. Since carbon is neither created nor destroyed but merely acts as a catalyst, the Q for this process is the same as for the proton–proton cycle. Although this process requires greater energy to overcome the Coulomb barrier, it is not limited by the rate of deuteron production, as is the proton–proton cycle. Hence, the CNO cycle is less likely at lower temperatures than the proton–proton cycle but becomes more probable at higher temperatures. This trend is indicated in figure 13.5. The internal temperature of the Sun, about 10^7 K, falls below the cross-over point indicating that the energy production in the Sun is dominated by the proton–proton cycle.

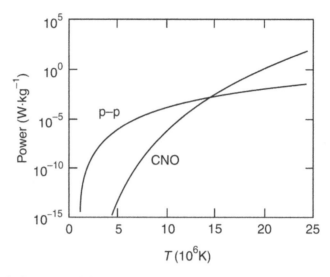

Figure 13.5. Relative importance of the proton–proton and CNO Cycles in stars as a function of their internal temperature. Curves are calculated from expressions for the temperature dependent reaction rates given in Schwarzschild (1958).

Stars that have depleted their supply of hydrogen can no longer produce energy by the proton–proton or CNO cycles. This frequently happens in the central region of a star where temperatures and hence hydrogen fusion rates are greater resulting in a core made up primarily of ^{4}He. If the temperature in this region is sufficiently high (greater than about 10^8 K) further energy can be produced by helium fusion,

$$^4\text{He} + {}^4\text{He} \rightarrow {}^8\text{Be} + \gamma.$$

The ^{8}Be is unstable and decays by α-decay,

$$^8\text{Be} \rightarrow {}^4\text{He} + {}^4\text{He}.$$

The lifetime for this process is 7×10^{-17} s. Although some (or even most) of the ^{8}Be that is formed decays back to ^{4}He, some will fuse with another ^{4}He to produce ^{12}C,

$$^8\text{Be} + {}^4\text{He} \rightarrow {}^{12}\text{C} + \gamma.$$

This process is known as the triple alpha process since it combines three ^{4}He or α-particles into a ^{12}C nucleus. It has a value of $Q = 7.27$ MeV. Heavier nuclei can be synthesized in the interior of stars by a variety of processes including neutron capture and α-particle capture. An important feature of all such processes is that fusion is no longer energetically favorable for nuclei heavier than around $A = 55$ (see figure 4.1). As a result, the relative abundance in the universe of elements heavier than Fe is much less than that of elements lighter than Fe. A good summary of nuclear astrophysics and, in particular, nucleosynthesis (that is, the methods by which heavier nuclei are synthesized in stellar interiors) can be found in Krane (1988) and references therein.

13.4 Fusion reactors

It is interesting to consider the energy release from various processes. Table 13.1 gives the typical energy produced per kilogram of fuel for different reactions. A chemical reaction corresponds to the burning of a fuel such as coal or oil. As a rough approximation for fission power, it has been assumed that all available fission energy can be extracted from the ^{235}U component of natural uranium and that there is no contribution from fission of ^{238}U. For fusion power it is assumed that the energy is produced by the indicated reaction from the naturally occurring hydrogen isotopes in water. It is clear in all cases that nuclear reactions produce more energy than chemical reactions. This is merely a result of the difference between electronic binding energies (a few eV) and nuclear binding energies (a few MeV). The substantial amount of energy produced by p–p fusion is the result of the large mass of protons contained in one kilogram of water combined with the small atomic weight of ^{1}H. The smaller amount of energy available from d–d fusion is primarily the result of the much lower natural abundance of ^{2}H.

On the basis of table 13.1 it would certainly be desirable to be able to extract energy from water by p–p fusion according to equation (13.2). However, as this reaction is dominated by the weak interaction its cross section is very small. Thus, although it is an important factor in the energy production in the Sun, it is unsuitable for a fusion reactor. The d–d fusion reactions given by equations (13.5) and (13.6) are possible candidates for a fusion reactor. These two reactions produce similar amounts of energy and natural water contains sufficient quantities of deuterium to make this a highly attractive energy source. The energy dependence of the reaction rate for d–d fusion is illustrated in figure 13.1. By comparison, the figure shows that for moderate energies the reaction rate for d–t fusion is about two orders of magnitude larger.

In order to gain energy from a sustained fusion reaction, it is essential that the fusion energy output is greater than the total energy input required to heat the fuel and the energy losses in the system. The energy gained per unit time per unit volume (power per unit volume) from fusion reactions in a plasma can be expressed as

$$P_f = RQ \qquad (13.14)$$

Table 13.1. Comparison of energy produced by chemical processes, fission of natural uranium, and fusion of the proton and deuteron components of natural water.

Fuel	Reaction	Mass (g) of reactive component per kg of fuel	Energy (J) per kg of fuel
Chemical	Chemical	–	6×10^6
Natural uranium	^{235}U fission	7.2	6×10^{11}
Natural water	p–p Fusion	110	7×10^{12}
Natural water	d–d Fusion	0.016	3×10^9

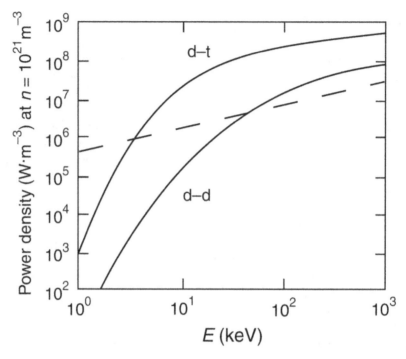

Figure 13.6. Power per unit volume produced by a plasma from d–d and d–t fusion and power loss per unit volume from bremsstrahlung (broken line) plotted as a function of energy.

where R is the reaction rate given by equation (13.11) and Q is the energy per fusion. The fusion power per unit volume can be determined from the values of $\langle \sigma v \rangle$ illustrated in figure 13.3 and is shown as a function of plasma temperature for d–d and d–t fusion in figure 13.6. Even in the ideal situation, a plasma will lose energy as a result of bremsstrahlung. This loss results primarily from the behavior of the electrons which, being much less massive than the ions, experience greater acceleration during electron–ion interactions. The power radiated per unit volume is

$$P_r = \frac{4\pi n n_e Z^2 e^6}{3(4\pi\varepsilon_0)^3 c^3 \hbar} \left(\frac{3k_B T}{m_e^3} \right)^{1/2}$$

where n and n_e are the ion and electron densities and Z is the ion charge. The bremsstrahlung losses are the same for d–d and d–t plasmas (since Z is the same for both ions). The temperature dependence of these losses is compared with the fusion power in figure 13.6. This makes clear the more demanding temperature requirements for d–d fusion compared with d–t fusion. In either case, it is necessary to operate the reactor at a temperature above the crossing point of these curves. It is now important to consider a comparison of the fusion energy produced and the thermal energy needed to heat the plasma. From equations (13.11) and (13.14) the fusion energy per unit volume can be expressed as

$$E_f = \frac{n^2 \langle \sigma v \rangle}{4} Q\tau \tag{13.15}$$

where τ is the time during which the plasma is held at a temperature and density compatible with the fusion rate of equation (13.11). For simplicity we have considered d–d fusion where $n_1 = n_2 = n/2$. The thermal energy needed to heat the plasma is the sum of the energy needed to heat the electrons and the energy needed to heat the ions. In the case where $n_e = n$ then the total thermal energy needed per unit volume is

$$E_{th} = 3nk_B T \tag{13.16}$$

For a reactor operating at a temperature above the bremsstrahlung crossing point, radiative losses can be ignored. Combining equations (13.15) and (13.16) gives the minimum operating conditions to gain energy from the fusion process,

$$n\tau > \frac{12k_B T}{\langle \sigma v \rangle Q} \tag{13.17}$$

This condition is referred to as the Lawson criterion and the quantity on the left-hand side of the equation is generally called the Lawson parameter. The value of the Lawson parameter for which the reactor produces net energy is a function of operating temperature. In recent years, it has become common to use the so-called triple product as a figure of merit for fusion reactors. The triple product is obtained from the Lawson criterion in equation (13.17) by multiplying both sides of the equation by temperature to give,

$$n\tau T > \frac{12k_B T^2}{\langle \sigma v \rangle Q} \tag{13.18}$$

In most reactors the operating temperature (in eV) will be in the range of 10–20 keV. Over this range of temperatures, the reactivity $\langle \sigma v \rangle$ is roughly proportional to T^2. This eliminates the temperature dependence on the right-hand side of equation (13.18) and, for the d–t reaction, allows us to write the condition on the triple product for which the Lawson criterion is satisfied as

$$n\tau T > 3 \times 10^{21} \text{ m}^{-3} \cdot \text{keV} \cdot \text{s}. \tag{13.19}$$

Another way of looking at the viability of a reactor is to look at energy gain. Obviously, it is necessary for a reactor to output more energy than it requires for input. The ratio of energy out to energy in is referred to as the reactor gain, Q. Of course, there are energy requirements for a reactor facility that involve energy that is not directly input into the reactor. These requirements include, pumps, material transport, building operations, etc as well as the energy losses that result from conversion of thermal energy to electrical output. A value of Q around 5 is generally considered to be the point at which there is a net energy gain. Viable reactor operating conditions are probably in the range of around $Q = 30$. Ignition, which is the ultimate goal of a reactor is the point at which the fusion energy that is produced

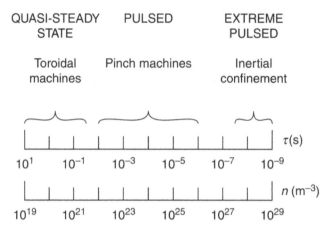

Figure 13.7. A comparison of the values of n and τ for different reactor designs that are necessary to achieve the Lawson criterion for d–t fusion.

is sufficient to maintain the plasma conditions. In this case, no further energy input into the plasma is required and excess output energy can be used for other reactor requirements. Ignition corresponds to $Q = \infty$.

A major difficulty in constructing a fusion reactor is the requirement for a means of confining the plasma. Certainly, the temperatures involved are sufficiently high that the interaction of the plasma with the walls of a containment vessel would readily melt any possible solid container material. It is also worth noting that any such interaction is also highly undesirable as it results in energy loss from the plasma.

Various approaches have been utilized to achieve conditions that satisfy the Lawson criterion. These conditions range from those where plasma densities are relatively low, but confinement times are long, to those where plasma densities are very high, but confinement times are very short. Figure 13.7 shows the relevant features of different types of potential fusion reactors. Most fusion research in recent years has concentrated on the two ends of the spectrum shown in the figure. In the two sections below, we consider in some detail the design and performance of these two approaches that are categorized by the plasma confinement method, that is, magnetic confinement and inertial confinement. A more detailed consideration of the current status of nuclear fusion reactors has been presented by Dunlap (2021).

13.5 Magnetic confinement reactors

Since the ions and electrons in a plasma are free to move independently their motion can be controlled by the application of a suitable magnetic field. Magnetic confinement reactors utilize magnetic fields to direct the particles in a plasma to prevent them from colliding with the walls of the containment vessel. There are two basic geometries that are used for these devices—a linear geometry and a toroidal geometry.

The linear geometry uses a plasma column that is pinched at the ends. The plasma is contained in a cylindrical chamber and an axial magnetic field is provided by coils around the outside of the chamber. Basically, the particles travel in a region of comparatively low field along the length of the cylinder and are reflected from the ends of the cylinder by regions of higher field. This is sometimes referred to as magnetic mirror confinement. In general, progress towards the conditions necessary for a sustained fusion reaction in a mirror confinement device has fallen short of that achieved in other reactor designs. As a result, most current fusion research is directed towards the toroidal reactors and inertial confinement reactors described below.

The plasma column may be closed in the form of a toroid, in which case the particles travel along toroidal field lines produced by poloidal currents in windings around the toroid. In this geometry the windings are closer together on the inside of the torus than on the outside resulting in a stronger magnetic field near the inside. The consequence of this is that the particles will slowly spiral outward, towards the region of weaker field and eventually strike the outer wall of the torus.

The usual approach to dealing with this problem is to introduce an additional magnetic field component so that the field lines in the torus follow a helical path. In this way the particles travel along a path that alternates between the inside and outside of the torus. Two different toroidal fusion reactor designs have been investigated in this respect. The *tokamak* induces a toroidal current in the plasma that generates a poloidal field that adds to the toroidal field, while the *stellarator* uses external field coils that are intrinsically non-axisymmetric to generate helical field lines. The details of these two types of reactors are discussed below.

13.5.1 Tokamak

Figure 13.8 shows the relationship between the currents and magnetic fields produced in a tokamak. The combination of the toroidal and poloidal magnetic fields results in the helical field lines inside the tokamak along which the charge

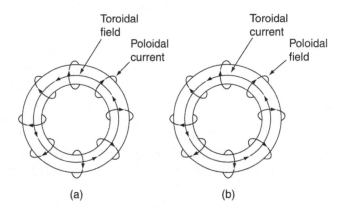

(a)   (b)

Figure 13.8. Geometry of currents and magnetic field lines in a tokamak, (a) toroidal field produced by poloidal currents and (b) poloidal field produced by a toroidal current.

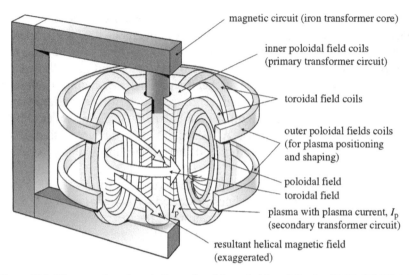

magnetic circuit (iron transformer core)

inner poloidal field coils
(primary transformer circuit)

toroidal field coils

outer poloidal fields coils
(for plasma positioning
and shaping)

poloidal field
toroidal field

plasma with plasma current, I_p
(secondary transformer circuit)

resultant helical magnetic field
(exaggerated)

Figure 13.9. Diagram of a tokamak. Reproduced from Smith and Cowley (2010) CC BY 4.0.

particles of the plasma travel. Figure 13.9 shows the design of a tokamak. The creation of the toroidal field from currents in the poloidal direction is straightforward using magnet windings wrapped around the torus, as shown in figure 13.8(a), and by the 'toroidal field coils' shown in figure 13.9. The creation of the poloidal field is not so straightforward, as a magnetic coil cannot be placed inside the torus to carry the toroidal current, as shown in figure 13.8(b). The poloidal field is produced by an induced current in the plasma itself. This is accomplished by means of a central solenoid shown as the 'inner poloidal field coils' in figure 13.9. The current in the solenoid is not constant but is ramped up and down at a frequency of the order of 1 Hz and this changing axial field induces the toroidal plasma current. The plasma current is related to the plasma resistance, R, and the rate of change of the magnetic flux density, Φ, as

$$I = -\frac{1}{R}\frac{d\Phi}{dt}.$$

The 'outer poloidal field coils' in figure 13.9 (green coils) are used to provide minor adjustments to the positioning and shape of the helical field lines inside the reactor.

A major factor in the operation of a tokamak is the means by which the plasma is heated. The motion of the electrons and ions through the plasma that constitutes the toroidal plasma current generates heat in the same way a current flowing through a resistor produces thermal energy. This, however, is not sufficient to heat the plasma to the necessary temperature. Additional energy must be input into the reactor to raise the plasma temperature. The most common method of doing this is by neutral beam injection. A collection of atoms (typically deuterium) is ionized and accelerated to high energy by a small linear accelerator. These ions are then recombined with electrons to produce a high energy beam of neutral atoms. These neutral atoms are injected into the plasma where they are quickly ionized to produce ions and electrons. This process has two important features: to heat the plasma by transferring kinetic energy to other ions and electrons through Coulomb interactions and to increase the density of the plasma. Both processes help to achieve the Lawson criterion. It is necessary to inject the particles as neutral particles because it would be difficult for charged particles to penetrate the magnetic fields surrounding the reactor.

The first tokamak (T-1) was operational at the Kurtchatov Institute in Moscow from 1957 to 1959. The tokamak name, in fact, is an acronym that is derived from the Russian description of the device, TOroidal'naya KAmera s MAgnitnymi Katushkami, meaning toroidal chamber with magnetic coils. Since that time more than 50 tokamaks have been constructed worldwide, of which about 30 are currently operational. The largest scale tokamak project to date is ITER located at the Cadarache research facility near the village Saint-Paul-lès-Durance in southern France (see figure 13.10). The name ITER (pronounced 'eater') is an acronym for

Figure 13.10. Concept model of the ITER exhibited at the International Fusion Energy Days in Monaco, December 2013. A person is shown (lower right) for scale. IAEA Imagebank (2014) Flickr CC BY SA 2.0.

International Thermonuclear Experimental Reactor and also means 'the way' in Latin. ITER has a plasma volume that is more than a factor of 10 larger than any other tokamak constructed thus far. It is interesting to note in the figure that the plasma chamber of ITER, like that of many modern tokamaks, is 'D' shaped rather than circular. The planning stage for ITER was complete around 2007, after which the construction phase began. As of 2022 construction is nearing completion and the first tests on d–d plasmas are expected by around the end of 2025. While ITER is a test reactor and is not designed to actually produce electrical output, it is anticipated that it will attain a Q value of around 10, close to the requirements for a commercial reactor. It is hoped that ITER will lead the way to a demonstration power reactor, designated DEMO, with a Q around 25 by somewhere around the mid-21st century.

13.5.2 Stellarator

A stellarator does not induce a plasma current to produce the helical magnetic fields lines but utilizes a non-axisymmetric arrangement of external field coils to generate a helical field. The simplest approach to this design, sometimes called a classical stellarator, uses a bi-directionally wound set of helical coils around the toroidal plasma chamber to introduce a helical twist to the magnetic field lines, as shown in figure 13.11. The resulting helical magnetic field is shown in the illustration.

Figure 13.11. Diagram showing the coil geometry of a classical stellarator. The bi-directionally wound helical coils are shown by the light blue lines and the resulting helical magnetic field is shown by the dark blue surface, where the surface of the last closed flux surface is illustrated. Reproduced from Suzuki (2020). Copyright IOP Publishing Ltd CC BY 4.0.

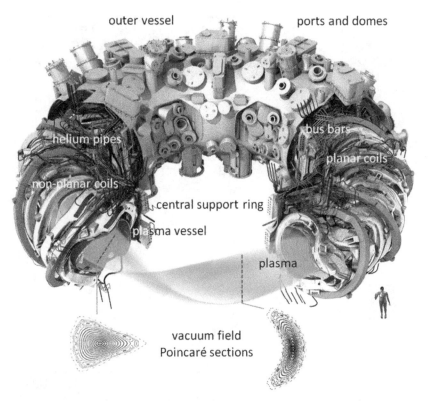

Figure 13.12. The Wendelstein 7-X stellarator. The design of the poloidal coils is seen in the image and the helical shape of the constant magnetic field surface is shown in pink. A person is shown (lower right) for scale. Reproduced from Klinger *et al* (2017). Copyright IOP Publishing Ltd CC BY 3.0.

The development of the stellarator predates that of the tokamak. The first stellarators were constructed in the early 1950s and this approach formed the basis for the majority of fusion reactor research until the late 1960s. At that time, the difficulties of developing viable fusion energy were becoming obvious and the newly developed tokamak offered hope that this new device would overcome some of these roadblocks. By the 1990s it had become obvious that the tokamak suffered from some of the same problems as the stellarator, as well as some different ones. Knowledge gained from tokamak experiments, as well as improved superconducting magnet technology and computer modelling techniques, led the way to the design of improved stellarators. To date, the Wendelstein 7-X is the most advanced stellarator to be constructed (see figure 13.12). This device, and most other modern stellarators, does not use the magnetic coil configuration shown in figure 13.11, but rather uses poloidal windings that are non-circular (and non-planar) to produce a helical magnetic field inside the toroidal chamber.

13.5.3 Progress in magnetic confinement fusion

The performance of various magnetic confinement fusion devices is illustrated by the plot of the triple product as a function of temperature, as shown in figure 13.13. It is

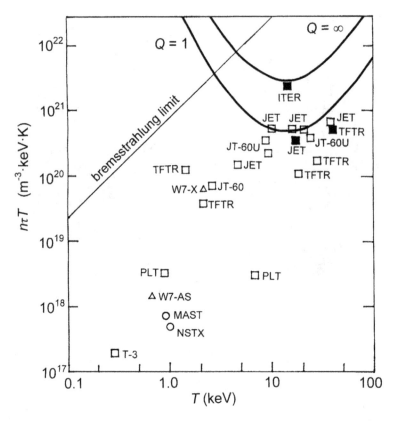

Figure 13.13. Triple product as a function of ion temperature showing results for some magnetic confinement fusion reactors. Expected performance for ITER is shown on the diagram. Tokamaks are shown as squares, spherical tokamaks as circles and stellarators as triangles. Closed symbols represent hydrogen or deuterium experiments and open symbols represent d–t experiments. Curves are shown for scientific breakeven ($Q = 1$) and ignition ($Q = \infty$). The bremsstrahlung limit is defined by the conditions for which bremsstrahlung losses exceed fusion power. From Dunlap (2021). Copyright IOP Publishing. Reproduced with permission. All rights reserved.

seen in the diagram that the Joint European Torus (JET) at the Culham Centre for Fusion Energy in the United Kingdom holds the record for the highest triple product obtained to date and is the only magnetic confinement fusion reactor to approach so-called scientific breakeven ($Q = 1$). Anticipated performance of ITER is shown and represents a significant advance in magnetic confinement fusion research. The results for Wendelstein 7-X (W7-X) shown on the diagram fall well below those of many recent tokamaks and represent the somewhat earlier stage of development of stellarators.

It is interesting to note that nearly all magnetic confinement fusion experiments have used plasmas consisting of hydrogen and/or deuterium. It is clear from the discussion in section 13.4 that a viable fusion reactor operating on a mixture of deuterium and tritium is much more probable than one operating without tritium in the fuel. However, deuterium is a naturally occurring isotope of hydrogen and is readily extracted from water. It is, therefore, plentiful, and relatively inexpensive. Tritium, on the other hand, is not naturally occurring, as it decays by β-decay with a lifetime of 17.7 years. Thus, all tritium must be bred artificially and is, at present rare and expensive. It is estimated that the current world stockpile of tritium contains about 20 kg.

Tritium can be bred from lithium using neutrons from the d–t fusion reaction in equation (13.7). This process would use natural lithium, which consists of about 7% ^{6}Li and 93% ^{7}Li, to produce tritium by the processes,

$$^6\text{Li} + \text{n} \rightarrow {}^3\text{H} + {}^4\text{He} \quad (Q = 4.78 \text{ MeV}) \tag{13.20}$$

$$^7\text{Li} + \text{n} \rightarrow {}^3\text{H} + {}^4\text{He} + \text{n} \, (Q = -2.47 \text{ MeV}). \tag{13.21}$$

The first reaction is exothermic and has a large cross section. The second reaction has a smaller cross section and, although endothermic, is possible because the kinetic energy carried by the fusion neutrons is above the reaction threshold. Tritium can be recovered from these reactions and utilized as reactor fuel. The only additional by-product is ^{4}He. Thus, in a sustainable d–t fusion reactor program tritium produced from lithium would be used to refuel the reactor.

13.6 Inertial confinement reactors

Inertial confinement refers to the situation where the fusion fuel is confined by inertial forces in the plasma itself. Most experiments that fall into this category are referred to as laser fusion experiments. A pellet of fuel (containing deuterium or a mixture of deuterium and tritium) contained in a capsule about a millimeter in diameter) is bombarded from several directions at once by pulsed high energy laser beams. The fuel pellet heats rapidly to a high temperature and is, at the same time, compressed to a high density. The actual processes that take place in the pellet when it is irradiated by an intense laser pulse are quite complex. Figure 13.14 shows a simplified description of these processes. In figure 13.14(a), the energy is absorbed by the fuel pellet, heating it from the outside. In figure 13.14(b) the heat propagates through the pellet transforming the outer portions into a plasma. In figure 13.14(c)

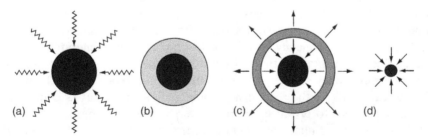

Figure 13.14. Processes that occur in a fuel pellet when energy from a laser is absorbed. (a) Absorption of energy from laser beam, (b) formation of plasma atmosphere, (c) ablation of plasma atmosphere, and (d) compression of pellet core by ablation shock wave.

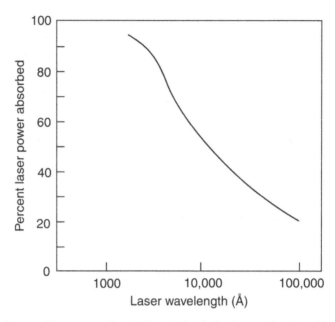

Figure 13.15. Percent of laser power absorbed by a fusion fuel pellet as a function of laser wavelength.

this outer plasma atmosphere is driven off as it heats and expands. This process is referred to as ablation. In figure 13.14(d) the remaining core of the pellet is compressed and heated by the inertial forces resulting from the expanding plasma atmosphere. The lasers used in such experiments are pulsed and the duration of the process shown in the figure is typically about 10^{-9} seconds. During this time the temperature of the fuel is very high because of the large amount of energy absorbed from the laser beam and the pellet core can be compressed to densities of several thousand times the density of water. Thus, despite the small values of τ, the values of n can be sufficiently large that there is a possibility of achieving the Lawson criterion.

A major factor in the operation of an inertial confinement fusion reaction is the ability of the fuel pellet to absorb energy from the laser beam. Figure 13.15 shows the percent power absorbed by a fuel pellet as a function of laser wavelength. High power lasers (for example, Nd-glass lasers) radiate in the infrared at wavelengths around 1050 nm. The figure shows that this is a disadvantageous situation. A crystal, known as a frequency doubler or second harmonic generator, can be used to reduce the wavelength of the radiation by a factor of two. This substantially increases the efficiency of energy absorption. From a quantum mechanical viewpoint one can think of the second harmonic generator crystal as combining two low energy photons to produce one high energy photon. From a classical standpoint, this property results from the nonlinear optical behavior of the crystal. The relationship between applied electric field and induced polarization is not linear, resulting in the production of higher harmonics in the radiated power spectrum. The greater the amplitude of the driving field the more intense the higher

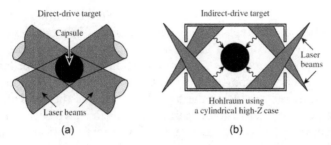

Figure 13.16. Comparison of (a) direct drive and (b) indirect drive laser fusion experiments. From Sangster *et al* (2007). Copyright IOP Publishing. Reproduced with permission. All rights reserved.

harmonics. Thus, the second harmonic generation efficiency is negligible for weak incident radiation but can be substantial for intense laser light. This approach can be taken one step further to produce radiation at the third harmonic. This is done by combining photons at the fundamental frequency with second harmonic photons to yield a photon with three times the fundamental frequency. For incident laser radiation at around 1050 nm the third harmonic will be at around 350 nm, in the near ultraviolet, and as shown in figure 13.15, energy absorption efficiency will be in excess of 90%. The most commonly used harmonic generation material is monopotassium phosphate (KDP) which can produce third harmonics with an efficiency of about 50%.

There are two different approaches to getting energy from the laser into the fuel pellet. One, referred to as direct drive, utilizes the laser beam to irradiate the pellet directly. The other approach, referred to as indirect drive, encloses the fuel pellet in a metal (typically gold) cylinder, called a hohlraum. The laser beams are incident on the hohlraum which vaporizes and in the process releases x-rays produced by electronic transitions in the metal atoms. These two approaches are illustrated in figure 13.16. In both cases the irradiated fuel pellet undergoes the procss shown in figure 13.14.

While it is true that there will be some energy loss in the conversion of near ultraviolet photons into x-rays, there are some distinct advantages of indirect drive. These advantages include the following:

- More uniform radiation on the surface of the pellet.
- Reduced preheating of the fuel by hot electrons.
- Elimination of crossbeam destructive interference.

In the first case, it is important to irradiate the fuel as uniformly as possible from as many different directions as possible to effectively compress the pellet. Typically, laser beams are split into multiple components and are incident on the fuel pellet simultaneously from many directions. Beams from existing high-power lasers (e.g., Nd-glass) are not as spatially uniform as would be desired. The x-rays produced by the hohlraum are capable of irradiating the surface of the fuel pellet more uniformly than the lasers in a direct drive design.

In the second case, electrons given off during the initial heating of the pellet cause the central portion of the fuel to heat prematurely (as in figure 13.14(b)) and this reduces the effectiveness of the ablation shock wave to compress the fuel.

Finally, refracted components from one portion of the incident laser beam can interfere destructively with portions of the beam that are incident on the fuel from other directions, thereby reducing the overall intensity of the incident beam.

The most extensive laser fusion experiments thus far have been conducted at the National Ignition Facility at Lawrence Livermore National Laboratory in California. This facility uses frequency tripled Nd-doped phosphate glass lasers that can provide up to a 2.05 MJ laser pulse with a peak power of around 500 TW. The fuel contained in a holhraum is irradiated using the indirect drive approach. To date the best results were obtained towards the end of 2022, where a laser pulse of 2.05 MJ produced 3.5 MJ of fusion energy (Kramer 2022). This represents a net energy gain due to fusion and meets the criteria for ignition. Although this is a major breakthrough for controlled fusion, it is still a long way from meeting the needs for a viable energy source. This is because Nd-lasers have a very low efficiency for converting electrical energy input into laser energy output. Currently, this conversion efficiency is only around 1%.

Research in laser fusion is progressing along several lines, but the development of new laser systems that can overcome the deficiencies of current approaches is one of the important directions. In this respect some goals include:

- development of near ultraviolet lasers that do not require harmonic generation;
- development of lasers with higher beam uniformity that can be used in direct drive systems;
- development of lasers with higher efficiency for converting electrical input into laser radiation.

The last point is of particular interest as it not only increases the overall efficiency of the system, but also decreases heating of the lasing medium, thereby increasing possible laser pulse repetition rates. Some lasers that have shown promise for future inertial fusion experiments include KrF excimer lasers and diode pumped solid state lasers.

Once a functioning experimental fusion reactor has been constructed it still remains to design and construct a viable reactor that can produce energy on a commercial scale. An important consideration in the design of an operational reactor is the means by which the fusion energy may be converted into electricity. A generic d–t laser fusion reactor is illustrated in figure 13.17. A similar design could be implemented for a magnetic confinement fusion reactor. Fusion neutrons incident on a lithium blanket surrounding the reactor vessel serve two purposes. Firstly, their kinetic energy heats the lithium, and this heat is used to produce steam to operate a turbine. Secondly, neutrons incident on lithium nuclei breed new tritium fuel through the reactions in equations (13.20) and (13.21).

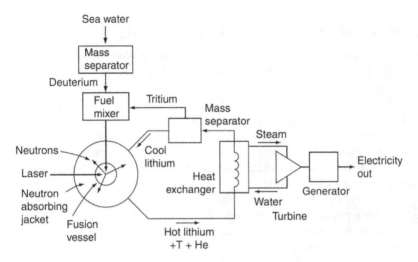

Figure 13.17. Design of a generic laser fusion reactor.

Problems

13.1. Justify the estimate of the energy produced by the combustion of 1 kg of fossil fuel as given in table 13.1.

13.2. Estimate the Coulomb barrier height for two nuclei that are just in contact for the following pairs of nuclei: (a) d–d, (b) ^{6}Li–^{6}Li, and (c) ^{20}Ne–^{20}Ne.

13.3.
   (a) For d–d fusion in a plasma of temperature (energy) of 10 keV calculate the minimum Lawson parameter for sustained fusion.
   (b) Repeat part (a) for d–t fusion.

13.4.
   (a) Calculate the Q for the process shown in equation (13.12).
   (b) Calculate the Q for each of the processes shown in equation (13.13) and compare the total energy with the result of part (a).

13.5. For the ^{15}N(p, α)^{12}C reaction in equation (13.13), calculate the kinetic energies of the reaction by-products.

13.6. The total energy output of the Sun is 3.86×10^{26} W. Assuming that the energy comes exclusively from the proton–proton cycle, calculate the number of hydrogen nuclei involved in fusion reactions per second. Note that each neutrino has an average energy of 0.26 MeV.

13.7.
   (a) The reaction in equation (13.7) is the most likely candidate for a commercial fusion reactor. Calculate the mass of tritium needed to fuel a reactor that produces 5 GW of power for a period of one year.
   (b) If tritium is produced by the reactions in equations (13.18) and (13.19) using natural lithium, calculate the mass of lithium required to produce the tritium mass calculated in part (a).

References and suggestions for further reading

Bobin J L 2014 *Controlled Thermonuclear Fusion* (Singapore: World Scientific Publishing)

Chen F F 2011 *An Indispensable Truth: How Fusion Power Can Save the Planet* (New York: Springer)

Dstrozzi 2009 *Plot of the Fusion Reactivity vs. Temperature for Three Common Reactions* (https://commons.wikimedia.org/wiki/File:Fusion_rxnrate.svg)

Dunlap R A 2021 *Energy from Nuclear Fusion* (Bristol: IOP Publishing)

IAEA Imagebank 2014 *ITER Exhibit* (https://flickr.com/photos/iaea_imagebank/12219071813/)

Klinger T *et al* 2017 Performance and properties of the first plasmas of Wendelstein 7-X *Plasma Phys. Control. Fusion* **59** 014018

Kramer D 2022 National Ignition Facility surpasses long-awaited fusion milestone *Phys. Today*

Krane K S 1988 *Introductory Nuclear Physics* (New York: Wiley)

Landen O L *et al* 2012 Progress in the indirect-drive National Ignition Campaign *Plasma Phys. Control. Fusion* **54** 124026

McCracken G and Stott P 2013 *Fusion: The Energy of the Universe* 2nd edn (New York: Academic)

Morse E 2018 *Nucl. Fusion* (Cham: Springer Nature Switzerland)

Parisi J and Ball J 2019 *The Future of Fusion Energy* (London: World Scientific)

Sangster T *et al* 2007 Overview of inertial fusion research in the United States *Nucl. Fusion* **47** S686–95

Schwarzschild M 1958 *Structure and Evolution of the Stars* (Princeton: Princeton University Press)

Smith C L and Cowley S 2010 The path to fusion power *Phil. Trans. Royal Soc.* A **368** 1091–108

Suzuki Y 2020 Effect of pressure profile on stochasticity of magnetic field in a conventional stellarator *Plasma Phys. Control Fusion* **62** 104001

Zohuri B 2016 *Plasma Physics and Controlled Thermonuclear Reactions Driven Fusion Energy* (Cham: Springer)

Zohuri B 2017 *Magnetic Confinement Fusion Driven Thermonuclear Energy* (Cham: Springer)

Zohuri B 2017 *Inertial Confinement Fusion Driven Thermonuclear Energy* (Cham: Springer)

Part IV

Particle physics

IOP Publishing

An Introduction to the Physics of Nuclei and Particles
(Second Edition)

Richard A Dunlap

Chapter 14

Particles and interactions

14.1 Classification of particles

Further to the discussion in chapter 2, we may divide all particles into two categories, fermions and bosons, as distinguished by their spins and the details of the statistics that describe their behavior. Figure 2.1 summarizes these two categories. Experimental evidence suggests that the leptons (electrons, neutrinos, etc) are fundamental particles with no internal structure. In fact, results indicate that these are truly point particles. The next section of this chapter describes some of the properties of leptons in detail. The baryons and mesons show related internal structure and are collectively referred to as hadrons. The hadrons are not fundamental particles but are comprised of quarks. This will be discussed further in chapters 15 and 16. A convenient way of categorizing particles for the purpose of the present discussion is illustrated in figure 14.1. It is the last row of particles (leptons, quarks, and gauge bosons) that are fundamental. It is important to distinguish between particles that are fundamental and those that are stable. An electron is both fundamental and stable. A muon is fundamental but is not stable, as discussed in the next section. The proton is stable (at least in the standard model—more on this in chapter 18) but is not fundamental, while the neutron is neither fundamental nor stable.

14.2 Properties of leptons

At present there are six known leptons and six corresponding antileptons, as summarized in table 14.1. It is seen that the leptons can be divided into three groups (called generations), each consisting of a lepton and an associated neutrino, as well as their antiparticles. It should be noted that the negatively charged leptons are referred to as particles, while the positively charged leptons are the antiparticles.

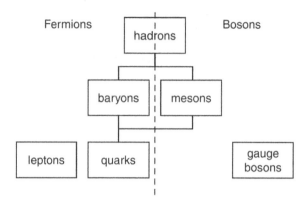

Figure 14.1. Classification of particles. The last row represents those particles that are believed to be fundamental.

Table 14.1. Properties of known leptons.

Lepton	Antiparticle	Generation	Mass (MeV c^{-2})	Lifetime (s)	Decays
e^-	e^+	e	0.511	∞	—
ν_e	$\bar{\nu}_e$	e	~ 0	∞	—
μ^-	μ^+	μ	105.7	3.2×10^{-6}	$\mu^- \to e^- + \bar{\nu}_e + \nu_\mu$
ν_μ	$\bar{\nu}_\mu$	μ	~ 0	∞	—
τ^-	τ^+	τ	1784	4.9×10^{-13}	$\tau^- \to e^- + \bar{\nu}_e + \nu_\tau$
					$\tau^- \to \mu^- + \bar{\nu}_\mu + \nu_\tau$
					$\tau^- \to \pi^- + \nu_\tau$
ν_τ	$\bar{\nu}_\tau$	τ	~ 0	∞	—

This results from the fact that the negatively charged electron is the particle that comprises normal matter. There may be some uncertainty as to whether the particles given in table 14.1 represent a comprehensive list of all possible leptons. Cosmological models of particle formation in the early universe indicate that the number of lepton generations is limited to four or less. Some experimental evidence suggests that the three generations listed in the table, in fact, represent all leptons. Further discussions in the present text will assume this to be true.

The decay of the unstable leptons is an important topic. The best known of these is the decay of a muon (or its antiparticle) into an electron (or a positron) and two neutrinos. This can be written as

$$\mu^- \to e^- + \bar{\nu}_e + \nu_\mu. \tag{14.1}$$

We have seen in the previous discussion of β-decay processes that lepton number must be conserved in all processes. Equation (14.1) indicates that this conservation law is even more strict. We can define three separate lepton numbers corresponding to the three lepton generations, L_e, L_μ, and L_τ. The decay of the muon shows that (at least) L_e and L_μ must be conserved, and it is this requirement that explains the existence of the two neutrinos on the right-hand side of the equation.

It is the proper identification of neutrino generation that is the basis of our belief that lepton generation number (as well as total lepton number) must be conserved. As an example, we can consider the β-decay of a proton,

$$p \rightarrow n + e^+ + \nu_e.$$

The electron neutrino can be moved to the left-hand side of the equation to represent the interaction of an electron antineutrino and a proton,

$$\bar{\nu}_e + p \rightarrow n + e^+. \tag{14.2}$$

The process in equation (14.2) is well known and has been extensively studied (although the cross section is very small). Experimental investigations of muon neutrinos sometimes involve the decay of pions. The basic properties of the pions (or π-mesons) are given in table 14.2. These particles will be discussed in further detail in the next chapter. The decay of a negative pion will produce muon antineutrinos according to the process,

$$\pi^- \rightarrow \mu^- + \bar{\nu}_\mu.$$

Experiments have utilized this process to look for reactions that violate conservation of lepton generation number, for example:

$$\bar{\nu}_\mu + p \rightarrow n + e^+. \tag{14.3}$$

In traditional experiments, no clear evidence for such reactions has been observed, although new observations will be discussed further in chapter 19.

It is also important to distinguish between neutrinos and antineutrinos. Neutrinos that have been discussed thus far are distinct from antineutrinos and are referred to as Dirac neutrinos. Neutrinos that are the same as their antiparticles are referred to as Majorana neutrinos. The failure to observe reactions that violate lepton number conservation, such as

$$\nu_e + p \rightarrow n + e^+ \tag{14.4}$$

demonstrates that neutrinos are Dirac particles and are distinct from their antiparticles. Another approach to distinguishing between Dirac and Majorana

Table 14.2. Properties of pions.

Particle	Charge (e)	Mass (MeV c^{-2})	Spin	Lifetime (s)	Decay products
π^-	−1	139.57	0	2.6×10^{-8}	$\mu^- + \nu_\mu$
π^0	0	134.98	0	8.5×10^{-17}	$\gamma + \gamma$
π^+	+1	139.57	0	2.6×10^{-8}	$\mu^+ + \nu_\mu$

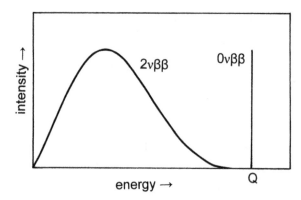

Figure 14.2. Electron energy spectra for two-neutrino (2ν) and neutrinoless (0ν) double beta decay. Q is the total decay energy. From Dunlap (2018). Copyright IOP Publishing. Reproduced with permission. All rights reserved.

neutrinos is thorough the study of double β-decay. The double β-decay process changes two neutrons to two protons,

$$2n \rightarrow 2p + 2e^- + 2\bar{\nu}_e$$

where the resulting kinetic energy is distributed between the electrons and antineutrinos on the right-hand side of the equation. This is referred to as a two-neutrino double β-decay ($2\nu\beta\beta$) and was discussed in detail in section 9.6. In the case where neutrinos and antineutrinos are not distinguishable, then the two antineutrinos on the right-hand side can annihilate leading to neutrinoless double β-decay ($0\nu\beta\beta$),

$$2n \rightarrow 2p + 2e^-.$$

In this case, the electrons would receive all of the kinetic energy from the process. The distinction between the energy distribution for the electrons for these two types of double β-decay is illustrated in figure 14.2. An inspection of the double β-decay electron spectrum in figure 9.11, shows no evidence for a neutrinoless contribution. In fact, all searches for neutrinoless double β-decay to date have not found any convincing evidence for its existence. This lends further support to the conclusion that neutrinos are Dirac particles.

14.3 Feynman diagrams

From a classical standpoint, the interaction between particles (or any objects) is the result of a field. From a macroscopic standpoint the fields that we can experience directly are gravitational fields and electromagnetic fields. An important feature of a classical field is the fact that it can transfer energy and momentum from one object to another. The acceleration of a mass in a gravitational field is a well-known example of this phenomenon. From a quantum mechanical standpoint, fields are quantized, and the field quanta are bosons. As an example, light can be considered classically in terms of electric and magnetic fields. However, from a quantum mechanical standpoint a quantum of electromagnetic radiation is described as a photon. Along similar lines the electromagnetic interaction between two charges is described quantum mechanically as the exchange of a photon. Bosons that take part in interactions are referred to as gauge bosons and are said to mediate the interaction. In addition to the photon, that mediates the electromagnetic interaction, there are gauge bosons that mediate the other interactions in nature. The four known interactions and the gauge bosons that mediate them are given in table 14.3.

Feynman diagrams are a convenient means of representing particle interactions. Although they can be used in a more quantitative sense to calculate properties such as cross sections, this approach is beyond the scope of the present book. Here we will use these diagrams to indicate the relationship between particles in interactions and decay processes. Table 14.4 describes the symbols commonly used to designate various types of particles in Feynman diagrams. As a simple example of a Feynman diagram, we can look at the Coulomb interaction between two electrons as shown in figure 14.3. The electrons are represented by straight lines with arrows indicating their direction of propagation. Time progresses from left to right in the diagram. The gauge boson that mediates the interaction (the photon) is represented by the wavy line. Since the photon is uncharged and has zero lepton and baryon numbers, it can propagate in either direction. The figure shows that the two electrons interact via the Coulomb interaction by exchanging a virtual photon. It should be noted that the arrows for antileptons (and other antifermions), as will be seen in some subsequent diagrams, are drawn in the opposite direction, as it is customary to view these particles as propagating backward in time.

Table 14.3. Gauge bosons associated with the four known interactions. The graviton is a hypothetical particle that mediates the gravitational interaction and has never been observed experimentally.

Interaction	Gauge boson	Mass (GeV c^{-2})	Spin	Acts on
Strong	Gluons	0	1	Hadrons
Electromagnetic	γ	0	1	Charges
Weak	W^+, W^-, Z^0	80.4, 80.4, 91.2	1	Leptons and hadrons
Gravity	Graviton	0	2	Masses

Table 14.4. Symbol convention for different particles in Feynman diagrams.

Particles	Symbol
Leptons, baryons, real mesons	————
Photons	∿∿∿
W^+, Z^0, W^-, virtual mesons	- - - -
Gluons	⦵⦵⦵

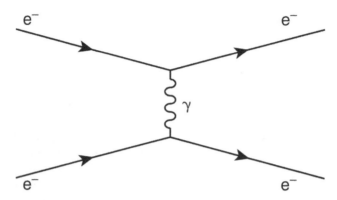

Figure 14.3. Feynman diagram for an electron–electron interaction illustrating the exchange of a virtual photon.

The conservation of mass/energy, linear momentum, angular momentum, lepton number, lepton generation number, baryon number, quark number and charge applies to all particles in a reaction or decay. These conservation laws can be applied to the vertices of the Feynman diagram and the implementation of these laws is, for the most part, straightforward. However, the conservation of mass/energy and momentum requires some specific comments. In general, it is not possible to satisfy these conservation laws if the mass of the mediating particle is identical to the mass of the same particle when it is a free particle. Such particles do not obey the Einstein relation,

$$E^2 = p^2c^2 + m^2c^4 \tag{14.5}$$

and are sometimes said to be 'off the mass shell'. Particles that satisfy equation (14.5) are referred to as real particles and those that do not are called virtual particles. In general, lines in Feynman diagrams that represent real particles have one free end and one vertex, while lines that represent virtual particles have vertices at both ends. The conservation of angular momentum also has some important implications for the construction of Feynman diagrams. The Feynman diagram shown in figure 14.3 has vertices involving two fermion lines (leptons) and one boson line. Conservation of angular momentum requires that the vector sum of the spins of the particles must be a conserved quantity. An even number of half integer spins (fermions) is always necessary to compensate for an integer spin boson.

The construction of proper Feynman diagrams may be demonstrated by considering some of the decays and reactions that have previously been discussed in this book and an inspection of these diagrams is beneficial to understanding the physics of these processes. The Feynman diagram for the decay of a negative muon as described by equation (14.1) is shown in figure 14.4. A W^- travelling downward (in the diagram) is equivalent to a W^+ travelling upward. Conservation of lepton generation at vertices in the diagram requires that the weak interaction converts a muon to its own neutrino and creates an electron antineutrino pair. It is, in fact, a

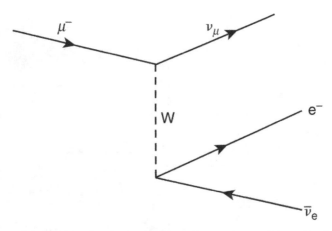

Figure 14.4. Feynman diagram for the decay of a negative muon (equation (14.1)).

general vertex rule that a charged weak gauge boson will convert a lepton into its own neutrino (or vice versa) or that it will create a lepton–antineutrino pair. The direction convention for antileptons can be seen in figure 14.4. Vertex rules for gauge bosons will be discussed in a more general context in chapter 16.

The Feynman diagrams for β-decay processes are illustrated in figure 14.5. For β⁻ and β⁺ decay the weak gauge boson results in the creation of a lepton–neutrino pair. Equivalent diagrams can be drawn for W⁺ and W⁻ with different directions of propagation. In each case, charge conservation must be upheld at the vertices. A comparison of the diagrams for β⁺ decay (equation (4.7)) and electron capture (equation (9.3)) illustrates an interesting feature of leptons. Moving a lepton from the right-hand side to the left-hand side of an equation converts it into its own antiparticle and results in a mirror reflection of the lepton line about the vertex in the Feynman diagram. An inspection of figure 14.5 shows that the relevant conservation laws are still obeyed. The interaction of the weak charged boson with a baryon, as shown in figure 14.5, is best dealt with in the context of the quark components of the baryon and is left for chapter 16. An attempt to construct valid Feynman diagrams for the processes given in equations (14.3) and (14.4) immediately demonstrates the relevant conservation law that is violated in each case.

Problems

14.1. The observation of the process given in equation (14.4) implies the existence of neutrinoless double β-decay. Explain how this is possible.

14.2. Following along the lines of figure 14.5, construct diagrams for:
   (a) electron neutrino capture by a neutron;
   (b) positron capture by a neutron;
   (c) electron antineutrino capture by a proton.

14.3. Calculate the Q values for the lepton decays shown in table 14.1.

14.4. Unstable particles are often created in high energy collisions and the particles themselves are highly relativistic. It is interesting to consider the

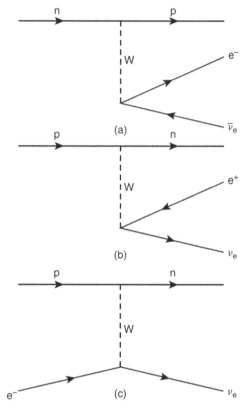

Figure 14.5. Feynman diagrams for β-decay processes. (a) β^- decay (equation (4.6)), (b) β^+ decay (equation (4.7)), and (c) electron capture [equation (9.3)].

distance such a particle travels prior to decay. For the unstable leptons in table 14.1 calculate the mean distance traveled by relativistic particles.

14.5. The lifetime of very short-lived states is sometimes given in energy units, that is the width in energy of the resonance corresponding to the state. Calculate the lifetime in time units of states with resonance widths of (a) 1 keV, (b) 1 MeV, and (c) 1 GeV.

14.6. Explain the observed trend for the pion lifetimes given in table 14.2.

Suggestions for further reading

Dunlap R A 2018 *Particle Physics* (San Rafael, CA: Morgan & Claypool)

IOP Publishing

An Introduction to the Physics of Nuclei and Particles
(Second Edition)

Richard A Dunlap

Chapter 15

The standard model

15.1 Evidence for quarks

The earliest versions of the quark model hypothesized the existence of three quarks (and three corresponding antiquarks). These were sufficient to explain the properties of the known hadrons at that time. Since both baryons and mesons are comprised of quarks, it is obvious that quarks must be fermions. Conservation of angular momentum implies that it is possible to construct either fermions (baryons) or bosons (mesons) from a combination of fermions, but it is not possible to construct fermions from a combination of bosons.

There is considerable experimental evidence that hadrons are made up of fundamental point-like particles that are, in many ways, analogous to the fundamental leptons. Although much of this evidence is rather indirect, taken as a whole it lends convincing support to the quark model. Here we review a small portion of the experimental evidence accumulated to date.

Neutral meson production—The interaction of high energy electrons with protons is known to produce neutral mesons according to reactions such as

$$e^- + p \rightarrow e^- + p + \pi^0.$$

These reactions are difficult (at best) to explain if it is assumed that the protons are, like electrons, fundamental structureless particles.

Excited states of the proton—The absorption cross section for photons on atoms shows anomalously large values at energies corresponding to those that allow for the population of excited states. The existence of excited states in atoms is a feature that results from the fact that an atom is a bound system consisting of more than one particle. High energy photon absorption cross sections for protons show similar

doi:10.1088/978-0-7503-6094-4ch15

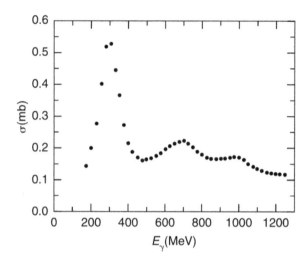

Figure 15.1. Cross section for photons incident on protons. Data are from Armstrong *et al* (1972).

features indicating that the proton is not a discrete particle but is a bound system of more than one particle. Some experimental results are illustrated in figure 15.1. The large peak at just below 300 MeV corresponds to the creation of the Δ^+ particle, a baryon with a mass of 1231 MeV c^{-2} (that is, the proton mass, 938 MeV c^{-2} plus the 293 MeV provided by the photon). Experimental results indicate that the Δ^+ is an excited state of the proton corresponding to a different spin state. Details of the distinction between the proton and the Δ^+ will be given in the next section.

Neutron magnetic moment—Although the neutron carries no charge, it does have a magnetic moment. This is a clear indication that the neutron has an internal structure involving a distribution of charges.

Deep inelastic scattering of electrons—The most conclusive and revealing evidence for the existence of quarks comes from deep inelastic scattering experiments. These most commonly involve the scattering of high energy electrons by protons. These experiments follow from Rutherford's original α-particle scattering experiments to study the structure of the atom. Subsequent experiments using higher energy incident particles were used to study the structure of the nucleus. The spatial resolution of a scattering experiment is given by the de Broglie wavelength of the incident particles,

$$\lambda = \frac{h}{p}.$$

Thus, higher energy particles can probe the structure of the scatterer on a smaller scale. Electrons at an energy of 10 GeV have a de Broglie wavelength of about 0.1 fm (about one-tenth the radius of a proton). Scattering experiments utilizing electrons with energies up to 100 GeV can, therefore, probe the internal structure of a proton in considerable detail. The results of such experiments can be summarized as follows:

1. The proton contains three point-like particles.
2. These particles carry charges of $-1/3$ or $+2/3$ of the electron charge.
3. The particles are spin 1/2 fermions.

These observations are consistent with the standard model of hadron structure as described in the remainder of this chapter.

15.2 Composition of light hadrons

The three quarks and three antiquarks of the original model are described in table 15.1. From a quantum mechanical standpoint, the identity of a quark (up, down, or strange) is referred to as the flavor of the quark. The baryons are comprised of three quarks while antibaryons are comprised of three antiquarks. The mesons are made up of a quark and an antiquark. It is, therefore, easy to see that the resulting baryons must be fermions and the mesons must be bosons. Even with only three quarks a substantial number of baryons and mesons can be constructed. This comes from the fact that quarks of different flavors can be combined and that these bound systems can exist in various quantum mechanical states. Some examples help to illustrate these points.

Table 15.2 gives the properties of light mesons consisting of up, down, and strange quarks (along with their antiquarks). It is customary to indicate the quark content of the particle as a quantum mechanical state vector, as given in the table. It is seen from the table that some particles (for example, π^0) are represented by a linear combination of quark flavor states.

The table illustrates some interesting relationships between the positively and negatively charged mesons and the neutral mesons. The process of *charge conjugation* interchanges quarks and antiquarks. Since the mass of a quark is the same as the mass of an antiquark of the same flavor, charge conjugation leaves the mass of a particle unchanged. The table indicates that charge conjugation changes positively charged mesons into corresponding negatively charged mesons of the same mass (or vice versa). These particles are the antiparticles of one another. The neutral mesons that are comprised of a quark and its own antiquark are said to be *self-conjugate* and are their own antiparticles.

Table 15.1. Properties of quarks. T_3, S, C, B' and T' represent isospin, strangeness, charm, bottom, and top, respectively (see section 15.4). the masses as shown are the current masses and are discussed further in the text.

Quark	Symbol	Charge (e)	Mass (GeV c^{-2})	T_3	S	C	B'	T'
Up	u	$+2/3$	0.0023	$+1/2$	0	0	0	0
Down	d	$-1/3$	0.0048	$-1/2$	0	0	0	0
Strange	s	$-1/3$	0.095	0	-1	0	0	0
Charm	c	$+2/3$	1.275	0	0	$+1$	0	0
Bottom	b	$-1/3$	4.18	0	0	0	-1	0
Top	t	$+2/3$	173.21	0	0	0	0	$+1$

Table 15.2. Properties of light mesons.

Quarks	Spin 0 ($^1S_0$, $J^\pi = 0^-$)			Spin 1 ($^3S_1$, $J^\pi = 1^-$)		
	Particle	Charge (e)	Mass (MeV c^{-2})	Particle	Charge (e)	Mass (MeV c^{-2})
$\lvert u\bar{d}\rangle$	π^+	$+1$	140	ρ^+	$+1$	770
$\frac{1}{\sqrt{2}}\lvert d\bar{d} - u\bar{u}\rangle$	π^0	0	135	ρ^0	0	770
$\lvert \bar{u}d\rangle$	π^-	-1	140	ρ^-	-1	770
$\frac{1}{\sqrt{2}}\lvert d\bar{d} + u\bar{u}\rangle$	η	0	549	ω	0	783
$\lvert u\bar{s}\rangle$	K^+	$+1$	494	K^{*+}	$+1$	892
$\lvert d\bar{s}\rangle$	K^0	0	498	K^{*0}	0	892
$\lvert \bar{u}s\rangle$	K^-	-1	494	K^{*-}	-1	892
$\lvert \bar{d}s\rangle$	$\overline{K}^0$	0	498	$\overline{K}^{*0}$	0	892
$\lvert s\bar{s}\rangle$	η'	0	958	ϕ	0	1020

Table 15.3. Meson states for $L = 0$ and 1.

Notation	L	S	J	π
$^1S_0$	0	0	0	$-$
$^3S_1$	0	1	1	$-$
$^1P_1$	1	0	1	$+$
$^3P_0$	1	1	0	$+$
$^3P_1$	1	1	1	$+$
$^3P_2$	1	1	2	$+$

Mesons have certain properties that are distinct from those of baryons that result directly from the fact that they are bosons. It is interesting to consider the properties of a bound quark–antiquark system in general. We can express the angular momentum of the quark–antiquark pair in terms of the three quantum numbers L, S, and J. In spectroscopic notation this is written as $^{2S+1}L_J$, where the usual convention S, P, D, ... is used for $L = 0, 1, 2, ...$ Since $\vec{J}$ is the vector sum of $\vec{L}$ and $\vec{S}$, the magnitude of J is constrained to take on values in the range

$$|L - S| \leqslant J \leqslant L + S. \tag{15.1}$$

Since the quark and the antiquark are spin 1/2 fermions the total meson spin will be 0 for an antiparallel alignment of the quark spins or 1 for a parallel alignment of the quark spins. Some examples of angular momentum states for mesons are given in table 15.3 where the values of L are 0 or 1, the values of S are 0 or 1, and the values of J are constrained by equation (15.1). Larger values of L are also possible, subject to the same constraints. These states are sometimes designated by their total angular momentum and their parity, π, as J^{π}. The question of the overall parity of a meson is quite interesting. The parity of a bound quark–antiquark pair with orbital angular momentum L is given by $(-1)^{L+1}$. The low-lying meson states are the least massive and those given in table 15.2 all correspond to $L = 0$. As indicated in table 15.3, the spin 0 mesons in table 15.2 are in the $^1S_0$ (or 0^-) state, while the spin 1 mesons are in the $^3S_1$ (or 1^-) state. Mesons with the same quark content but different spins states are sometimes distinguished by appending a superscript asterisk to the name of the higher spin state particle to indicate an excited state, for example, K*° versus K°. Another scheme is to append the mass of the particle in MeV c^{-2} to the name of the particle, for example, $\rho(1450)$ versus $\rho(770)$. The obvious increase in mass associated with increasing spin state as illustrated in table 15.2 is evidence for a spin–spin interaction term in the potential.

The properties of baryons that can be formed from up, down, and strange quarks are given in table 15.4. The three spin 1/2 quarks can yield a total spin of 1/2 (for two parallel and one antiparallel) or 3/2 (for three parallel). Following the discussion above, the possible total angular momentum values of the three-quark bound state are related to the spin and orbital components and some examples are illustrated in table 15.5. The parity of an L state baryon is given by $(-1)^L$. The particles described in table 15.4 are in the $L = 0$ state meaning that the spin 1/2 baryons are in the $^2S_{1/2}$ (or $1/2^+$) state and the spin 3/2 baryons are in the $^4S_{3/2}$ (or $3/2^+$) state. Finally, one should note that certain quark flavor combinations are not allowed in certain spin states, for example, the spin 1/2 state of uuu.

Antibaryons are formed from three antiquarks and can be viewed in terms of the application of charge conjugation to the baryons given in table 15.4. In all cases the masses of the antibaryons are the same as the masses of the corresponding baryons, while the charges are opposite (in the case of charged baryons). The parity of an

Table 15.4. Properties of light baryons.

Quarks	Spin 1/2 ($^2S_{1/2}$, $J^{\pi} = 1/2^+$)			Spin 3/2 ($^4S_{3/2}$, $J^{\pi} = 3/2^+$)		
	Particle	Charge (e)	Mass (MeV c^{-2})	Particle	Charge (e)	Mass (MeV c^{-2})
$\lvert uuu\rangle$	—	—	—	Δ^{++}	+2	1230
$\lvert uud\rangle$	p	+1	938	Δ^+	+1	1231
$\lvert udd\rangle$	n	0	940	Δ^0	0	1232
$\lvert ddd\rangle$	—	—	—	Δ^-	−1	1234
$\frac{1}{\sqrt{2}}\lvert(ud - du)s\rangle$	Λ	0	1116	—	—	—
$\lvert uus\rangle$	Σ^+	+1	1189	Σ^{*+}	+1	1383
$\frac{1}{\sqrt{2}}\lvert(ud + du)s\rangle$	Σ^0	0	1192	Σ^{*0}	0	1384
$\lvert dds\rangle$	Σ^-	−1	1197	Σ^{*-}	−1	1387
$\lvert uss\rangle$	Ξ^0	0	1315	Ξ^{*0}	0	1532
$\lvert dss\rangle$	Ξ^-	−1	1321	Ξ^{*-}	−1	1535
$\lvert sss\rangle$	—	—	—	Ω^-	−1	1672

Table 15.5. Baryon angular momentum states for $L = 0$ and 1.

Notation	L	S	J	π
$^2S_{1/2}$	0	1/2	1/2	+
$^4S_{3/2}$	0	3/2	3/2	+
$^2P_{1/2}$	1	1/2	1/2	−
$^2P_{3/2}$	1	1/2	3/2	−
$^4P_{1/2}$	1	3/2	1/2	−
$^4P_{3/2}$	1	3/2	3/2	−
$^4P_{5/2}$	1	3/2	5/2	−

antibaryon in an orbital angular momentum state L is given by $(-1)^{L+1}$. The neutral baryons are not self-conjugate and there are positive and negative varieties of some baryons that are not antiparticles of one another that have similar (but not identical) masses.

15.3 Composition of heavy hadrons

The description of the heavier mesons and baryons requires the introduction of additional, heavier quarks. Since the original version of the quark model, three additional quarks have been added: charm, bottom, and top, as indicated in table 15.1.

A number of mesons can be formed that involve one or more heavy quarks (or antiquarks). Some of those that have been observed experimentally are summarized in table 15.6. The masses given in the table are the spin 0 ground state masses. In most cases, more massive excited states are also observed. The $c\bar{c}$ state, usually referred to as charmonium, was first observed in 1974. Almost simultaneously electron–positron collider experiments at the Stanford Linear Accelerator and synchrotron experiments involving the collision of protons with light nuclei at Brookhaven National Laboratory yielded evidence for charmonium. The former researchers named the new meson, ψ, while the latter referred to it as J. Although the name controversy persisted for a number of years, current convention seems to be to call it J/ψ as indicated in the table. Details of the operation of the Stanford Linear Accelerator and the physical principles of synchrotrons are discussed in Appendix C.

Mesons containing the most massive of the quarks, top, were first reported from a series of experiments using the Tevatron synchrotron at Fermilab in 1994. However, theoretical prediction for the existence of the top quark predated these observations. The Fermilab experiments utilized collisions between high energy protons and antiprotons to produce a variety of massive hadrons. In a very small number of instances, quark pairs are formed. The top quark decays to the bottom quark by the weak interaction on a time scale of 10^{-24} s. This decay produces mesons with bottom and lepton–neutrino pairs that can be observed experimentally. A careful analysis of the energy and momentum of decay by-products allows for a reconstruction of the

Table 15.6. Properties of mesons with charm and bottom. Masses are for ground state $^1S_0$ particles.

Quarks	Particle	Charge (e)	Mass (MeV c^{-2})
$\lvert \bar{u}c \rangle$	D^0	0	1865
$\lvert u\bar{c} \rangle$	$\bar{D}^0$	0	1865
$\lvert \bar{d}c \rangle$	D^+	+1	1869
$\lvert d\bar{c} \rangle$	D^-	−1	1869
$\lvert \bar{s}c \rangle$	$D_s{}^+$	+1	1969
$\lvert s\bar{c} \rangle$	$D_s{}^-$	−1	1969
$\lvert c\bar{c} \rangle$	J/ψ	0	3097
$\lvert u\bar{b} \rangle$	B^+	+1	5279
$\lvert \bar{u}b \rangle$	B^-	−1	5279
$\lvert d\bar{b} \rangle$	$B_d{}^0$	0	5279
$\lvert \bar{d}b \rangle$	$\bar{B}_d{}^0$	0	5279
$\lvert s\bar{b} \rangle$	$B_s{}^0$	0	5369
$\lvert \bar{s}b \rangle$	$\bar{B}_s{}^0$	0	5369
$\lvert c\bar{b} \rangle$	$B_c{}^+$	+1	6400
$\lvert \bar{c}b \rangle$	$B_c{}^-$	−1	6400
$\lvert b\bar{b} \rangle$	Υ	0	9460

original top quark mass. The meaning of quark masses will be discussed further in the next section. Figure 15.2 shows the reconstructed top quark mass from the relevant events observed in experiments at the Large Hadron Collider.

The most recently discovered family of mesons contains both charm and bottom quarks. These were first reported in 1998 in experiments involving proton–antiproton collisions at Fermilab. These B_c mesons, which have a lifetime of 0.46 ps, were observed via the decay mode,

$$B_c{}^+ \rightarrow J/\psi + \ell^+ + \nu$$

where ℓ^+ is a positron or a positive muon. Similar decays are seen for $B_c{}^-$. Other possible decays for B_c include $J/\psi + \pi$, $B_s + \ell + \nu$, $B_s + \pi$ and a direct decay to leptons, $\tau + \nu_\tau$. In all cases, these decays are the result of the weak interaction.

15.4 More about quarks

The standard model as described above assumes the existence of six quarks (and six corresponding antiquarks). These are divided into three generations and are, in many ways, analogous to the six known leptons. These relationships are summarized in table 15.7. The leptons are either charge $-e$ or charge zero while the quarks are either charge $-e/3$ or charge $+2e/3$. Previous discussions have shown that lepton number as well as lepton generation must be conserved in all processes. Baryon number conservation can be viewed in terms of the quark content of the various

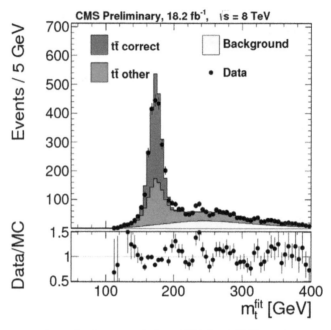

Figure 15.2. Reconstructed mass for the top quark from ATLAS and CMS experiments at the Large Hadron Collider (see appendix C). Reproduced from Mirman (2015) CC BY 4.0.

Table 15.7. Relationship of the generations of leptons and quarks. Antileptons and antiquarks are charge conjugates of the particles in the table.

	Leptons		Quarks	
Generation	Charge $(-e)$	Charge (0)	Charge $(-e/3)$	Charge $(+2e/3)$
1	e^-	ν_e	d	u
2	μ^-	ν_μ	s	c
3	τ^-	ν_τ	b	t

hadrons. If we assign a baryon number of $+1/3$ to quarks and $-1/3$ to antiquarks, then the quark content of baryons and mesons explains the assignment of baryon numbers to these particles. In a more fundamental sense, baryon number conservation should be viewed as quark conservation. Quark flavor is clearly not conserved within a generation as β-decay processes represent the change of an up quark to a down quark, or vice versa. Changes in quark flavor are the result of the weak interaction and, as will be discussed further in the next chapter, changes in quark flavor between generations can occur.

Experimentally quarks have not been observed to exist as free particles and are said to be confined. *Confinement* results from the nature of the strong interaction. Although the strong interaction between nucleons in a nucleus is a short-ranged interaction, this is not a fundamental interaction between fundamental particles. The strong interaction between quarks (or between quarks and antiquarks), as mediated by gluons, actually increases with increasing distance. This is rather like stretching a spring. To get quarks further and further apart requires more and more energy. It has been speculated that quarks and gluons may have existed in the very early universe (for the first 10^{-5} s or so) in the form of a quark–gluon plasma rather than in the form of hadrons. It is also possible that a quark–gluon plasma may exist at the center of a neutron star and experiments are underway to create this form of matter (for very short periods of time) in high energy particle collisions.

Because quarks are not observed as free particles, the concept of mass cannot be viewed in the same way as it is for particles such as electrons or protons. In the simplest approach, the quark mass may be determined in terms of the mass of the hadrons that it comprises. The mass in this context is called the *effective mass* or *constituent mass*. This type of analysis would lead us to the conclusion that the masses of the up quark and the down quark were both about 310 MeV c^{-2}, with the down quark being more massive than the up quark by a small number of MeV c^{-2}. This simplistic approach is, more or less, consistent with the masses of many other hadrons, although it does not provide a measure of the mass of a quark if it existed as a free particle. Another approach is to view the quark masses as parameters in the theory of quark interactions (see section 15.5 for more on this topic). Masses determined in such a way are referred to as *bare masses* (as they are, more or less, related to the intrinsic mass of the quark) or *current masses* (as the mass parameter appears in terms for currents in the theory). In the case of the light quarks, at least, the current masses are substantially less than the constituent masses. Some estimates of current masses for quarks are given in table 15.1.

The relationships between the properties of the various hadrons can be understood in terms of quantum numbers associated with the flavors of their constituent quarks. Table 15.1 shows that quantum numbers for strangeness, charm, bottom, and top are defined for the respective quarks. Quantum numbers for the antiquarks are conjugates of those listed in the table. The up and down quarks are dealt with differently than other quarks. Table 15.1 indicates that these quarks are much less massive than the other quarks and as a result there are groups (or *multiplets*) of hadrons with very nearly the same masses that can be formed by interchanging up and down quarks. Tables 15.2 and 15.4 illustrate some examples of these multiplets; for example (n, p), (Σ^-, Σ^0, Σ^+), (π^-, π^0, π^+), etc. One view is to consider the up and down quarks as indistinguishable except for their difference in charge. As an analogy we can consider an atomic electron. The electron has an intrinsic spin of 1/2. In the presence of a magnetic field electrons are designated as spin up or spin down on the basis of the z-component of their spin, +1/2 or −1/2, respectively. The flavor quantum number of the up and down quarks is defined as a pseudo-spin referred to as *isospin* with an intrinsic value of $T = 1/2$. In the presence of electromagnetic interactions, the up and down quarks are distinguished on the basis of the *third component* of isospin, T_3, where the up quark is defined to have $T_3 = +1/2$ and the

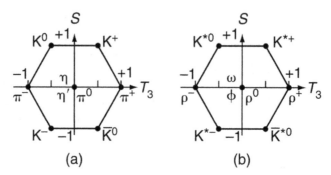

Figure 15.3. Eight-fold way diagrams for light mesons with (a) spin 0 and (b) spin 1.

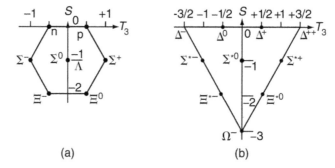

Figure 15.4. Eight-fold way diagrams for light baryons with (a) spin 1/2 and (b) spin 3/2.

down quark is defined to have $T_3 = -1/2$. These can be related to the usual definitions of isospin of $+1/2$ and $-1/2$ for the proton and neutron, respectively.

The relationships between different hadrons can be seen schematically in what are referred to as eight-fold way diagrams by plotting strangeness as a function of isospin. Examples for light mesons with spin 0 and spin 1 are illustrated in figure 15.3. Nearly mass degenerate multiplets are indicated by families of particles on lines with constant strangeness. Eight-fold way diagrams for light baryons are illustrated in figure 15.4. Similar diagrams that include heavier hadrons can also be constructed. For example, mesons and baryons with charm can be illustrated in a three-dimensional diagram with the charm of the particle plotted along the third axis. Because individual quarks have well-defined values for their various quantum numbers, as well as charge, the hadrons must also have certain relationships between these numbers. These relationships are defined by the Gell–Mann–Nishijima formula,

$$Q = T_3 + \frac{B + \Sigma(\text{flavour})}{2}$$

where Q is the particle charge in units of e, B is the baryon number and Σ(flavor) is the sum of flavor quantum numbers as defined in table 15.1. Since individual quarks (and antiquarks) satisfy this relationship, all mesons and baryons must also obey the Gell–Mann–Nishijima formula. This is verified for all but the heaviest known hadrons by direct experimental evidence.

15.5 Color and gluons

Since quarks, like electrons, are fermions they obey Fermi–Dirac statistics and are subject to the Pauli exclusion principle. A careful consideration of the properties of some baryons indicates a problem with the quark model as described above. For example, the Δ^{++} baryon is composed of three up quarks in the same spin and orbital states, and this violates the Pauli exclusion principle.

This difficulty is not encountered for mesons because a quark and an antiquark are always distinguishable particles. To resolve the problem of the Δ^{++} baryon, each of the three identical quarks in the system must be distinguished in some way. This is done by assigning an additional quantum number to each quark in such a way that they are distinguishable. This new quantum number is referred to as *color*, although this is in no way related to the conventional concept of color. The quantum number for the color of a quark can take on one of three values referred to as *red* (R), *green* (G), or *blue* (B). Antiquarks are assigned anticolor, which can have values of *anti*red ($\overline{\mathrm{R}}$), *anti*green, ($\overline{\mathrm{G}}$) or *anti*blue ($\overline{\mathrm{B}}$).

Baryons are made up of three quarks of three different colors, that is, one red, one green, and one blue. Mesons are made of a quark of a given color and an antiquark of the same anticolor. This scheme has the net result that all hadrons are colorless. This is a requirement of the model since no manifestation of color is observed in the measured physical properties of mesons or baryons. This is in contrast to the flavor quantum numbers of the quark constituents of hadrons. A hadron can exhibit overall strangeness, charm and other characteristics resulting from the sum of the quantum numbers of the component quarks associated with these quantities. The physical manifestation of strangeness or charm is the mass of the hadron since quarks of different flavors have different masses. The theory of interactions involving color is referred to as *quantum chromodynamics* (QCD) and will be dealt with only briefly in this text. However, a very phenomenological interpretation of color interactions can be made by analogy with electrostatic interactions. In the same way that like charges repel and unlike charges attract, we can view the strong force between quarks as color dependent with like colors repelling and unlike colors attracting one another. Thus, a meson, which contains a quark of a certain color, must contain an antiquark of the same anticolor in order to leave the meson as a color-neutral particle.

Quantum mechanically the properties of hadrons can be considered in terms of their wave functions and this approach illustrates the necessity for color. The total wave function of a particle is comprised of four components (space, spin, flavor, and color),

$$\psi = \psi_{space}\psi_{spin}\psi_{flavor}\psi_{color}.$$

For baryons (which are fermions) the total wave function is required to be antisymmetric under interchange of any two quarks. In this case it can be shown that the space–spin–flavor wave function is symmetric. The wave function for the color state is defined by a linear combination of all possible color states:

$$\psi_{color} = \frac{1}{\sqrt{6}}[RGB + GBR + BRG - RBG - BGR - GRB].$$

It can be seen that this function is antisymmetric under interchange of any two quarks. When combined with the baryon space–spin–flavor wave function this restores the antisymmetric properties of the total wave function.

It is now possible to consider the details of the strong interaction between quarks. Virtual pion exchange, as illustrated in figure 15.5(a), is sometimes used to describe the interactions between nucleons in a nucleus. However, neither the hadrons nor the virtual pions in figure 15.5(a) are fundamental particles, so the description of the interaction is not fundamental. An appropriate description of the interaction between hadrons in terms of the massless gluons that mediate the strong interaction is illustrated in figure 15.5(b). As mediators of the strong interaction, gluons cannot change the flavor of a quark. Thus, the only allowed quark–gluon (q–g) vertices in Feynman diagrams are of the form $qg \rightarrow q$, $q \rightarrow qg$, $q\bar{q} \rightarrow g$ or $g \rightarrow q\bar{q}$ and their charge conjugates, where the quarks must be of the same flavor. Figure 15.5(b)

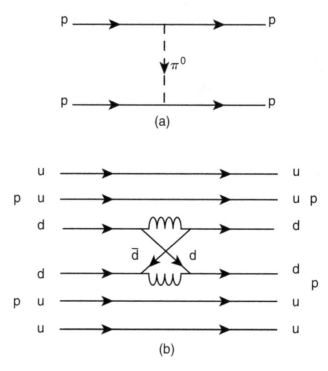

(a)

(b)

Figure 15.5. Proton–proton scattering viewed in terms of (a) virtual pion exchange between baryons and (b) gluon exchange between quarks.

shows that in the quark description of the process, the virtual pion is represented by its quark components and results from the creation and annihilation of a $d\bar{d}$ pair.

The gluons form an octet with color state vectors given by,

$$|R\overline{G}\rangle$$

$$|R\overline{G}\rangle$$

$$|R\overline{G}\rangle$$

$$|R\overline{G}\rangle$$

$$|R\overline{G}\rangle$$

$$|R\overline{G}\rangle$$

$$\frac{1}{\sqrt{2}}|R\overline{R} - G\overline{G}\rangle$$

$$\frac{1}{\sqrt{6}}|R\overline{R} + G\overline{G} - 2B\overline{B}\rangle.$$

Examples of the simplest cases of color exchange between quarks or between quarks and antiquarks of the same flavor are illustrated in figure 15.6. Color is conserved at

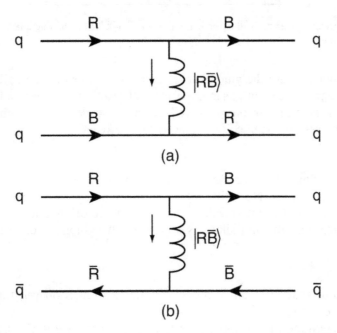

Figure 15.6. Examples of simple gluon color exchange interaction (a) between quarks and (b) between a quark and an antiquark.

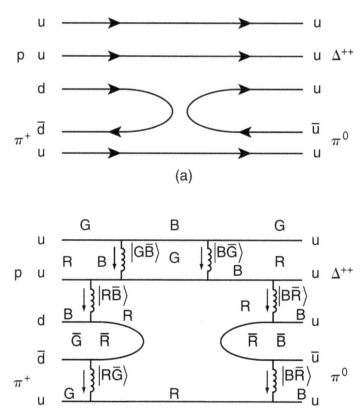

Figure 15.7. Feynman diagram for the reaction $p + \pi^+ \to \pi^0 + \Delta^{++}$; (a) showing quark relationships and (b) showing colored gluons.

all vertices but because the gluon carries color, the color of the quark is changed. Color exchange between quarks in the context of reactions between hadrons requires that all mesons and baryons involved remain colorless. An example of color exchange process involving colored gluons for the reaction

$$p + \pi^+ \to \pi^0 + \Delta^{++}$$

is illustrated in figure 15.7. This is, of course, only one possible valid way of coloring the quarks and gluons. Note that all initial and final hadron states are colorless. Note as well, that the intermediate Δ^{++} state is colorless and that the quark–antiquark creations and annihilations do not involve gluons or color exchange.

Problems

15.1. Show that the Gell–Mann–Nishijima relation is obeyed for the mesons given in table 15.6.

15.2. Determine the isospin of the following particles: (a) π^-, (b) K^+, (c) D^0, (d) J, (e) Ξ^-, (f) n and (g) Δ^{++}.

15.3. Draw a charm space eight-fold way diagram for spin 0 non-strange mesons.

15.4. Draw Feynman diagrams showing quarks and gluons for the exchange of positive and negative pions between a neutron and a proton.

15.5. Draw Feynman diagrams for the τ^- lepton decays given in table 14.1.

References and suggestions for further reading

Abe F *et al* 1995 Observation of top quark production in $\bar{p}p$ collisions with the collider detector at Fermilab *Phys. Rev. Lett.* **74** 2626–31

Armstrong T A *et al* 1972 Total hadronic cross section of γ rays in hydrogen in the energy range 0.265–4.215 GeV *Phys. Rev.* D **5** 1640–52

Burcham W E and Jobes M 1995 *Nuclear and Particle Physics* (Harlow, Essex: Pearson)

Dunlap R A 2018 *Particle Physics* (San Rafael, CA: Morgan & Claypool)

Henley E M and García 2007 *Subatomic Physics* 3rd edn (Singapore: World Scientific)

Krane K S 1988 *Introductory Nuclear Physics* (New York: Wiley)

Mirman N 2015 Measurements of the top quark mass at ATLAS and CMS *CMS Report CMS-CR-2015-306* (https://cds.cern.ch/record/2105512?ln=en)

Martin B R 2006 *Nuclear and Particle Physics—An Introduction* (Chichester: Wiley)

Williams W S C 1991 *Nuclear and Particle Physics* (Oxford: Oxford University Press)

IOP Publishing

An Introduction to the Physics of Nuclei and Particles
(Second Edition)

Richard A Dunlap

Chapter 16

Particle reactions and decays

16.1 Reactions and decays in the context of the quark model

With a knowledge of the quark content of baryons and mesons it is possible to view particle decays and reactions in terms of the fundamental quarks and leptons and the relevant interactions. As an example, we can consider the Feynman diagram of the β^- decay process in figure 14.4 in terms of the quark components of the neutron and proton. This is shown in figure 16.1 and illustrates the manner in which the weak interaction acts on quarks and on leptons. As we have seen previously, the weak interaction changes a lepton to a neutrino or vice versa, but (in the context of our previous discussions) only within the same generation. Lepton–antineutrino creation or annihilation is an analogous process. Figure 16.1 shows that the weak interaction (unlike the strong interaction) can change the flavor of a quark. In fact, the W^+ and W^- bosons must change the flavor of a quark. (We will discuss the properties of the Z^0 Boson in more detail in the next section.) This is obvious, at least, from charge conservation, as the charged W boson must change a $+2e/3$ quark to a $-e/3$ quark (or vice versa), and similarly for the antiquarks. In the case of β-decay processes, the weak interaction changes a down quark to an up quark (as in figure 16.1) or vice versa (for β^+ decay).

Another example of a decay process is illustrated in figure 16.2. This corresponds to the decay,

$$D^0 \to K^- + \pi^+ \tag{16.1}$$

Here the weak $W^\pm$ boson couples two quarks and two flavor changes are necessary. In this case, the $u\bar{u}$ quark pair is formed by the strong interaction. An alternate decay of the D^0 meson, illustrated in figure 16.3, is given by,

$$D^0 \to K^- + e^+ + \nu_e \tag{16.2}$$

doi:10.1088/978-0-7503-6094-4ch16

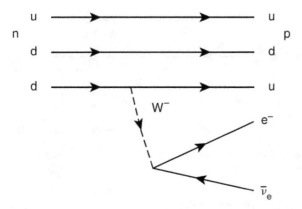

Figure 16.1. Feynman diagram for β^- decay showing quark relationships.

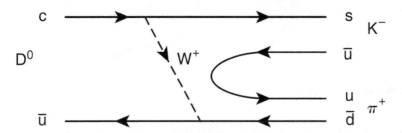

Figure 16.2. Feynman diagram of the weak decay of a D^0 meson to hadrons (equation (16.1)). In this diagram, and in all subsequent diagrams shown in this chapter that involve the strong interaction, the gluons are not shown.

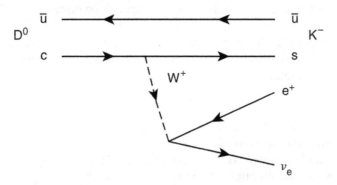

Figure 16.3. Feynman diagram of the weak decay of a D^0 meson to a hadron and leptons (equation (16.2)).

and shows the coupling of the weak boson to leptons, as in the case of β-decay. Figure 16.4 illustrates the decay,

$$D^+ \rightarrow \overline{K}^0 + \pi^+ \qquad (16.3)$$

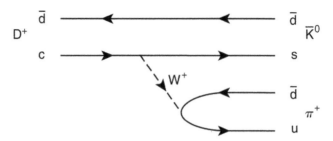

Figure 16.4. Feynman diagram of the weak decay of a D^+ meson (equation (16.3)).

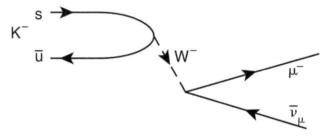

Figure 16.5. Feynman diagram of the weak decay of a K^- meson to leptons (equation (16.4)).

and shows that the weak W^+ boson must form quark–antiquark pairs of different flavors. In the examples given above, there is no mixing of quark generation. Figure 16.5 shows an example of quark generation mixing in the K^- meson decay,

$$K^- \rightarrow \mu^- + \bar{\nu}_\mu. \tag{16.4}$$

The implications of changes in quark generation will be discussed further in section 16.3.

The above examples illustrate decays in which the weak interaction plays a role. Most interactions involving hadrons proceed dominantly by means of the strong interaction. The reaction

$$p + \pi^+ \rightarrow \Delta^{++} + \pi^0,$$

as illustrated in figure 15.7, falls into this category. Here the reaction requires only the creation and annihilation of quark–antiquark pairs of the same flavor. In this sense, quark number as well as quark flavor is conserved, as required for the strong interaction. A similar situation occurs in reactions such as

$$p + \pi^- + \rightarrow n + \pi^0. \tag{16.5}$$

This is illustrated in figure 16.6 and shows that there is merely an exchange of quarks between hadrons. An example of a process that proceeds via the electromagnetic interaction, as illustrated in figure 16.7, is

$$\Delta^+ \rightarrow p + \gamma. \tag{16.6}$$

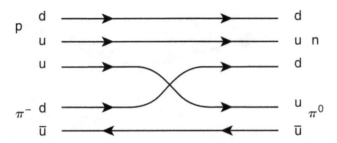

Figure 16.6. Feynman diagram of an example of the strong interaction between hadrons showing quark exchange (see equation (16.5)).

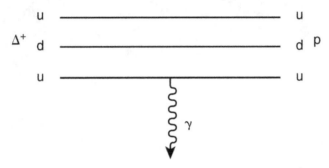

Figure 16.7. Feynman diagram of the electromagnetic decay of a Δ^+ baryon (equation (16.6)).

Here a Δ^+ baryon decays to a proton by emitting a real photon. The quark content of the two baryons is the same but the masses are different. The emission of the photon conserves mass/energy and corresponds to the flipping of one of the quark spins to change the $^4S_{3/2}$ state to a $^2S_{1/2}$ state. Electromagnetic interactions can also be responsible for the creation or annihilation of particle–antiparticle pairs, as shown in figure 16.8. The particles may be leptons, as in e^-e^+ annihilation (figure 16.8(a)), or quarks, as in the electromagnetic decay of a neutral pion (figure 16.8(b)). In either case two (or more) real photons are produced, as is necessary in order to conserve momentum.

16.2 $W^{\pm}$ and Z^0 bosons

A variety of weak processes involving charged weak bosons have been discussed in previous sections. There is, as well, a neutral weak boson, the Z^0. Since the Z^0 boson does not carry charge, it cannot mediate the kinds of processes that we have seen the $W^{\pm}$ bosons involved in. Examples of processes involving the Z^0 boson are the scattering of neutrinos by electrons, as shown in figure 16.9(a), and neutrino–hadron scattering is shown in figure 16.9(b). Scattering involving only charged leptons and/ or quarks (for example, electron–electron or electron–proton scattering) may be mediated by the Z^0 boson, although at low energies the electromagnetic interaction would be dominant. The Z^0 can also mediate particle–antiparticle creation or

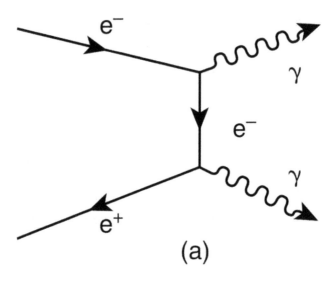

(a)

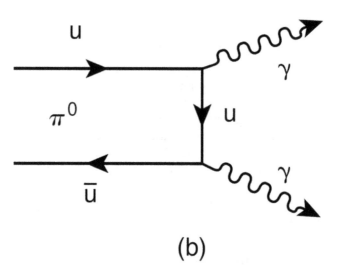

(b)

Figure 16.8. Feynman diagrams for (a) electron–positron annihilation to photons and (b) electromagnetic decay of a π^0 meson.

annihilation. Some examples are shown in figure 16.10. Electron–positron annihilation as shown in figure 16.10(a),

$$e^- + e^+ \rightarrow \nu_e + \bar{\nu}_e \tag{16.7}$$

requires the weak neutral boson since the neutrinos cannot couple electromagnetically. However, at low energies the cross section for annihilation to real photons as

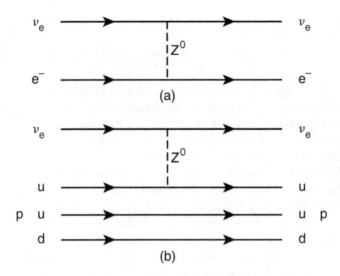

Figure 16.9. Feynman diagrams showing neutral weak bosons in (a) neutrino–electron scattering and (b) neutrino–proton scattering.

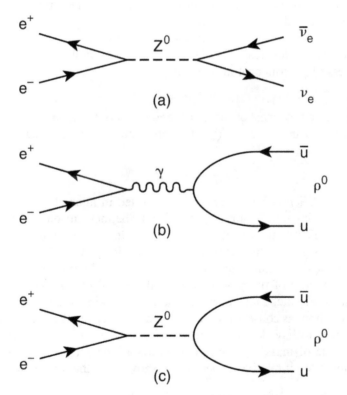

Figure 16.10. Feynman diagrams for (a) electron–positron annihilation to neutrinos as mediated by the Z^0 boson (equation (16.7)), to ρ^0 as mediated by the photon and (c) to ρ^0 as mediated by the Z^0 boson (equation (16.8)).

shown in figure 16.8(a) is greater. Electron–positron annihilation to hadrons, for example,

$$e^- + e^+ \rightarrow \rho^0 \qquad (16.8)$$

(see figure 16.10(b) and (c)) are generally mediated at low energies by the photon (figure 16.10(b)) and at higher energies by the neutral weak boson (figure 16.10(c)).

The above discussion shows that there are a number of similarities and some differences between the electromagnetic interaction and the weak interaction. The similarities are more obvious at high energies than at low energies. In fact, at sufficiently high energies the two interactions become completely equivalent. This feature is the basis of the *electroweak* theory developed by Sheldon Glashow (b. 1932), Steven Weinberg (1933–2021) and Abdus Salam (1926–96) that unifies the electromagnetic and weak interactions. At high energy the electroweak interaction is mediated by four massless gauge bosons consisting of a singlet (uncharged) and a triplet (with charges $-e$, 0, and $+e$). At lower energy the symmetry between the two interactions is broken and the singlet gauge boson remains massless and becomes the photon. The triplet particles acquire mass and become the weak $W^\pm$ and Z^0 bosons. This theory is able to predict the masses of the $W^\pm$ and Z^0 with a reasonable degree of accuracy and, in fact, made this prediction before the weak bosons were observed experimentally as free particles.

Thus far, we have discussed the mediating bosons as virtual particles. The possibility of producing real particles in high energy collisions can be illustrated by the process of proton–proton scattering,

$$p + p \rightarrow p + p.$$

This can be viewed in terms of the quark components of the protons and the virtual pion as illustrated in figure 15.5(b). The production of a real neutral pion by the reaction

$$p + p \rightarrow p + p + \pi^0 \qquad (16.9)$$

can occur at sufficiently high energy, as illustrated in figure 16.11. In a traditional accelerator experiment, a beam of protons may be incident on a target containing protons at rest. These target protons are often in the form of liquid hydrogen atoms that also serve as part of the bubble chamber used for particle detection. The kinetic energy of the incident beam required for real pion production substantially exceeds the rest mass energy of the pion that is produced. In the laboratory frame this is readily seen by the fact that not all of the incident kinetic energy is available for particle production, as conservation of momentum requires the scattered proton and the created pion to have kinetic energy. For a reaction $m_1 + m_2 \rightarrow m_3 + m_4 + m_5$, where a particle of mass m_1 is incident on a stationary (in the laboratory frame) particle of mass m_2, a relativistic derivation shows that the threshold kinetic energy is given by

$$K_{th} = \frac{(m_3 + m_4 + m_5)^2 c^2 - (m_1 + m_2)^2 c^2}{2m_2}. \qquad (16.10)$$

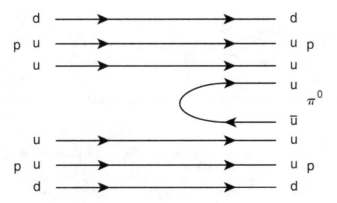

Figure 16.11. Feynman diagram for the production of a real π^0 meson during proton–proton scattering (equation (16.9)).

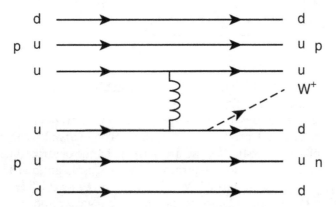

Figure 16.12. Feynman diagram for real W^+ production during a proton–proton collision (equation (16.11)).

Utilizing the masses of the particles in equation (16.10), it is found that the production of a 135 MeV c^{-2} neutral pion requires about 280 MeV of incident laboratory frame energy. Many contemporary particle experiments use a colliding beam geometry where conservation of momentum does not require that the particles on the right-hand side of the reaction have kinetic energy. As a result, the total kinetic energy of both beams becomes available for particle production and the advantages of this geometry become immediately obvious.

At sufficiently high energy proton–proton collisions can release a real weak boson by a process such as (see figure 16.12),

$$p + p \rightarrow p + n + W^+. \tag{16.11}$$

Proton–antiproton collisions can produce real $W^\pm$ by quark–antiquark annihilation such as

$$p + \bar{p} \rightarrow \pi^- + \pi^0 + W^+ \tag{16.12}$$

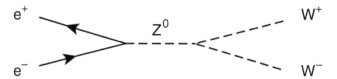

Figure 16.13. Feynman diagram for electron–positron annihilation to real weak bosons.

Table 16.1. Decay modes for the real W^+ boson decay to branching ratio (%) partial width (GeV).

Decay to	Branching ratio (%)	Partial width (GeV)
$e^+\nu_e$	11	0.23
$\mu^+\nu_\mu$	11	0.23
$\tau^+\nu_\tau$	11	0.23
$u\bar{d}$	34	0.72
$\bar{s}c$	34	0.72

and lepton collisions at sufficiently high energy can also lead to real $W^\pm$ production. An example is illustrated in figure 16.13.

The first experimental observation of real weak boson production was in proton–antiproton collisions that produced W^+ bosons. The charged weak bosons decay with a lifetime of about 3×10^{-25} s and are detected by the observation of their decay products. These can be lepton–neutrino pairs or quark–antiquark pairs as are observed for the decay of virtual $W^\pm$ in weak processes. These processes are summarized in table 16.1. Conservation of lepton generation requires that the lepton and neutrino that are formed are from the same generation. This is not a strict requirement for quark–antiquark pair production. However, mixed quark generation decays have virtually zero branching ratios because of the very small cross sections for these processes (see more on this in section 16.3). Decay to $t\bar{b}$ is not allowed for reasons of mass/energy conservation.

16.3 Quark generation mixing

Experimentally it is known that the proton is the only stable baryon and there are no stable mesons. These facts can be readily explained by the existence of weak decays such as

$$\Sigma^- \rightarrow n + e^- + \bar{\nu}_e$$

where there is a change of strangeness ($|\Delta S| = 1$). Similar generation mixing decays involving heavy flavors are also known. It is also possible for the Σ^- to decay by the strangeness-conserving process

$$\Sigma^- \rightarrow \Lambda + e^- + \bar{\nu}_e$$

where $|\Delta S| = 0$. On the basis of mass differences, and hence different Q values, we would expect that the branching ratios for these two processes would be different. However, these differences can be accounted for by following the discussion in section 9.4 and we would expect that the two decays would have similar $f\tau_{1/2}$ values. Experimentally, however, it is found that,

$$\frac{f\tau_{1/2}(|\Delta S| = 1)}{f\tau_{1/2}(|\Delta S| = 0)} \approx 12$$

indicating that the strangeness-changing process can occur but is strongly suppressed. These observations are explained in the context of the standard model by the following theory as originally proposed by Nicola Cabibbo (1935–2010).

The branching of decays is, in many ways, analogous to the flow of current from a node in a circuit as described by Kirchhoff's laws and the term current is often used in connection with this aspect of particle interactions. The total current, J, can be expressed as a linear combination of the currents representing the different decay modes. In the case of strangeness-conserving (J_0) and strangeness changing (J_1) decays we can write

$$J = aJ_0 + bJ_1.$$

Since the transition rates are proportional to the square of the coefficients, normalization requires

$$a^2 + b^2 = 1. \tag{16.13}$$

It is customary to assume the form for the coefficients,

$$a = \cos \theta_C$$
$$b = \sin \theta_C$$

where θ_C is the Cabibbo angle, and the condition given in equation (16.13) is satisfied. The Cabibbo angle can then be found to be,

$$\theta_C \approx \cot^{-1} \sqrt{\frac{f\tau_{1/2}(|\Delta S| = 1)}{f\tau_{1/2}(|\Delta S| = 0)}}.$$

Following the example given above for Σ^- decay we obtain $\theta_C \approx 16°$.

Strangeness-changing decays exist because the weak interaction couples the up quark (u) to a linear combination of down (d) and strange (s) quarks, given by d', rather than to just the down quark. Cabibbo theory gives,

$$d' = d \cdot \cos \theta_C + s \cdot \sin \theta_C. \tag{16.14}$$

The charm quark (c) can be viewed as coupling to the linear combination,

$$s' = -d \cdot \sin \theta_C + s \cdot \cos \theta_C. \tag{16.15}$$

This approach can be extended to include heavy flavors and, in general, explains the existence of generation mixing in weak processes.

16.4 Conservation laws and vertex rules

On the basis of the information given in chapters 14 and 15 and the previous two sections, it is possible to summarize the relevant conservation laws that apply at vertices in Feynman diagrams. In this section we summarize the types of vertices that are allowed for each type of interaction. We have seen that we can describe four kinds of gauge bosons: gluons (for strong interactions), photons (for electromagnetic interactions), $W^{\pm}$ (for charged current weak interactions), and Z^0 (for neutral current weak interactions). The weak bosons are divided into two categories for the purpose of the present discussion since different Feynman diagram vertex rules apply to the two types. The relevance of different conservation laws for the various gauge bosons acting on leptons and quarks are summarized in table 16.2. As examples of allowed three-vertices in Feynman diagrams we can consider processes of the form:

$$\text{fermion} \leftrightarrow \text{gauge boson} + \text{fermion}$$

or those that correspond to creation/annihilation processes:

$$\text{fermion} + \text{fermion} \leftrightarrow \text{gauge boson}.$$

Figure 16.14 illustrates examples of these vertices.

16.5 Classification of interactions

In the previous sections we have seen a wide variety of processes involving leptons and/or quarks and various gauge bosons. It is clear that processes involving leptons cannot be the result of the strong interaction as gluons do not couple to leptons (see figure 16.14). However, a wide variety of other combinations of particle types and interactions are possible. We can categorize decay processes as described in table 16.3. The relevance of the various interactions for each of these decays is best seen by examining the relevant Feynman diagram. Figures 16.15 and 16.16 show some examples. These examples should allow for the immediate classification of decays according to the type of initial and final particles as well as the type of gauge boson(s) involved.

Table 16.2. Summary of conservation laws for various gauge bosons acting on leptons and quarks.

Interaction	Gauge boson	Lepton $\leftrightarrow$ neutrino	Quark flavor change	Quark color change
Strong	Gluons	No interaction	No change	Yes
Electromagnetic	Photon	No change	No change	No change
Weak	$W^{\pm}$	Yes	Yes	No change
Weak	Z^0	No change	No change	No change

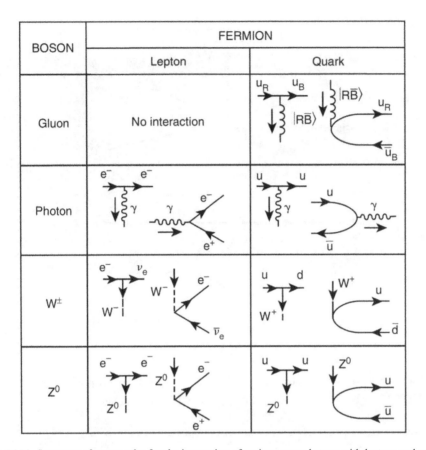

Figure 16.14. Summary of vertex rules for the interaction of various gauge bosons with leptons and quarks.

Table 16.3. Classification of particle decays.

No.	Type of decay	Particles	Interactions	Example
1	Leptonic decay	Lepton→leptons	Weak	$\tau^+ \to e^+ + \bar{\nu}_\tau + \nu_e$
2	Hadronic lepton decay	Lepton→lepton+hadron	Weak	$\tau^+ \to \pi^+ + \bar{\nu}_\tau$
3	Hadronic decay	Hadron→hadrons	Strong	$\Delta^{++} \to \pi^+ + p$
4	Leptonic hadron decay	Hadron→leptons	Weak	$\pi^+ \to \mu^+ + \nu_\mu$
5	Semileptonic hadron decay	Hadron→hadron+leptons	Weak	$n \to p + e^- + \bar{\nu}_e$
6	Nonleptonic hadron decay	Hadron→hadrons	Strong+weak	$K^+ \to \pi^+ + \pi^+ + \pi^-$
7	Electromagnetic	Hadron→hadron+photon	Electromagnetic	$\Delta^+ \to p + \gamma$

16.6 Transition probabilities and Feynman diagrams

Although a detailed analysis of reaction cross sections on the basis of Feynman diagrams is not possible here, we can make some approximate estimates of relative transition rates based on a qualitative analysis of the diagrams. In general, processes

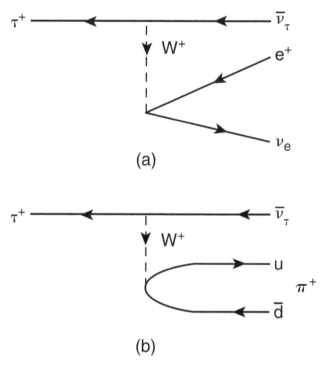

Figure 16.15. Examples of lepton decays as given in table 16.3. (a) and (b) correspond to numbers 1 and 2 in the table, respectively.

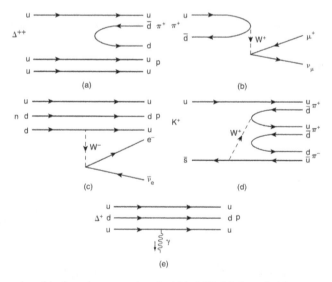

Figure 16.16. Examples of hadron decays as given in table 16.3. (a) through (e) correspond to numbers 3 through 7 in the table, respectively.

that are dominated by strong interactions proceed more rapidly. In situations such as figure 16.16(d), where both the strong interaction and the weak interaction are required, the lifetime is determined by the weak interaction, as this determines the maximum rate at which the process can proceed. We have already mentioned that electron–positron annihilation via the electromagnetic interaction (figure 16.8(a)) is more likely than via the weak interaction (figure 16.13) at low energies. Along similar lines, we can compare neutral pion decay via the electromagnetic interaction (figure 16.8(b)) and charged pion decay via the weak interaction (figure 16.16(b)). The former decay has a lifetime of 8.7×10^{-17} s and the latter decay has a lifetime of 2.6×10^{-8} s.

A careful inspection of Feynman diagrams allows for a more detailed analysis of relative transition rates for a variety of hadronic processes that involve the strong and/or weak interactions. We begin with a consideration of weak processes. Figure 16.1 shows a well-known weak process, β-decay, which involves both leptons and quarks. Figures 16.2 and 16.4 show weak D meson decays that do not involve leptons. In all of these cases, there is no mixing of quark generation, and the decays are not suppressed. As we saw in section 16.3, processes involving quark generation changes are suppressed relative to generation conserving processes. An example of a process that involves leptons and quark generation mixing, the decay of the K^- meson, is shown in figure 16.5. The decay of a K^+ meson, as shown in figure 16.16(d), does not involve leptons but has a quark generation change. In some cases, a weak boson that couples to two quarks can result in two generation changes and is highly suppressed. An example is the decay,

$$A^+ \rightarrow \pi^+ + n$$

as illustrated in figure 16.17. From these trends we can establish an approximate hierarchy of transition probabilities for the weak decay of hadrons based on the number of quark generation changes that are involved.

Other suppressed situations occur, such as

$$\Sigma^+ \rightarrow n + e^+ + \nu_e$$

(see figure 16.18), where two weak bosons are required in order to create the lepton–neutrino pair and to produce the necessary quark flavor changes.

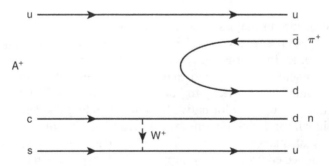

Figure 16.17. The decay of the A^+ baryon showing generation mixing for two quarks.

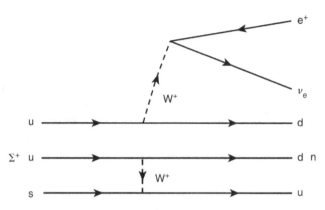

Figure 16.18. The decay of the Σ^+ baryon showing two weak bosons.

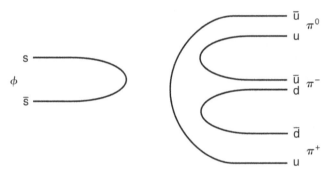

Figure 16.19. The decay of a ϕ meson to pions. Gluons are not shown but are discussed in the text.

A similar analysis can be considered for strong processes. As an example, we consider the decay for the ϕ meson (s$\bar{\text{s}}$). The two principal decay modes for ϕ are decay to pions,

$$\phi \rightarrow \pi^+ + \pi^- + \pi^0$$

and decay to K mesons (kaons),

$$\phi \rightarrow K^+ + K^-.$$

An inspection of meson masses indicates that the Q for the former process is 605 MeV, while for the latter process it is 32 MeV. Our previous discussions suggest that the branching ratio for these decays would significantly favor decay to pions. However, experimentally it is found that the branching ratios are about 17% for decay to pions and 83% for decay to kaons. The Feynman diagrams, as shown in figures 16.19 and 16.20, are informative for understanding this behavior. Diagrams such as that shown for decay to pions are referred to as disconnected diagrams since there is no flow of quarks from the left to right sides of the diagram. *Zweig's rule* states that disconnected processes are suppressed relative to connected processes. To understand the reasons for this we must consider the transfer of color from the initial

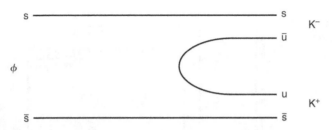

Figure 16.20. The decay of a ϕ meson to kaons. Gluons are not shown but are discussed in the text.

to final quarks, realizing that all meson states must be colorless. In figure 16.20 quark color is transferred by the quarks on the left side of the diagram to the strange quarks on the right side of the diagram. Color changes can be associated with individual gluon interactions along the lines of those shown in figure 15.7. In figure 16.19, however, individual quarks cannot carry color from the left side to the right side of the diagram. Since individual gluons do carry color, a colorless interaction requires two (or more) gluons. This is analogous to the suppression of weak processes that require two (or more) weak bosons.

16.7 Meson production and fragmentation

The study of meson production in high energy collisions provides important insight into the physics of particle interactions. Let us consider the production of particles in electron–positron collisions. One possible process is the production of lepton–antilepton pairs. At moderate energies this is most likely muon production,

$$e^- + e^+ \rightarrow \mu^- + \mu^+.$$

Another possible process is hadron production, that is, the production of quark–antiquark pairs (of the same flavor) as shown in figure 16.10(b). We would expect that when the total center of mass energy before the collision is equal to the rest mass energy of the quark–antiquark pair after the collision, then there would be a preference for hadron production relative to lepton production.

A common experimental technique is to plot the ratio of cross sections for hadron production and muon production, as shown in figure 16.21. Resonances corresponding to the formation of several meson states are illustrated in the figure. The lowest energy states correspond to various $u\bar{u}$ and $d\bar{d}$ states followed by the ϕ resonance corresponding to $s\bar{s}$ and, at higher energies, resonances involving heavy flavors, such as J/ψ and Υ, are observed. The ordering of states in this figure and their description in terms of the quark components of the corresponding mesons is generally considered to be confirmation of the quark model.

In this type of experiment, we would expect quark–antiquark states to be formed from quarks of the same flavor. However, in practice a wide variety of meson states of mixed flavor are typically observed. The reasons for this behavior have important implications for our understanding of the fundamental properties of gluons and

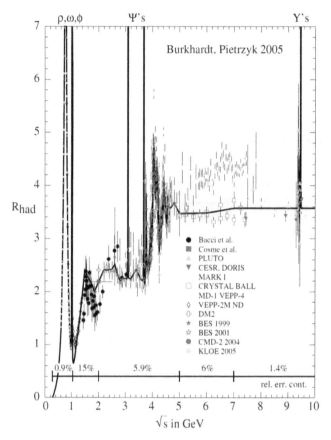

Figure 16.21. The ratio, R_{had} = cross section for hadrons/cross section for muons in electron–positron collisions as a function of center of mass energy. Reprinted from Pietrzyk (2006), Copyright (2006), with permission from Elsevier.

quarks. Let us consider, for example, the formation of $c\bar{c}$ during an electron–positron collision. A possible observation might show the process,

$$e^- + e^+ \rightarrow D^0 + \pi^0 + \pi^+ + D^-.$$

We can understand how this can come about by looking at the Feynman diagram as shown in figure 16.22. The gauge boson, shown as a photon, can also be a Z^0. Particle formation results from quark–antiquark pair formation. In this case, there is no flavor mixing and the process on the right side of the gauge boson is the result of the strong interaction. We can view this process, called *fragmentation*, in terms of the breaking of gluons and the formation of new particle pairs, as illustrated in figure 16.23. Breaking a gluon bond produces new quarks and antiquarks at the ends of the broken bonds, as shown in the figure. In this way, a large variety of mixed flavor mesons can be formed.

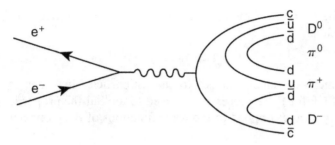

Figure 16.22. Feynman diagram for the process $e^- + e^+ \rightarrow D^0 + \pi^0 + \pi^+ + D^-$.

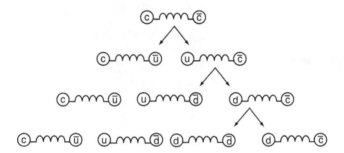

Figure 16.23. Representation of the physical processes involved in the breaking of gluon bonds in the reaction shown in figure 16.22.

16.8 CP violation in neutral meson decays

We saw in chapter 9 that β-decay can violate conservation of parity. This is because the parity operation (P) does not leave the system invariant. If the parity operation is combined with charge conjugation (C), then it can be shown that the β-decay process is in fact invariant. Thus, although β-decay can violate parity conservation it leaves CP unchanged.

The decay of certain neutral mesons demonstrates that the weak interaction can violate CP conservation. This has been observed in the decay of K^0 ($d\bar{s}$) and its antiparticle $\overline{K}^0(\bar{d}s)$ as discussed below and should apply as well to the decay of D^0, B_d^0, and B_s^0. The K^0 and $\overline{K}^0$ mesons can decay by several modes. These modes are to two pions, to three pions, and to a pion and leptons. The two pion modes are

$$K^0 \rightarrow \pi^0 + \pi^0$$

and

$$K^0 \rightarrow \pi^+ + \pi^-$$

and similarly, for $\overline{K}^0$

$$\overline{K}^0 \rightarrow \pi^0 + \pi^0$$

and

$$\overline{K}^0 \rightarrow \pi^+ + \pi^-.$$

Since both the K^0 and $\overline{K}^0$ mesons have the same decay products, it is possible for processes such as

$$K^0 \rightarrow 2\pi \rightarrow \overline{K}^0 \qquad (16.16)$$

to occur, which will change a K^0 to its antiparticle. In order to consider the invariance of CP for these processes, we need to look at the properties of the pions. Charge conjugation as applied to the wave functions of the pions has the following properties:

$$C\psi(\pi^0) \rightarrow \psi(\pi^0)$$

$$C\psi(\pi^+) \rightarrow \psi(\pi^-)$$

$$C\psi(\pi^-) \rightarrow \psi(\pi^+)$$

and the parity operation is written as

$$P\psi(\pi^0) \rightarrow \psi(\pi^0)$$

$$P\psi(\pi^+) \rightarrow \psi(\pi^-)$$

$$P\psi(\pi^-) \rightarrow \psi(\pi^+)$$

For a two-pion state, $\pi^0 + \pi^0$ or $\pi^+ + \pi^-$, with $L = 0$ it can be shown that the CP operation leaves the wave function invariant. That is,

$$CP\ \psi(\pi^0, \pi^0) \rightarrow \psi(\pi^0, \pi^0)$$

$$CP\ \psi(\pi^+, \pi^-) \rightarrow \psi(\pi^+, \pi^-).$$

From equation (16.16) it follows that if the intermediate pion state is CP invariant, then the initial and final meson states must also be CP invariant. However, applying the CP operation we find

$$CP\ \psi(K^0) \rightarrow \psi(\overline{K}^0)$$

and

$$CP\ \psi(\overline{K}^0) \rightarrow \psi(K^0)$$

and CP conservation is violated. Because, according to equation (16.16), K^0 can change into $\overline{K}^0$ and vice versa, the K^0 states are never pure but are admixtures of K^0 and $\overline{K}^0$. If we write these mixed states as

$$\psi(K_s) = \frac{1}{\sqrt{2}}[\psi(K^0) + \psi(\overline{K}^0)]$$

and

$$\psi(K_s) = \frac{1}{\sqrt{2}}[\psi(K^0) - \psi(\overline{K}^0)] \qquad (16.17)$$

it can be shown that

$$CP\ \psi(K_S) \rightarrow \psi(K_S)$$

and

$$CP\ \psi(K_L) \rightarrow -\psi(K_L).$$

The meaning of the subscripts S and L will be discussed shortly. It is now clear that the mixture of states given by equation (16.17) for K_S is consistent with CP conservation for the two-pion decay mode. In order to understand the properties of the K_L state, we can look at the three-pion decay mode. This is

$$K^0 \rightarrow \pi^0 + \pi^0 + \pi^0$$

or

$$K^0 \rightarrow \pi^0 + \pi^+ + \pi^-$$

and similarly, for the $\overline{K}^0$. It can be shown that the CP operation as applied to the three-pion state gives,

$$CP\ \psi(3\pi) \rightarrow -\psi(3\pi).$$

Thus, the three-pion decay mode for K_L is consistent with CP invariance. Experimentally K^0 mesons (which are a mixture of K^0 and $\overline{K}^0$ states) are found to have two well-defined lifetimes, 8.9×10^{-10} s and 5.2×10^{-8} s, corresponding to the decay of K_S and K_L. The subscripts, therefore, refer to short-lived and long-lived states. It is also observed that the decay modes for these states are $K_S \rightarrow 2\pi$ and $K_L \rightarrow 3\pi$, consistent with CP invariance. K_L can also decay to a pion plus leptons. From a consideration of masses, the decay to three pions will have a smaller value of Q than the decay to two pions and it is, therefore, apparent that the three-pion decay will have a longer lifetime.

In 1964 an experiment conducted by James Cronin (1931–2016), Val Fitch (1923–2015) and coworkers looked for the decay $K_L \rightarrow 2\pi$, which would violate CP invariance. A careful analysis of their data indicated that, indeed, about 0.3% of the K_L decays were to two-pion states. These results provided clear evidence that weak processes can violate CP conservation.

Just as the inclusion of charge conjugation in the β-decay experiments described in chapter 9 restored the conservation laws, there is common belief that inclusion of time reversal (T) for weak processes will restore the conservation laws, that is, these processes are invariant under CPT operations. Experimental evidence thus far does not contradict this hypothesis.

Problems

16.1. Some of the weak charmed decay modes of the bottom meson, B^+ are
   (a) $B^+ \rightarrow +\overline{D}^0 + \mu^+ + \nu_\mu$
   (b) $B^+ \rightarrow +\overline{D}^0 + \pi^+$
   (c) $B^+ \rightarrow D^- + \pi^+ + \pi^+$
   (d) $B^+ \rightarrow D_s^+ + \overline{K}^0$

Draw Feynman diagrams for these decays.

16.2. The decay

$$\overline{D}^0 = K^- + \mu^+ + \nu_\mu$$

is suppressed, while the decay

$$\overline{D}^0 = K^+ + \mu^+ + \nu_\mu$$

is not. Explain.

16.3. In addition to the decay modes of the τ^- lepton given in table 14.1, modes involving other combinations of light and strange mesons are possible. Discuss these and show Feynman diagrams as appropriate.

16.4. Discuss the possibility of decay modes of the W^+ boson other than those shown in table 16.1. Consider, in particular, decays to $u\overline{s}$, $u\overline{b}$, $c\overline{d}$, $c\overline{b}$, $t\overline{d}$, $t\overline{s}$ and $t\overline{b}$.

16.5. Draw a Feynman diagram for the process given in equation (16.12).

16.6. Draw figure 16.2 indicating a valid coloring scheme for the quarks. Show gluons as necessary.

16.7. Discuss the possible decay modes of a real Z^0 boson.

16.8. Determine if each of the following decays is allowed. If it is allowed determine the interaction by which it proceeds. If it is not allowed determine which conservation law(s) is/are violated.

(a) $\pi^+ \rightarrow \pi^0 + e^+ + \nu_e$
(b) $\tau^+ \rightarrow \pi^+ + \overline{\nu}_\tau$
(c) $\Xi^- \rightarrow \Lambda + \pi^-$
(d) $\mu^- \rightarrow \pi^- + \nu_\mu$
(e) $\pi^+ \rightarrow e^+ + \nu_e$
(f) $\pi^0 \rightarrow e^+ + e^-$
(g) $\phi \rightarrow K^+ + K^-$
(h) $\Omega^- \rightarrow \Xi^0 + K^-$
(i) $K^- \rightarrow e^- + \nu_e$
(j) $D^0 \rightarrow \Delta^- + e^+ + \nu_e$

16.9. Determine if each of the following reactions is allowed. If it is allowed, determine the interaction by which it proceeds. If it is not allowed, determine which conservation law(s) is/are violated.

(a) $\Delta^+ + p \rightarrow p + p + \gamma$
(b) $\nu_e + \tau^- \rightarrow e^- + \nu_\tau$
(c) $K^+ + n \rightarrow p + \pi^0$
(d) $\Delta^{++} + \pi^+ \rightarrow K^+ + K^+$
(e) $p + \overline{p} \rightarrow \pi^+ + \pi^- + \pi^0$
(f) $\nu_\mu + \mu^- \rightarrow e^- + \nu_e$
(g) $\nu_\tau + e^- \rightarrow \nu_\tau + e^- + e^+ + e^-$
(h) $n + p \rightarrow \pi^0 + \pi^+$
(i) $\mu^+ + \mu^- \rightarrow \nu_e + \overline{\nu}_e$
(j) $\pi^- + p \rightarrow K^- + \Sigma^0$

References and suggestions for further reading

Dunlap R A 2018 *Particle Physics* (San Rafael, CA: Morgan & Claypool)

Halzen F and Martin A D 1991 *Quarks and Leptons—An Introductory Course in Modern Particle Physics* (New York: Wiley)

Henley E M and García 2007 *Subatomic Physics* 3rd edn (Singapore: World Scientific)

Krane K S 1988 *Introductory Nuclear Physics* (New York: Wiley)

Pietrzyk B 2006 Tests of the standard model and $\alpha(m_z^2)$ *Nucl. Phys. B (Proc. Suppl.)* **162** 18–21

IOP Publishing

An Introduction to the Physics of Nuclei and Particles
(Second Edition)

Richard A Dunlap

Chapter 17

The Higgs boson

17.1 Yukawa theory and the mass of the weak boson

While the standard model provides insight into many particle properties such as spin, it does not specifically provide a means by which particles acquire mass. Composite particles, such as baryons, have mass associated with the binding energy of the component quarks, but an additional mechanism is needed to explain the mass of fundamental particles. The so-called Higgs mechanism was proposed in the 1960s shortly after the quark hypothesis was made and is a means by which the symmetry of interactions between particles, which prevents them from acquiring mass, is broken. The Higgs mechanism deals specifically with the mass associated with the weak bosons and explains the short range of the weak interaction. François Englert (b. 1932) and Peter Higgs (b. 1929) were awarded the 2013 Nobel Prize in Physics for their theoretical work on this subject.

The weak boson mass as given in table 14.3 may be related to its range as given in table 2.2 by Yukawa theory, which was originally proposed by Hideki Yukawa (1907–81), to describe the mediation of the strong interaction by virtual mesons. A simple formulation of Yukawa theory is based on the Einstein relation,

$$E^2 = p^2 c^2 + m^2 c^4.$$

Replacing energy and momentum by their respective quantum representations

$$E = i\hbar \frac{\partial}{\partial t} \text{ and } p = -i\hbar \nabla$$

gives the time dependent wave equation

$$-\hbar^2 \frac{\partial^2 \psi}{\partial t^2} + \hbar^2 c^2 \nabla^2 \psi = m^2 c^4 \psi.$$

doi:10.1088/978-0-7503-6094-4ch17

The corresponding time independent form

$$\nabla^2 \psi = \left[\frac{mc}{\hbar}\right]^2 \psi$$

has the exponentially damped solution

$$\psi(r) = \frac{1}{r} \exp\left(-\frac{mc}{\hbar} r\right). \tag{17.1}$$

The characteristic length scale for the wavefunction in equation (17.1) is the reduced Compton wavelength

$$\lambda = \frac{\hbar}{mc} \tag{17.2}$$

and gives the range of an interaction mediated by a virtual particle whose mass as a real particle is given in the equation as m. It is often convenient to express the range in fm and the mass in MeV c^{-2} in which case equation (17.2) may be written as

$$\lambda \text{ (fm)} = (197.3 \text{ MeV} \cdot \text{fm})\frac{1}{mc^2}. \tag{17.3}$$

In the case of the weak interaction, a range of 2×10^{-3} fm in equation (17.3) gives a weak boson mass of about 90 GeV c^{-2}, as shown in table 14.3. The limited range of the weak interaction is, therefore, a direct consequence of the mass associated with the weak bosons.

17.2 Spontaneous symmetry breaking and the Higgs field

In order to relate the symmetry of a system to its physical properties, we can look at the simple example of a magnetic material, specifically the transition from a paramagnetic state at high temperatures to a ferromagnetic state at low temperatures. As shown in the two-dimensional diagram of the paramagnetic state illustrated in figure 17.1(a), the magnetic moments are randomly oriented. This means that the system is symmetric, as all directions are equivalent. As the

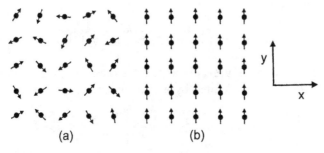

(a)　　　　　(b)

Figure 17.1. Orientation of magnetic moments (a) in the high temperature paramagnetic regime and (b) in the low temperature ferromagnetic regime. Adapted from Dunlap (2019). Copyright IOP Publishing. Reproduced with permission. All rights reserved.

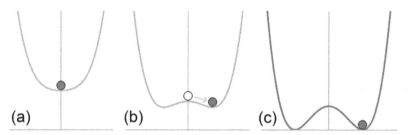

Figure 17.2. One-dimensional representation of spontaneous symmetry breaking in a magnetic system as a function of temperature; (a) high temperature, (b) medium temperature and (c) low temperature. This [explanatory diagram showing how symmetry breaking works] image has been obtained by the author from the Wikimedia website where it was made available by FT2 (2012) under a CC BY-SA 3.0 licence. It is included within this chapter on that basis. It is attributed to FT2.

temperature is lowered below the transition (Curie) temperature into the ferromagnetic regime, the magnetic moments align, as in figure 17.1(b), and the system has a preferred orientation. Thus, lowering the temperature has broken the symmetry of the system.

The effects of lowering the temperature (and hence the energy) of a magnetic system, along with the associated breaking of the system symmetry, is graphically illustrated in the one-dimensional example shown in figure 17.2. In figure 17.2(a) the system is in a symmetric state where both directions are equivalent. As the energy is lowered, the system shows the behavior in figures 17.2(b) and 17.2(c) and the lowest energy state, which is no longer symmetric, is occupied.

We can look at the magnetic example from a mathematical perspective. We begin by considering a two-dimensional potential which is a function of the x and y spatial coordinates as

$$V(x, y) = \frac{k_1}{4}[x^2 + y^2]^2$$

where k_1 is a constant to be discussed further below. The ground state of this potential is at $x_{min} = y_{min} = 0$ and the potential is symmetric in x and y around this minimum. We can break this symmetry by adding an additional x and y dependent term (and constant k_0),

$$V(x, y) = -\frac{k_0}{2}[x^2 + y^2] + \frac{k_1}{4}[x^2 + y^2]^2. \tag{17.4}$$

This potential is shown in figure 17.3. It is clear that the ground state is no longer symmetric in x and y and is defined by a circle where

$$x^2_{min} + y^2_{min} = \left[\frac{k_0}{k_1}\right]^{1/2}. \tag{17.5}$$

In the magnetic analogy, x and y may be viewed as the x and y coordinates of the magnetization, M_x and M_y and the potential represents the thermodynamic

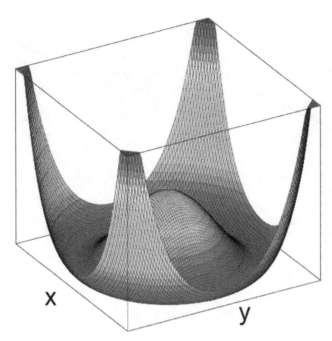

Figure 17.3. Potential function given by equation (17.4). This [the potential for the Higgs field] image has been obtained by the author from the Wikimedia website where it was made available by Gonis (2007) under a CC BY-SA 3.0 licence. It is included within this chapter on that basis. It is attributed to Gonis (axis labels changed).

potential of the system. The magnitude of the magnetization may be found from equation (17.5) as

$$M = \left[M_x^2 + M_y^2 \right]^{1/2} = \frac{1}{\sqrt{\lambda}} [T_C - T]^{1/2}$$

and is represented by the radius of the circle in figure 17.3 that defines the minimum in energy. This result is the so-called mean field approximation. Here we have taken k_0 to be the reduced temperature $(T_C - T)$, where T_C is the Curie temperature. For the magnetic system, $M \rightarrow 0$ as the reduced temperature $(T_C - T) \rightarrow 0$, or as the Curie temperature is approached from below. This means that the symmetry that is characteristic of the randomly oriented magnetic moments, at high temperature (energy) is spontaneously broken at low temperature (energy).

The analogy of spontaneous symmetry breaking at low energy can be applied to particle physics. Figure 16.14 provides some insight into how symmetry breaking applies to particle interactions. An inspection of the vertex rules that apply to interactions mediated by the photon and those mediated by the Z^0 boson, shows that these rules are identical. The difference between interactions mediated by photons and those mediated by Z^0 bosons is that the former are prevalent at low energies and the latter are prevalent at high energies. As an example, we can look at the e^-e^+ interaction

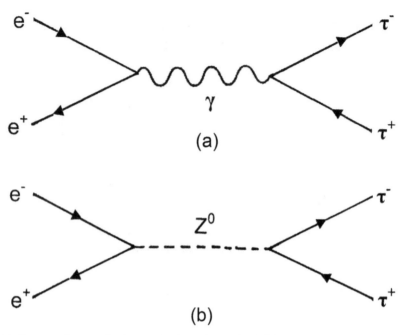

Figure 17.4. Feynman diagrams for the reaction shown in equation (17.6), (a) as mediated by a photon and (b) as mediated by a Z^0 boson. From Dunlap (2018). Copyright IOP Publishing. Reproduced with permission. All rights reserved.

$$e^- + e^+ \rightarrow \tau^- + \tau^+. \tag{17.6}$$

Based on the vertex rules from figure 16.14 we can construct two valid Feynman diagrams, as shown in figure 17.4. Thus, the reaction can be the result of the electromagnetic interaction as mediated by a photon or it can be viewed in terms of the weak interaction as mediated by a Z^0 boson. In either case, there must be sufficient kinetic energy available on the left-hand side of the reaction to create the τ-lepton pair on the right-hand side. While this can occur via a mediating photon at low energies, at sufficiently high energies the processes shown in figures 17.4(a) and 17.4(b) become equivalent. As noted in the previous chapter, the concept of a unified electroweak interaction was developed in the 1960s means that at a sufficiently high energy, the electromagnetic interaction and the weak interaction become unified and there is no distinction between the photon and the Z^0 boson. The energy at which this unification occurs is related to the rest mass energy of the weak bosons, ~80 GeV, although depending on context, it is sometimes taken to be the calculated energy for spontaneous electroweak symmetry breaking, ~160 GeV or sometimes the vacuum expectation value of the scalar Higgs potential, 246 GeV (see below). This symmetric high energy state is analogous to the symmetric magnetic system at high temperature. As the energy is lowered, the symmetry is spontaneously broken, thereby creating the photon to mediate the electromagnetic interaction and the three weak bosons to mediate the weak interaction. The Higgs mechanism as described below endows mass on the three weak bosons, W^+, W^- and Z^0, while the photon remains massless.

It is important to look at the physical reason for spontaneous symmetry breaking. In the case of magnetic ordering, it is the spin–spin interaction which can be expressed in terms of a coupling constant, K, as

$$E = -K\left[\vec{s_i} \cdot \vec{s_j}\right]$$

that is responsible for the alignment of the magnetic moments and the breaking of the symmetry. In the case of the spontaneous symmetry breaking of the electroweak interaction at low energies, it is the coupling of the weak bosons, W^+, W^- and Z^0, to the Higgs field that is responsible. The Higgs field is a scalar field that is present throughout space. It can be described by figure 17.3 where x and y represent the real and imaginary components of the Higgs field, ϕ_{re} and ϕ_{im}, respectively. The magnitude of the Higgs field is given by the radius of the circle representing the minimum energy.

A scalar field has a magnitude at all points in space but is distinguished from a vector field which has both magnitude and direction. The magnitude and direction of a vector field can easily be measured by observing the motion of an object that couples to the field. The magnitude of a scalar field, on the other hand, is most easily measured by observing changes in potential energy as a function of location. In general, adding a constant to the magnitude of a scalar field often does not alter our perception of the properties of the field. Gravitational potential is an example. For practical purposes, we generally assign a value of zero to the magnitude of the gravitational potential at the surface of the earth. This is not correct in absolute terms, which is easily demonstrated by dropping a mass into a deep hole. If the magnitude of a scalar field is constant throughout space, it is difficult to determine that value.

If a vector field has a non-zero value when averaged over all space, then this would imply that space is anisotropic. We have no observational evidence that this is the case. A scalar field, however, can have a non-zero value when averaged over all space and this will not alter our perception of the universe. Theoretically, the Higgs field is expected to have a constant magnitude throughout space with a value of 246 GeV. As might be expected on the basis of the above discussion, confirming the existence of the Higgs field is very challenging from an experimental standpoint. The next section describes how this has been undertaken.

17.3 The Higgs boson

Since the theorized Higgs field is a scalar field with a constant value everywhere in space, it is very difficult to observe experimentally. However, if the Higgs field exists, then the Higgs boson will be a scalar (spin 0) boson that is a quantum manifestation of the field. Thus, the experimental observation of the Higgs boson is a clear indication that the Higgs field exists, and that the Higgs mechanism is responsible for endowing mass on the weak bosons. Nearly all of the predictions of the standard model had been confirmed by experimental observations by the mid-1990s.

These included the observation of the weak bosons in 1983 and the observation of the top quark in 1995. The sole exception to the experimental confirmation of the standard model was the Higgs boson.

While the standard model makes certain predictions concerning many of the properties of the Higgs boson, including zero spin, even parity and no charge, it failed to make detailed predictions concerning its mass. In fact, the standard model could accommodate a Higgs boson mass anywhere from around 100–1000 GeV c^{-2}. So, although theory gave relatively little information about where to look for the Higgs boson, it did provide some guidance on what to look for in terms of what reactions could produce a real Higgs boson and what it would ultimately decay into. Higgs boson production occurs in high energy particle collisions and, from a practical standpoint, experimental searches for the Higgs boson were ultimately limited by the energies of suitable particles that were available in existing accelerators.

We begin with an overview of the reactions that could be used to produce a Higgs boson. The three major production routes for the Higgs boson are shown in figures 17.5–17.7 and are described below. The various Feynman diagram vertices seen in this figure can be related to those that are shown in figure 16.14.

Gluon fusion—In high energy hadron collisions such as proton–antiproton collisions, the gluons binding the hadrons together can interact, thereby forming a virtual quark loop as shown in figure 17.5. This quark loop can produce a Higgs boson as illustrated. Since the Higgs boson couples to mass, Higgs production is most likely when the quarks are the most massive. Therefore, only top quark or bottom quark loops are relevant to this process.

Higgs-strahlung—Fermion–antifermion collisions, such as electron–positron and quark–antiquark collisions, can produce a weak boson as shown in figure 17.6. The

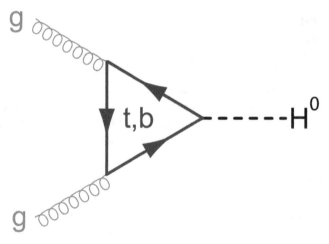

Figure 17.5. Higgs boson production by gluon–gluon fusion. This [Feynman diagram of Higgs production through gluon fusion] image has been obtained by the author from the Wikimedia website where it was made available by Rias (2012a) under a CC BY-SA 3.0 licence. It is included within this chapter on that basis. It is attributed to TimothyRias. (H changed to H^0.)

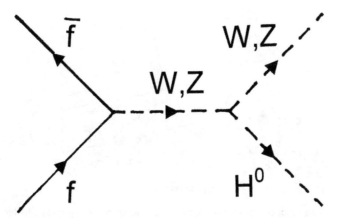

Figure 17.6. Higgs boson production by Higgs-strahlung.

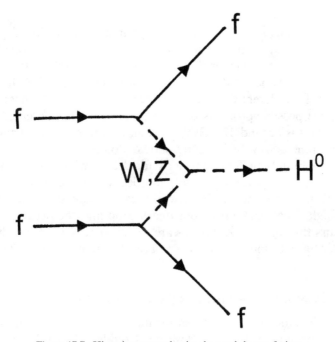

Figure 17.7. Higgs boson production by weak boson fusion.

charge of the weak boson will be compatible with charges of the incident fermions. The weak boson that is created can emit a Higgs boson as shown in the figure.

Weak boson fusion—The collision of a fermion and an antifermion can create a virtual weak boson, as shown in figure 17.7, and this weak boson can emit a real Higgs boson.

The relative cross sections for each of the processes given above depend on the particles involved in the collision and their energies. This will be discussed further below in the context of experimental searches for the Higgs boson.

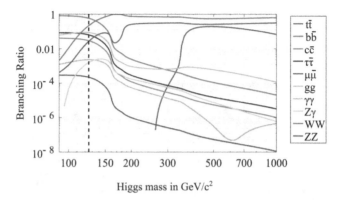

Figure 17.8. Branching ratios for Higgs boson decay as a function mass as predicted by the standard model. This [plot of the Branching Ratios for different decay modes of the Higgs boson] image has been obtained by the author from the Wikimedia website where it was made available by Rias (2012b) under a CC BY-SA 3.0 licence. It is included within this chapter on that basis. It is attributed to TimothyRias.

Since the Higgs boson is unstable, its experimental detection depends on the observation of its decay products in combination with other properties that are characteristic of this particle, including charge, spin and lifetime. The decay products for the Higgs boson depend on its mass and figure 17.8 shows predicted branching ratios for different decays. The Higgs boson couples to mass and decay to the most massive possible products is generally favored. The accepted value of the Higgs boson mass is around 125 GeV c^{-2} and, according to the figure, decays to bottom–antibottom and W^+–W^- pairs should dominate. From an experimental standpoint it is certainly preferable to look for decays with large branching ratios. However, it is also important to look for decays that can unambiguously be associated with the Higgs boson. In view of these considerations, the decay to two photons through a bottom–antibottom quark loop and the decay to two lepton–antilepton pairs through two Z^0 bosons are the most reasonable candidates for an experimental Higgs boson search. These are shown in figures 17.9 and 17.10, respectively.

17.4 Experimental observation of the Higgs boson

In the 1990s the first extensive experimental search for the Higgs boson was conducted at the Large Electron-Positron Collider (LEP) located at CERN (Conseil Européen pour la Recherche Nucléaire) in Geneva, Switzerland. Since the LEP collided electrons and positrons, the principal route to the production of the Higgs boson was by Higgs-strahlung and the secondary route was by weak boson fusion. At the time of its decommissioning (in 2000) to make way for the construction of the Large Hadron Collider (LHC) (see below) the LEP did not observe the Higgs boson but was able to put a limit on its minimum mass of 114.4 GeV c^{-2}.

After the LEP experiments, Higgs boson searches continued at the Tevatron, located at the Fermi National Accelerator Laboratory (generally known as

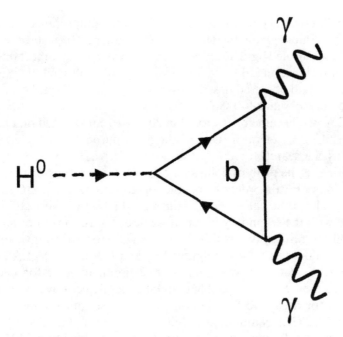

Figure 17.9. Higgs boson decay to two photons through a quark loop.

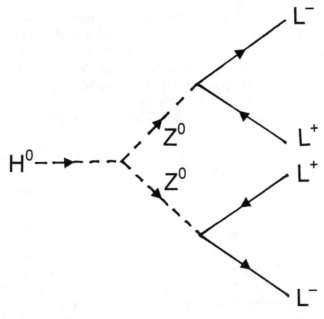

Figure 17.10. Higgs boson decay to four leptons.

Fermilab) in Illinois. The Tevatron utilized proton–antiproton collisions in these investigations. Gluon–gluon fusion was the dominant Higgs producing reaction investigated here, while Higgs-strahlung was the second most important. At the time it was decommissioned in 2011, the Tevatron had not confirmed the existence of the Higgs boson but had observed possible events suggesting a Higgs boson with a mass between 115 GeV c^{-2} and 140 GeV c^{-2}.

The LHC which became operational in 2010 was constructed on the CERN site of the former LEP accelerator. One of the main purposes for the construction of the LHC was the search for the Higgs boson. The LHC utilizes proton–proton collisions either in the form of collisions between two protons or between protons and heavy ions. Further details of the operation of the LHC are given in Appendix C. The principal mode of Higgs production in the LHC is by gluon–gluon fusion with the secondary mode of weak boson fusion. In order to optimize confidence in the results, two research groups were formed to investigate the two possible decay modes as shown in figures 17.9 and 17.10. The ATLAS (A Toroidal LHC ApparatuS) detector was used by one group to look for decays to two photons, while the CMS (Compact Muon Solenoid) detector was used by the other group to look for decays to four leptons (in this case four muons, i.e., two pairs of μ^-–μ^+). In 2012 CERN announced that both groups had confirmed the existence of the Higgs boson. Its mass and lifetime were determined to be 125.09 GeV c^{-2} and 1.56×10^{-22} s, respectively. The Higgs boson is the final particle to be observed from those predicted by the standard model of particle physics that was developed in the 1960s.

A final note concerning the mass of fermions is appropriate. The Higgs field gives mass to the weak bosons as described above. It is also responsible for giving mass to the charged fermions, that is the leptons and quarks, by a somewhat different mechanism. In this case, the Yukawa interaction (see discussion above) couples the lepton to the Higgs field and gives the lepton a mass of

$$m_f = \frac{g_f V}{\sqrt{2}}.$$

Here g_f is coupling constant, and V is the vacuum expectation value of the Higgs field (i.e., 246 GeV). Thus, each charged fermion acquires a mass related to the appropriate coupling constant. The question of neutrino mass is considered further in chapter 19.

Problems

17.1. In Yukawa's original theory, the pion was viewed as a mediator of the strong interaction between nucleons in the nucleus. Calculate the range of the strong interaction on the basis of this approach.

17.2. Discuss possible by-products of Higgs boson decay to W$^+$W$^-$.

References and suggestions for further reading

Bernstein J 1974 Spontaneous symmetry breaking, gauge theories, the Higgs mechanism and all that *Rev. Mod. Phys.* **46** 7–48

Dunlap R A 2018 *Particle Physics* (San Rafael, CA: Morgan & Claypool)

Dunlap R A 2019 *The Mössbauer Effect* (San Rafael, CA: Morgan & Claypool)

FT2 2012 *Explanatory Diagram Showing How Symmetry Breaking Works* (https://commons. wikimedia.org/wiki/File:Spontaneous_symmetry_breaking_(explanatory_diagram).png)

Gonis 2007 *The Potential for the Higgs Field* (https://commons.wikimedia.org/wiki/File: Mecanismo_de_Higgs_PH.png)

Gunion J, Haber H, Kane G and Dawson S 1990 *The Higgs Hunter's Guide* (New York: Basic Books)

Henley E M and García 2007 *Subatomic Physics* 3rd edn (Singapore: World Scientific)

Lederman L and Teresi D 2006 *The God Particle: If the Universe is the Answer, What is the Question?* (Boston: Mariner)

Rias 2012a *Feynman Diagram of Higgs Production Through Gluon Fusion* (https://commons. wikimedia.org/wiki/File:Higgs-gluon-fusion.svg)

Rias 2012b *Plot of the Branching Ratios for Different Decay Modes of the Higgs Boson* (https:// commons.wikimedia.org/wiki/File:HiggsBR.svg)

IOP Publishing

An Introduction to the Physics of Nuclei and Particles
(Second Edition)

Richard A Dunlap

Chapter 18

Proton decay

18.1 Grand unified theories

The standard model, as described in the last four chapters, is based on the fundamental assumptions that hadrons are comprised of quarks and quark (or baryon) number and lepton number are conserved. There is considerable experimental evidence in favor of the quark model and no convincing experimental observations have provided direct evidence of non-conservation of quark or lepton number. However, there is a desire to describe particle physics with a theory that unifies all interactions. This follows along the lines of electroweak unification as described in previous chapters, and theories that take this approach are referred to as grand unified theories (GUTs). The basic assumption is that at sufficiently high energies, the electroweak interaction becomes unified with the strong interaction. At such a unification energy, estimated to be of the order of 10^{15} GeV, all interactions are equivalent, or symmetric. At lower energies the symmetry is spontaneously broken, and the electroweak interaction becomes distinct from the strong interaction, and at even lower energies the electroweak symmetry is broken.

The simplest (and first) grand unified theory, referred to as $SU(5)$ was developed in 1974 by Howard Georgi (b. 1947) and Sheldon Glashow (b. 1932). The basic concept of unification can be viewed phenomenologically in the following way. The strong interaction decreases in strength with decreasing distance (recall the arguments for confinement) and at sufficiently small distances will become comparable to the electromagnetic and weak interactions. This situation is shown in figure 18.1 where, at some distance, the magnitude of the three interactions will become comparable and this defines a mass/energy scale on which the interactions become unified. In $SU(5)$, it is expected that this unification will take place on a distance scale of about 10^{-31} m and this corresponds to the unification energy of 10^{15} GeV.

doi:10.1088/978-0-7503-6094-4ch18

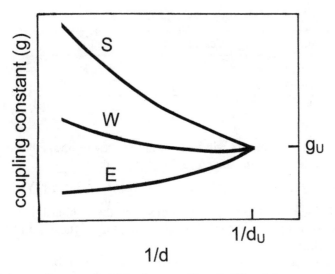

Figure 18.1. Relative coupling constants (g) for the strong (S), weak (W) and electromagnetic (E) interactions as a function of inverse distance ($1/d$) showing unification values (U).

This energy is 10 or more orders of magnitude greater than that which can be obtained by an accelerator using current technology.

Grand unified theories make predictions that fall outside of the framework of the standard model. One prediction that is common to most grand unified theories is the existence of massive magnetic monopoles. While there have been a number of intensive experimental searches for these particles, no conclusive evidence for their existence has been reported. Of particular relevance to the present discussion, however, is the prediction that baryon number, B, and lepton number, L, are not conserved. $SU(5)$ predicts that the quantity $B - L$ is conserved. Feynman diagram vertex rules for the 12 gauge bosons discussed thus far (γ, W^+, W^-, Z^0 and the eight colored gluons), show that these mediating particles interact in such a way that baryon number and lepton number are always conserved (see figure 16.14). Thus, grand unified theories that allow for non-conservation of these quantities include the existence of other gauge bosons. Specifically, $SU(5)$ includes a total of 24 gauge bosons; 12 bosons as included in the standard model and 12 additional bosons that mediate interactions that do not conserve either baryon number or lepton number.

The 12 additional gauge bosons have two charge states, 4/3 and 1/3, which are referred to as X-bosons and Y-bosons, respectively. These two bosons carry color, and each charge state contains three bosons colored R, G and B, leading to six different bosons. In addition, each of the six bosons has a charge conjugate (or antiparticle), bringing the total to 12 bosons. These bosons are sometimes referred to as X and Y for the particles, and as $\overline{X}$ and $\overline{Y}$ for the antiparticles. However, by analogy with the W^+ and W^- bosons, it is less ambiguous to refer to them as X^+, Y^+,

Table 18.1. Properties of the X^+ boson and Y^+ boson.

Boson	Charge (e)	Spin	$B - L$
X^+ boson	$+4/3$	1	2/3
Y^+ boson	$+1/3$	1	2/3

X^- and Y, and this convention will be used in the present book. The complete set of X-bosons and Y-bosons can, therefore, be written as

$$X_R{}^+, X_B{}^+, X_G{}^+, Y_R{}^+, Y_B{}^+, Y_G{}^+, X_R{}^-, X_B{}^-, X_G{}^-, Y_R{}^-, Y_B{}^-, Y_G{}^-$$

and these combined with the 12 gauge bosons from the standard model, make up the 24 gauge bosons of the $SU(5)$ model. Basic properties of the X^+ and Y^+ bosons are given in table 18.1.

The properties of X-bosons and Y-bosons as given in table 18.1, allows for the construction of valid Feynman diagrams for the interaction of these gauge bosons with known fermions (i.e. quarks and leptons). The possible interactions are shown by the allowed vertices in the Feynman diagrams in figure 18.2. It is seen that these interactions fall into two general categories. The X- and Y-bosons can convert a quark to a lepton, or they can convert a quark to an antiquark, and these processes occur in such a way as to conserve $B - L$.

From the allowed Feynman diagram vertices in figure 18.2, it is easy to see the decay modes of the X^+ boson and Y^+ boson. For the X^+ boson, the decay modes are

$$X^+ \rightarrow e^+ + \bar{d} \tag{18.1}$$

and

$$X^+ \rightarrow u + u. \tag{18.2}$$

For the Y^+ boson the decay modes are

$$Y^+ \rightarrow \ + \bar{d} \tag{18.3}$$

$$Y^+ \rightarrow e^+ + \bar{u} \tag{18.4}$$

and

$$Y^+ \rightarrow d + u. \tag{18.5}$$

Validating grand unified theories is not straightforward. Creating real X-bosons and Y-bosons and observing their decay products (e.g., equations (18.1) through (18.5)), as has recently been done for the Higgs boson, is not possible because of the large energies that are required. Thus, at present, the confirmation of predictions of grand unified theories is only possible through the observation of naturally occurring processes that are predicted by grand unified theories but are not allowed in the standard model. Two important implications of grand unified theories are the decay of the proton and the decay of the bound neutron. Certainly, in the standard model,

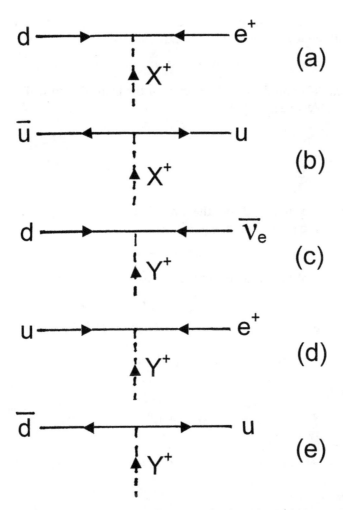

Figure 18.2. Allowed (first generation) Feynman diagram vertices involving X^+ bosons and Y^+ bosons; (a) $d + X^+ \rightarrow e^+$, (b) $\bar{u} + X^+ \rightarrow u$, (c) $d + Y^+ \rightarrow \bar{\nu}_e$, (d) $u + Y^+ \rightarrow e^+$, (e) $\bar{d} + Y^+ \rightarrow u$. Vertices involving charge conjugates of the particles shown can also be constructed for X^- and Y^-.

free protons are stable and neutrons which are bound in a stable nucleus are also stable, however, both can decay through X-boson and/or Y-boson mediated processes as allowed in $SU(5)$. Experimental searches for evidence for the validity of grand unified theories have concentrated on proton decay. The basic physics of this process in the context of $SU(5)$ is described in the next section.

18.2 Proton decay

One important consequence of grand unified theories concerns the stability of the proton. If quark number and lepton number are not conserved, then the proton can decay and this serves as, perhaps, the most convenient test of grand unified theories.

Here we look at the most common modes of proton decay that are allowed in $SU(5)$. The principal mode of proton decay is expected to be

$$p \rightarrow e^+ + \pi^0. \tag{18.6}$$

This decay can also produce excited states of the neutral pion with the same quark content (see table 15.2), such as

$$p \rightarrow e^+ + \rho^0 \tag{18.7}$$

and

$$p \rightarrow e^+ + \omega. \tag{18.8}$$

These decays are followed by the annihilation of the quark and antiquark components of the neutral pion, e.g.,

$$\pi^0 \rightarrow \gamma + \gamma$$

and annihilation of the positron with electrons in the environment,

$$e^+ + e^- \rightarrow \gamma + \gamma.$$

Other possible decay modes follow from equation (18.6), which has the following form

baryon $\rightarrow$ charged antilepton + neutral meson

and include, for example,

$$p \rightarrow \mu^+ + K^0. \tag{18.9}$$

Additional modes may be described as

baryon $\rightarrow$ neutral antilepton + charged meson

and include, for example,

$$p \rightarrow \bar{\nu}_e + \pi^+. \tag{18.10}$$

Table 18.2 gives the branching fractions for the three most likely proton decay modes. Different modes of decay may be observed experimentally by the detection of the relevant by-products with appropriate geometric and temporal relationships.

Table 18.2. Branching fractions for the principal proton decay modes. Data adapted from Burcham and Jobes (1995).

Decay mode	Branching fraction
$p \rightarrow e^+ + \pi^0$	0.40
$p \rightarrow e^+ + \rho^0$ and $p \rightarrow e^+ + \omega$, combined	0.30
$p \rightarrow \bar{\nu}_e + \pi^+$	0.16

In order to understand how proton decay fits into grand unified theories, we consider the principal decay mode as given in table 18.2 in further detail. It is clear that proton decay as described above is not consistent with the standard model as it requires the non-conservation of quark and lepton number. The decay process as given in equation (18.6) cannot, therefore, be mediated by a lepton-conserving weak boson or a quark-conserving gluon. An X-boson or Y-boson, as described in the previous section, is required. Feynman diagrams showing X^+ boson and Y^+ boson mediated proton decay as given in equations (18.6) to (18.8), are illustrated in figures 18.3 and 18.4, respectively. In these processes, the up quark that does not couple to the mediating boson is referred to as a spectator quark. Figure 18.5 shows the Feynman diagram for proton decay by equation (18.10). It is interesting to note that proton decay to a charged lepton and a neutral meson can be mediated by either

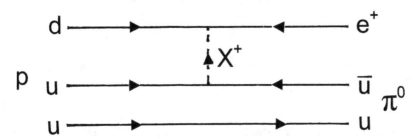

Figure 18.3. Feynman diagram for X^+ boson mediated proton decay as given in equation (18.6).

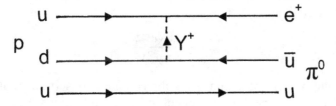

Figure 18.4. Feynman diagram for Y^+ boson mediated proton decay as given in equation (18.6).

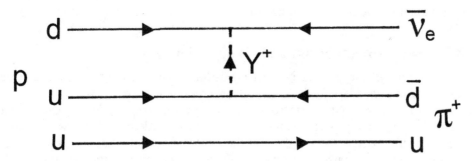

Figure 18.5. Feynman diagram for Y^+ boson mediated proton decay as given in equation (18.10).

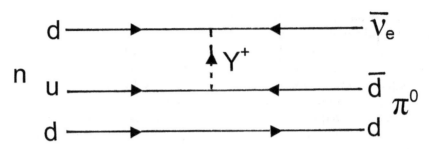

Figure 18.6. Feynman diagram for the decay of a bound neutron as mediated by a Y^+ boson [equation (18.11)].

an X-bason or a Y-boson, while decay to a neutral lepton and a charged meson can only be mediated by a Y-boson. Interactions involving mediating X^+ bosons and Y^+ bosons in figures 18.3 to 18.5 can be related to the allowed vertices as shown in figure 18.1.

A final point of interest is the decay of a bound neutron as mentioned above. In the context of $SU(5)$ this decay is

$$n \to \bar{\nu}_e + \pi^0. \tag{18.11}$$

The Feynman diagram for this decay is shown in figure 18.6. This decay is mediated by the Y-boson and is similar to the proton decay shown in figure 18.5, except that the spectator quark is a down quark rather than an up quark.

An important aspect of any theoretical prediction of, for example, proton decay, is the ability of the theory to accurately predict the decay lifetime. It can be shown that the proton lifetime, τ_p, is related to the mass of the X-boson, M_X, as

$$\tau_p \approx \frac{1}{g_U{}^4} \frac{(M_X c^2)^4}{(m_p c^2)^5} \tag{18.12}$$

where m_p is the proton mass and g_U is the coupling constant at unification. The values of M_X and g_U depend on the details of the grand unified theory and specific values for the proton lifetime, as predicted by equation (18.12) will be discussed in detail in section 18.5. In the next two sections, we consider the details of the experimental methods by which proton decay searches have been conducted.

18.3 Cherenkov radiation and its detection

As the lifetime of the proton, if it decays, is long, the decay rate is small and the number of expected events in a typical experiment will be small. As such, it is essential to use a very large detector containing protons to provide as great a signal as possible. Inevitably, large water-filled detectors, as described below, are used. Such water-filled detectors serve both as the source of protons to study, as well as the detecting medium. Charged particles produce Cherenkov radiation as they travel through water and are used as a means of detecting the presence of the charged particles. γ-rays resulting from the annihilation of decay by-products may be detected by means of Cherenkov radiation produced by electrons liberated during Compton scattering events in the detector medium.

Cherenkov radiation is electromagnetic radiation that is produced by charged particles travelling through a dielectric medium at a velocity that is greater than the phase velocity of light in the medium. Cherenkov radiation was first observed by Pavel Cherenkov (1904–90) in 1934 and the theoretical framework for the phenomenon was developed shortly thereafter, in 1937, by Igor Tamm (1895–1971) and Ilya Frank (1908–90).

In order to produce Cherenkov radiation a charged particle, travelling at a velocity, v, in a dielectric with an index of refraction, n, must satisfy the relationship,

$$v > \frac{c}{n}.$$

(18.13)

Of specific relevance to the observation of proton decay, Cherenkov radiation is produced by positrons. For many other applications, such as those described in the next chapter, the Cherenkov radiation is produced by electrons. For water, which has an index of refraction of about 1.33, the velocity condition in equation (18.13) gives a threshold velocity of 0.75 c. The energy of the positron (or electron) is obtained from this velocity in terms of the relativistic energy,

$$E = \frac{m_e c^2}{\left[1 - \frac{v^2}{c^2}\right]^{1/2}}.$$

(18.14)

For the rest mass energy of the positron (electron), $m_e c^2 = 0.511$ MeV c^{-2}, and equation (18.14) gives $E = 0.77$ MeV.

The basic phenomenon of Cherenkov radiation can be explained in terms of the details of the interaction of the charged particle with the atoms in the surrounding dielectric medium. As a charged particle travels through a dielectric medium, the electric field generated by the charge polarizes the electric dipole moments of the nearby atoms. As the charged particle moves away from the polarized atom, the polarization decreases as a result of the decreasing interaction between the atom's electric dipole moment and the particle's charge. If the velocity of the particle in the dielectric medium exceeds the phase velocity of light in the medium, then the atom is left in a polarized state and the energy associated with the polarized state is given off as photons when the atom depolarizes.

Figure 18.7 shows the geometric relationship between the charged particles velocity and the emitted Cherenkov radiation. If the charge polarizes an atom at $t = 0$, then at a time t later then, as shown in the figure, the emitted photon will have travelled a distance, ct/n, while the particle will have travelled a distance βct, where

$$\beta = \frac{v}{c}.$$

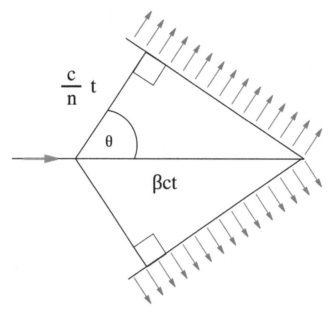

Figure 18.7. Geometry of Cherenkov radiation. This [the geometry of the Cherenkov radiation] image has been obtained by the author from the Wikimedia website where it was made available by Horvath (2006) under a CC BY-SA 2.5 licence. It is included within this chapter on that basis. It is attributed to Arpad Horvath.

Since the wavefront is perpendicular to the direction of propagation of the photon, then it is easy to see from the figure that the radiation will be emitted from the charged particle at an angle θ in the forward direction where

$$\theta = \sin^{-1}\left[\frac{1}{n\beta}\right].$$

Cherenkov radiation is responsible for the characteristic blue glow around the core of nuclear fission reactors as shown in figure 18.8. The Cherenkov radiation, in this case, results from the relativistic electrons that are emitted by the β^- decay from the fission products, as described in chapter 12. The blue color of Cherenkov radiation can be explained in terms of the Frank–Tamm formula which gives the energy spectrum of the photons in terms of the number of photons emitted per unit distance, dx, per unit wavelength, $d\lambda$, as

$$\frac{d^2N}{dxd\lambda} = \frac{q^2}{2\varepsilon_0\hbar c} \cdot \frac{1}{\lambda^2}\left[1 - \frac{c^2}{v^2n^2}\right] \tag{18.15}$$

where q is the particle charge and, in general, the index of refraction is a function of wavelength, $n(\lambda)$. It is often customary to express the power spectrum of the emitted photons. Since the photon energy is related to the inverse of the wavelength, $E \propto 1/\lambda$, then the power per unit wavelength is

Figure 18.8. Cherenkov radiation in the core of the High Flux Isotope Reactor at Oak Ridge National Laboratory, Oak Ridge, TN. Reproduced from Oak Ridge National Laboratory (2017) Flickr CC BY 2.0.

$$\frac{dP}{d\lambda} \propto \frac{1}{\lambda^3}.$$

A typical power spectrum of Cherenkov radiation is illustrated in figure 18.9. The spectrum shows the predicted $1/\lambda^3$ dependence. Also shown in the figure is the effect of passage of the Cherenkov radiation through various distances in water. It is the increasing intensity of Cherenkov radiation with decreasing wavelength (towards the blue end of the spectrum) that gives the radiation its characteristic blue color.

If we look at the total power radiated, that is the integral of curve in figure 18.9 over all wavelengths, we see a potential concern. As the wavelength approaches zero, $\lambda \rightarrow 0$, (i.e. the high frequency limit) then the power per unit wavelength will diverge as $1/\lambda^3$. In order to understand why the total power does not diverge, we have to go back to the details of equation (18.15). The index of refraction is a function of wavelength for all materials. In all cases, $n \rightarrow 0$ as $\lambda \rightarrow 0$. This means that for all media, the intensity of the Cherenkov radiation will become zero at some point, while λ is non-zero because the relevant condition in equation (18.13) will no longer be satisfied by the value of $n(\lambda)$.

The detection of the photons that are produced when Cherenkov radiation is emitted requires appropriate instrumentation. While a variety of semiconducting devices are in common use for light intensity measurements. These include, for example, photodiodes used in dvd players and charge coupled devices used in digital cameras. However, for the purpose of collecting meaningful data for proton decay studies or neutrino studies, as described in the next chapter, suitable photon sensors must be sufficiently sensitive to low level of light, have sufficiently low background

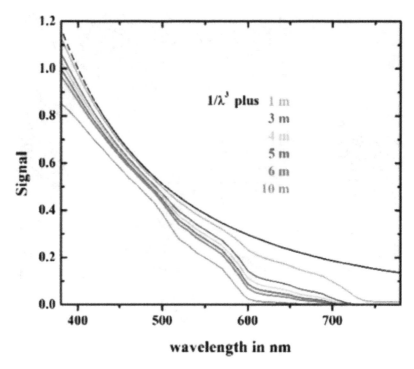

Figure 18.9. Wavelength dependence of Cherenkov radiation over the visible portion of the electromagnetic spectrum, as emitted and after passing through various lengths of water. From Vollmer (2020). Copyright 2020 European Physical Society. All rights reserved.

noise and have suffiuciently fast time response. For such purposes, photomultiplier tubes are the standard, and probably only, approach.

A photomultiplier tube (commonly referred to as a PMT) is a type of vacuum tube that contains a photoemissive cell (the photocathode) and a current amplifier. A diagram of a simple photomultiplioer tube is shown in figure 18.10. The photon is detected by the photocathode, which is made from a material that has a suitably low work function, as described below, and liberates an electron by means of the photoelectric effect when the photon is incident upon it. The photocathode is connected electrically to a series (typically of the order of 10 or so) dynodes, each of which is at a slightly higher positive potential. The dynode voltages are provided by a high voltage DC power supply connected to a voltage divider network, as shown in the figure. The dynodes are coated with an active material that, when struck with an electron, will liberate more than one electron. Electrons are directed through the set of dynodes by the increasingly positive electric potential until they arrive at the anode (the most positive electrode) and the charge which is collected there is converted to a voltage for processing.

The total gain of the photomultiplier tube may be determined by the properties of the dynodes. If each dynode emits δ electrons in response to one incident electron and there are N dynodes, then the total multiplication factor for one electron emitted by the photocathode will be

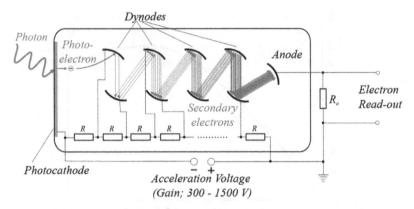

Figure 18.10. Diagram of a photomultiplier tube. This [scheme of a photomultiplier tube (PMT) in English] image has been obtained by the author from the Wikimedia website https://commons.wikimedia.org/wiki/File: Photomultiplier_schema_en.png (Krieger 2007), where it is stated to have been released into the public domain. It is included within this chapter on that basis included within this chapter on that basis.

$$M = \delta^N \tag{18.16}$$

The performance of a photomultiplier tube will depend on the design of the tube, in particular the materials that are used for the photocathode and the dynode, as well as the potential difference between dynodes. Here we look at some general characteristics of photomultiplier tubes.

We begin with the photocathode. The object of the photocathode is to release an electron when a photon is incident upon it. For this purpose, we would like to use a material in which the electrons are readily released. We define the work function of the photocathode material as ϕ. An electron can be released if the energy of the photon exceeds the work function,

$$\hbar\omega \geqslant \phi.$$

Any energy carried by the photon in excess of the work function will be given to the electron as kinetic energy. Most common photocathode materials are alkali metal compounds, such as Sb–Na–K–Cs, and these materials can have quantum efficiencies as high as 40%. An important consideration for the photocathode is thermal energy. At room temperature, where most photomultiplier tubes are used, thermal energy can result in thermionic emission from the photocathode. This results from electrons near the Fermi energy escaping from the material's surface and causes a signal to be detected at the anode when no photon was incident on the photocathode. This contribution to the photocurrent is referred to as dark current and is a component of the noise that is observed in the output of the tube. Dark current can be reduced by lowering the temperature of the photomultiplier, but this is generally not convenient or possible, as in the case of proton decay experiments.

Similarly, the surface of the dynodes needs to be from a material that will readily release electrons. Silver–magnesium alloys are commonly used. The number of secondary electrons or secondary emission ratio, δ in equation (18.16), depends not

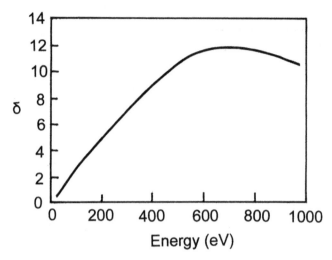

Figure 18.11. Typical secondary electron emission ratio as a function of primary electron energy.

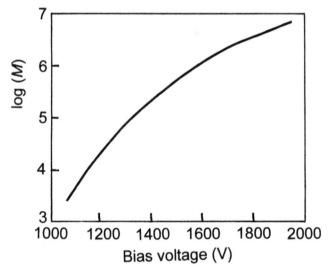

Figure 18.12. Total multiplication factor, as in equation (18.16), for a typical photomultiplier tube as a function of total bias voltage.

only on the properties of the dynode surface material, but also on the energy of the primary electrons striking it. The electron energy can be increased by increasing the potential difference between dynodes. Figure 18.11 shows a typical relationship between primary electron energy and secondary electron emission ratio. The total gain of the photomultiplier is, therefore, a sensitive function of the voltage between dynodes. The gain can be related to the total bias voltage applied to the tube and typical photomultiplier tube characteristics are illustrated in figure 18.12.

A final important factor in the operation of a photomultiplier tube deals with time response. We can get an idea of time response by looking at the physics of the electrons as they propagate through the series of dynodes in the tube. Assuming that secondary electrons are emitted at low energy and that the electric field between dynodes is constant,

$$E = \frac{V}{d}$$

where V is the potential difference between the dynodes and d is their spacing. The electron's acceleration, a, is given by the Lorentz force,

$$a = \frac{eV}{m_e d}$$

and the time, t, for the electron to travel between dynodes is

$$t = \sqrt{\frac{2d^2 m_e}{eV}}.$$

Using typical values of $V = 200$ V and $d = 0.01$ m, we find $t = 2.4 \times 10^{-9}$ s. If there are 10 dynodes, then the total time for the electron pulse to travel from the photocathode to the anode will be about 2.4×10^{-8} s. While this result provides an estimate of the timing characteristics of the photomultiplier tube it is not necessarily a precise measure of the time response of a light measurement. If all pulses were delayed by the exact same amount, then it would have little effect on the overall measurement. However, as the electron pulse progresses through the dynodes the pulse becomes spread out in time. This is largely due to the fact that the electric field between the dynodes is not actually constant, so that electrons that travel along different paths require different lengths of time. This phenomenon is referred to as transit time spread. The ability to measure the time that a pulse of light is detected by the photocathode depends on the ability of the electronics to trigger consistently on the rising edge of the voltage pulse from the anode. In this respect, timing accuracy can be significantly better than our transit time estimate above. Another factor of particular relevance is the effect of the location on the photocathode where the photon is detected. This is particularly relevant for large diameter photomultiplier tubes, as it affects the time delay between the production of the electron by the photocathode and its arrival at the first dynode. This is an important consideration for the present discussion, as large diameter photomultiplier tubes are required for proton decay experiments, as well as the neutrino experiments discussed in the next chapter, in order to improve sensitivity.

18.4 The Kamioka observatory

The most extensive investigation of proton decay has been conducted with the Kamiokande detector located at the Kamioka Observatory, Institute for Cosmic Ray Research, operated by the University of Tokyo. The facility is located about 1000 m underground in the Mozumi Mine in order to shield it from cosmic rays. The

Table 18.3. Properties of Cherenkov detectors constructed (or under construction) at the Kamioka Observatory. In some cases, the number of photomultiplier tubes varied over time. The table gives the maximum number installed.

Detector	Operational dates	Dimensions (h × d)	Water mass (kg)	Number of PMTs
KamiokaNDE	1983–5	16.0 m × 15.6 m	3.058×10^6	~1000
Super-Kamiokande	1986–Present	41.4 m × 39.3 m	5×10^7	11 146
Hyper-Kamiokande	Proposed 2027	60 m × 74 m	2.6×10^8	40 000

KamiokaNDE detector (for Kamioka Nucleon Decay Experiment) was designed specifically for the purpose of investigating proton decay. The first Kamiokande detector became operational in 1983 and has subsequently been replaced with a larger version, Super-Kamiokande. Super-Kamiokande has continued the search for proton decay but has also been used as a neutrino observatory to study neutrinos from a variety of sources such as solar neutrinos, cosmic ray generated neutrinos in the atmosphere and accelerator neutrinos. A larger version of Super-Kamiokande, Hyper-Kamiokande, is currently in the construction stage. Table 18.3 gives some basic information for the Cherenkov detectors that have been constructed or are under construction at the Kimioka Observatory.

18.4.1 Super-Kamiokande

At present, Super-Kamiokande has provided the most detailed results for proton decay and its design and operation are summarized here. Figure 18.13 shows a diagram of the detector. The detector is divided into an inner detector of dimensions 33.8 m diameter by 36.2 m high, which is surrounded by an outer detector, as shown in the figure. The two sections of the detector are separated by an opaque barrier, so that events that create Cherenkov radiation in one section are not seen in the other. The inner detector contains 32 000 tonnes of water and Cherenkov radiation is detected by 11 146 inward facing photomultiplier tube. The tubes are 50 cm diameter Hamamatsu R3600 photomultipliers, as shown in figure 18.14. Cherenkov radiation produced in the outer detector is detected by 1885 outward facing Hamamatsu R1408 20 cm diameter photomultiplier tubes. The outer detector serves as a veto detector to eliminate events that are caused by cosmic rays and radioactivity in the surrounding rocks. Only events that occur within a fiducial volume of about 22 500 tonnes of water, defined by a volume separated from the interior walls of the inner detector, and are not vetoed by an event from the outer detector are considered valid.

A major factor in the correct interpretation of observed events in the Super-Kamiokande detector is that of calibration. Calibration falls into several categories including optical attenuation of the water, timing calibration and energy calibration. These are discussed briefly.

Optical attenuation—The Cherenkov radiation produced by charged particles in the detector is attenuated as it travels from the location of the event to the

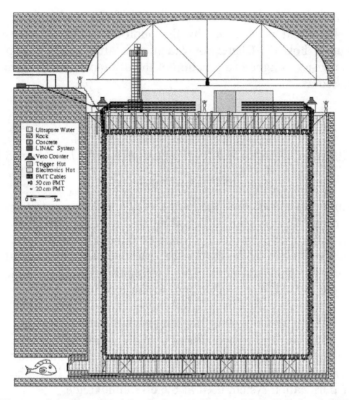

Figure 18.13. Diagram of the Super-Kamiokande detector. Reprinted from Fukuda *et al* (2003). Copyright (2003). With permission from Elsevier.

Figure 18.14. 50 cm diameter photomultiplier tube as used in Super-Kamiokande. This [Photomultiplier Tube (PMT)] image has been obtained by the author from the Wikimedia website where it was made available by Motokoka (2020) under a CC BY-SA 4.0 licence. It is included within this chapter on that basis. It is attributed to Motokoka.

photomultiplier tube where it is detected. Therefore, the interpretation of the intensity of light that is observed by the photomultiplier tubes depends on an understanding of how the Cherenkov radiation is attenuated as it passes through water. The intensity of light as a function of the distance it travels through water, $I(d)$, is expressed as

$$I(d) = \frac{I_0}{d^2} \exp\left[-\frac{d}{L}\right] \tag{18.17}$$

where I_0 is its initial intensity and the attenuation length, L, is defined by

$$L = [\alpha_{abs} + \alpha_{scat}]^{-1}.$$

Here α_{abs} represents attenuation due to absorption and α_{scat} represents attenuation due to scattering. The most direct method of measuring the attenuation length is by means of a tunable laser. Light from a tunable titanium–sapphire laser is introduced into the detector through a fiber optic cable connected to a diffused ball. The ball is positioned at different locations within the detector and the light intensity is monitored using a CCD camera located on the top of the detector as a function of the depth of the water. By this means the attenuation length can be obtained and equation (18.17) can be used in the interpretation of the measured intensity.

As the attenuation length is a function of the water quality (i.e., purity) the attenuation length must be measured frequently, and the water must be regularly purified to maintain its quality and optical properties. Figure 18.15 shows the details of the water purification system used at Super-Kamiokande.

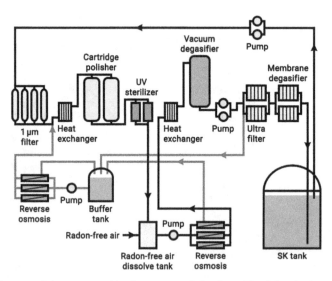

Figure 18.15. Diagram of the water purification system of the Super-Kamiokande detector. This [Super-Kamiokande water purification system] image has been obtained by the author from the Wikimedia website where it was made available by Haythornthwaite (2021) under a CC BY-SA 4.0 licence. It is included within this chapter on that basis. It is attributed to nagualdesign.

Timing calibration—As the detection of events in the Super-Kamiokande detector is in real-time, it is essential to know the timing characteristics for each photo-multiplier tube. Using a short pulse (< 3 ns) from a N_2 laser which is introduced into the detector through an optical fiber and the diffuser ball, as for the attenuation studies, the time response for each individual photomultiplier tube can be established. The individual timing characteristics for each photomultiplier tube can then be used in the analysis of the data.

Energy calibration—For some measurements, particularly solar neutrino measurements as discussed in detail in the next chapter, it is important to have an accurate energy calibration, that is, the relationship between the observed signal from the photomultiplier tubes and the initial energy of the electron that produced the Cherenkov radiation. The primary approach to energy calibration for the Super-Kamiokande detector is through the use of a linear electron accelerator (LINAC). Details of the operation of a linear electron accelerator is presented in appendix C. The LINAC system used in the Super-Kamiokamde detector is seen in figure 18.13. The electron beam from the linear accelerator is injected into the detector through a beam pipe. The end of the beam pipe can be positioned within the detector volume at specific appropriate positions. Electrons of controlled energy enter the fiducial volume of the detector at desired locations and produce Cherenkov radiation with specific energy parameters. Figure 18.16 shows a three-dimensional representation of the Super-Kamiokande detector where the tower for lowering the beam pipe into the detector. Measurements of photomultiplier output as a function of known electron energy allows for energy calibration of Super-Kamiokande, as well as a measure of the energy resolution of the system.

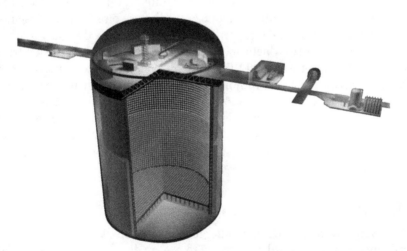

Figure 18.16. Three-dimensional cut-away image of the Super-Kamiokande detector. The inner and outer chambers are visible, and the photomultiplier tubes are shown by the array of dots on the detector inner surface. The detector dome is shown on top of the cylindrical detector. This dome contains electronics as well as the beam pipe for the electron linear accelerator (LINAC) used for calibration (the tower just to the left of center). CC BY 4.0 Reprinted from Kajita *et al* (2018). Copyright (2018). With permission from Elsevier.

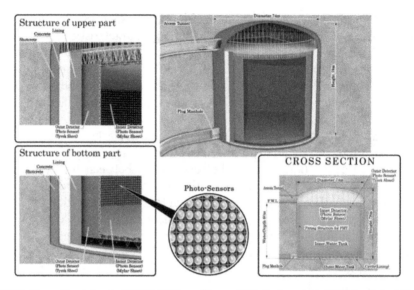

Figure 18.17. Hyper-Kamiokande CC BY 4.0 From Abe *et al* (2018). Copyright (2018) by Oxford University Press.

18.4.2 Hyper-Kamiokande

Planning for the construction of Hyper-Kamiokande began in 2010. The design was finalized in 2019 and construction started in 2022. It is scheduled for operation in 2027. As indicated in table 18.3, Hyper-Kamiokande will have a total volume of water that is about five times that of Super-Kamiokande and a fiducial volume approximately ten times greater. As well, it will incorporate almost four times the number of photomultiplier tubes. Figure 18.17 shows some general design specifications for Hyper-Kamiokande. Hyper-Kamiokande is being designed to investigate proton decay, as well as solar neutrinos, supernova neutrinos and the neutrino mass hierarchy. It will also provide data that may provide evidence concerning whether neutrinos are Dirac particles or Majorana particles. These neutrino experiments will be considered further in the next chapter. The increased fiducial volume, along with the increased number of photomultiplier tubes, compared to Super-Kamiokande, will vastly increase data production rates. It is estimated that ten years of data collection on Hyper-Kamiokande would be equivalent to 100 years of data collection of Super-Kamiokande.

18.5 Experimental limits to proton decay

Thus far, Super-Kamiokande and all previous proton decay searches have not observed any events that could be ascribed to proton decay based on the signature of the signal. The fact that proton decay has not been observed does not mean that proton decay does not exist. It does, however, allow us to place a lower limit on the lifetime of such a decay based on the experimental sensitivity and the duration of the measurements and to relate this limit to the predictions of specific grand unified theories.

Current results from Super-Kamiokande have placed lower limits on the partial lifetimes for proton decay to positrons [equation (18.6)] and to antimuons [equation (18.9)] of

$$\tau_p(e^+\rho^0) > 1.67 \times 10^{34} \text{ years}$$

and

$$\tau_p(\mu^+K^0) > 1.08 \times 10^{34} \text{ years},$$

respectively. These lower limits can be compared to the lifetimes that are predicted by grand unified theories. Following from the basic $SU(5)$ theory as formulated by Georgi and Glashow, equation (18.12) can be used to estimate the lifetime of proton decay. If reasonable uncertainties in the values of the input parameters are included, then the predicted lifetime for proton decay is in the range of 10^{30}–10^{31} years. Clearly, the results from Super-Kamiokande have eliminated the existence of proton decay as predicted by Georgi–Glashow $SU(5)$ theory. However, more contemporary grand unified theories typically predict longer proton lifetimes than Georgi–Glashow. Table 18.4 gives the predicted proton lifetimes for some grand unified theories. Details of these theories are beyond the scope of the present book, but results are presented in the table to emphasize the range of theoretically predicted proton lifetimes for the purpose of a comparison with experimental studies.

The lower limit placed on the proton lifetime by the experimental results from Super-Kamiokande are sufficient to rule out the validity of the simplest of grand unified theories. The validity of theories that predict lifetimes with an upper limit of around 10^{34} years is borderline. $SUSY\ SO(10)$, $SUSY\ SU(5)$ (5-dimensions) and (particularly) Flipped $SU(5)$ ($MSSM$) cannot be ruled out on the basis of Super-Kamiokande results. Hyper-Kamiokande, which should be operational by the late 2020s, will have a sensitivity that is 5–10 times that of Super-Kamiokande. That means that by the late 2030s, it should be able to push the experimental lower limit of the proton lifetime to around 10^{35} years. If proton decay is not observed within this timeframe, it would rule out, or at least seriously question all grand unified

Table 18.4. Ranges of lifetimes that are consistent with different grand unified theories.

Theory*	Proton lifetime (y)
$SU(5)$ (Georgi–Glashow)	10^{30}–10^{31}
$SUSY\ SU(5)$	10^{28}–10^{32}
$SUGRA\ SU(5)$	10^{32}–10^{34}
$SUSY\ SO(10)$	10^{32}–10^{35}
$SUSY\ SU(5)$ ($MSSM$)	$\sim 10^{34}$
$SUSY\ SU(5)$ (5-dimensions)	10^{34}–10^{35}
$SUSY\ SO(10)$ ($MSSM$)	2×10^{34}
Flipped $SU(5)$ ($MSSM$)	10^{35}–10^{36}

*$SUSY$ = Supersymmetry, $SUGRA$ = Supergravity, $MSSM$ = Minimal Supersymmetric Standard Model.

theories in table 18.4 except Flipped $SU(5)$. Non-$SUSY$ grand unified theories such as Flipped $SU(5)$ have an upper limit for proton lifetimes of around 1.4×10^{36} years. It is conceivable that future experiments could confirm or refute proton decay on this timescale by using a combination of even larger detectors and/or longer search times.

Problems

18.1. Draw possible Feynman diagrams for proton decay to an antimuon.

18.2. Calculate the Cherenkov threshold for the following particles in water:
    (a) proton (p);
    (b) antimuon (μ^+);
    (c) positive kaon (K^+);
    (d) positive pion (π^+).

18.3. 500 photons per second are incident on the photocathode of a 12-dynode photomultiplier tube with a quantum efficiency of 30%. Calculate the average number of electrons emitted per dynode that is required to produce an anode current of 1.5 μA.

References and suggestions for further reading

Abe K *et al* 2018 Physics potentials with the second Hyper-Kamiokande detector in Korea *Prog. Theor. Exp. Phys.* **2018** 063C01

Bajca B, Hisano J, Kuwahara T and Omura Y 2016 Threshold corrections to dimension-six proton decay operators in non-minimal SUSY SU(5) GUTs *Nucl. Phys.* B **910** 1–22

Burcham W E and Jobes M 1995 *Nuclear and Particle Physics* (Harlow, Essex: Pearson)

Dunlap R A 1988 *Experimental Physics—Modern Methods* (New York: Oxford University Press)

Fukuda S *et al* 2003 The Super-Kamiokande detector *Nucl. Instrum. Methods* **501** 418–62

Horvath A 2006 *The Geometry of the Cherenkov Radiation* (https://commons.wikimedia.org/wiki/File:Cherenkov.svg)

Haythornthwaite J 2021 *Super-Kamiokande Water Purification System* (https://commons.wikimedia.org/wiki/File:Super-Kamiokande_water_purification_system.png)

Jelley J V 1958 *Cerenkov Radiation and Its Applications* (Oxford: Pergamon)

Kajita T *et al* 2018 Establishing atmospheric neutrino oscillations with Super-Kamiokande *Nucl. Phys.* B **908** 14–29

Krieger J 2007 *Scheme of a Photomultiplier Tube (PMT) in English* (https://commons.wikimedia.org/wiki/File:Photomultiplier_schema_en.png)

Motokoka 2020 *Photomultiplier Tube (PMT)* (https://commons.wikimedia.org/wiki/File:Photomultiplier_Tube_(PMT).jpg)

Nishino H *et al* 2009 Search for proton decay via p→e$^+\pi^0$ and p→$\mu^+\pi^0$ in a large water Cherenkov detector *Phys. Rev. Lett.* **102** 141801

Oak Ridge National Laboratory 2017 *Beautiful Refueling* (https://flickr.com/photos/oakridgelab/34300813941/)

Sreekantan B V 1984 Searches for proton decay and superheavy magnetic monopoles *J. Astrophys. Astr.* **5** 251–71

Vollmer M 2020 Cherenkov radiation: why is it perceived as blue? *Eur. J. Phys.* **41** 065304

Weinberg S 1981 The decay of the proton *Sci. Am.* **1981** 64–75

IOP Publishing

An Introduction to the Physics of Nuclei and Particles
(Second Edition)

Richard A Dunlap

Chapter 19

Neutrino oscillations and masses

19.1 Solar neutrinos

Neutrinos are among the most difficult subatomic particles to investigate because of their extremely small reaction cross section with other particles. The earliest extensive studies of neutrinos dealt with the neutrinos that are emitted by the Sun. The production of energy in the Sun has been discussed previously in chapter 13. Most of this energy (over 98%) is produced by the proton–proton cycle as illustrated in figure 13.4. Electron neutrinos are produced in two kinds of solar processes: (1) β^+ decay processes such as

$$p + p \rightarrow d + e^+ + \nu_e$$

and (2) electron capture processes such as

$$^7\text{Be} + e^- \rightarrow {}^7\text{Li} + \nu_e.$$

In both cases, the neutrinos are neutrinos (rather than antineutrinos). In the former processes, the neutrinos have a continuous distribution of energies up to the end point energy given by the value of Q. In the latter processes, the neutrinos have discrete energies. The overall energy spectrum of the neutrinos that are emitted from the Sun depends on the branching ratios of the various processes involved. These ratios depend on the internal temperature and composition of the Sun. The expected energy spectrum of the neutrinos from the processes shown in figure 13.4 is illustrated in figure 19.1. End point energies for these decays are given in table 19.1. Since the 1960s there has been substantial experimental work on the detection of solar neutrinos. The cross section for such reactions is very small because neutrinos only interact via the weak interaction. The most common method of neutrino detection is by the reaction

$$\nu_e + n \rightarrow p + e^-.$$

doi:10.1088/978-0-7503-6094-4ch19

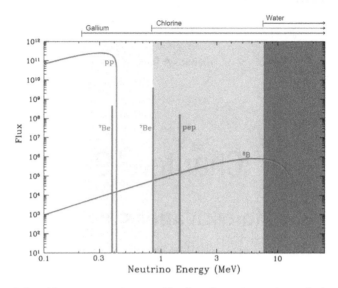

Figure 19.1. Expected neutrino energy spectrum resulting from the proton–proton cycle shown in figure 13.4. The neutrinos from electron capture by ^{7}Be have two discrete energies because the decay can go to either the ground state or an excited state of ^{7}Li. The threshold energies for the experiments described in the text are indicated. The units for flux are $cm^{-2} \cdot s^{-1} \cdot MeV^{-1}$ at 1 AU (astronomical unit) for β^+ decay processes and $cm^{-2} \cdot s^{-1}$ for electron capture processes. Reproduced from Bahcall (1996). The American Astronomical Society. All rights reserved.

Table 19.1. End point energies, E_{max}, for β-decay processes in the proton–proton cycle.

Decay	E_{max} (MeV)
$p + p \rightarrow d + e^+ + \nu_e$	0.42
$p + e^- + p \rightarrow d + \nu_e$	1.44
$^7Be + e^- \rightarrow {^7}Li + \nu_e$	0.383, 0.861
$^8B \rightarrow {^8}Be + e^+ + \nu_e$	~15.0

The by-products of this reaction can be observed by standard charged particle detection methods or by radiochemical methods as described below.

The most extensive solar neutrino studies were initiated in the late 1960s by Raymond Davis, Jr (1914–2006) in the Homestake Mine in South Dakota where a 380 m^3 container of C_2Cl_4 (perchlorethylene) was used as a detector (see figure 19.2). The chlorine consists of about 24% naturally occurring ^{37}Cl. Neutrinos with sufficient energy can produce the reaction

$$\nu_e + {^{37}}Cl \rightarrow {^{37}}Ar + e^-. \tag{19.1}$$

The threshold energy for this process is 0.814 MeV. An inspection of figure 19.1 and table 19.1 shows that the most significant contribution to this process will be from

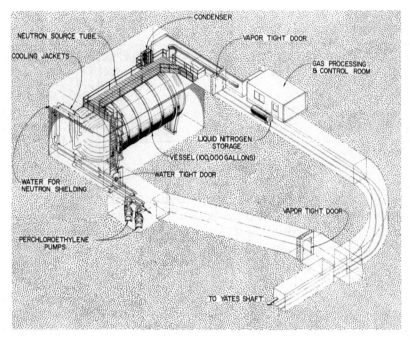

Figure 19.2. Diagram of the Homestake Mine neutrino experiment. Public domain by U.S. Department of Energy (2014).

the ^{8}B β-decay neutrinos. The p–e–p process neutrinos and the ^{7}Be electron capture neutrinos will also contribute, but to a lesser extent. Since the process in equation (19.1) is endothermic, ^{37}Ar is not β-stable and will spontaneously decay back to ^{37}Cl by electron capture,

$$^{37}Ar + e^- \rightarrow {}^{37}Cl + \nu_e. \tag{19.2}$$

The lifetime for this decay is about 50 days. On a timescale of 1–2 months the argon is removed from the detector fluid by chemical methods. The amount of argon present is determined by observing the decay products from equation (19.2). When electron capture occurs, a vacancy is left in one of the inner electron shells of the daughter ^{37}Cl atom. This vacancy is filled by another atomic electron and the difference in binding energies can liberate an outer shell electron. This is the so-called *Auger process*, and the liberated electron is an *Auger electron*. The energy spectrum of the Auger electrons is characteristic of the particular decay process and is a convenient means for determining the quantity of ^{37}Ar atoms present. This, in turn, can be related to the number of neutrinos detected and is a function of: (1) the flux of solar neutrinos; (2) the number of ^{37}Cl nuclei in the detector; and (3) the cross section of the reaction in equation (19.1). In the present case the first two factors are large, while the third factor is very small.

After collecting data for approximately 30 years, the Homestake experiment has measured a solar neutrino flux of 2.55 ± 0.25 SNU. The number of detected

Table 19.2. Specifications and results of some solar neutrino experiments. Data for SAGE are from the period 1990–2007.

Experiment	Detector nuclei	Mass of detector nuclei (kg)	Threshold (MeV)	Predicted flux (SNU)	Measured flux (SNU)
Homestake	^{37}Cl	1.3×10^5	0.814	8.0 ± 1.0	2.55 ± 0.25
SAGE	^{71}Ga	2.3×10^4	0.233	132 ± 8	65.4 ± 8.5
GALLEX/GNO	^{71}Ga	1.2×10^4	0.233	132 ± 8	67.5 ± 5.1

neutrinos is generally expressed in solar neutrino units (SNU) where 1 SNU corresponds to one neutrino detected per 10^{36} detector nuclei per second. Based on the solar model described in figure 19.1 and the threshold energy for this experiment, the standard solar model predicts a neutrino flux of 8.0 ± 1.0 SNU. Information about the Homestake experiment is summarized in table 19.2. Clearly the agreement between theory and experiment is less than ideal. There are three possible sources of uncertainty that could account for this discrepancy; (1) the Sun does not produce the flux of neutrinos as predicted by the standard solar model, (2) the performance of the detector is different than expected or (3) we do not properly understand the behavior of neutrinos. Solar models are very well developed and are accurate in predicting all other aspects of energy production in the Sun. It is, therefore, reasonable to begin by considering the other two possibilities. To this end, additional experiments that take a different approach can be beneficial. The most significant of these are described below, and later in the chapter.

A major problem with the Homestake experiment is that it is sensitive to only a small fraction of the total solar neutrino flux. It is not sensitive to neutrinos from the p–p process which is responsible for the majority of solar neutrinos and for which the flux can be predicted to an accuracy of better than 1% by solar models. The neutrino flux from other solar processes in the proton–proton chain is a very sensitive function of solar model parameters (such as temperature) and can be predicted less accurately. It would, therefore, be highly desirable to undertake an experiment that would be sensitive to at least some of the p–p neutrinos below their 0.42 MeV end point energy. A reaction that is useful in this respect is

$$\nu_e + {}^{71}\text{Ga} \rightarrow {}^{71}\text{Ge} + e^-.$$

This reaction has a threshold energy of 0.23 MeV and is sensitive to at least some of the neutrinos from each of the proton–proton chain processes. Two experiments along these lines have been undertaken in recent years. One experiment, SAGE, utilizes a detector containing metallic Ga. This detector is located in Baksan, Russia and has been in operation since 1989. A second experiment, GALLEX, is located outside of Rome and utilizes a solution of $GaCl_3$ as the detector. This experiment collected data from 1991 to 1997 and from 1998 to 2003 (under its new name, GNO for Gallium Neutrino Observatory). In all cases neutrino detection is by radiochemical methods. The resulting ^{71}Ge decays back to ^{71}Ga by β^+ decay with a lifetime of 16.5 days and periodic Ge extraction and analysis provides a measure of the number of neutrinos detected. Data obtained from SAGE and GALLEX/GNO are shown in figures 19.3 and 19.4, respectively, and the results are summarized in table 19.2.

The two Ga experiments are consistent with one another, but these experimental measurements are not in agreement with the solar model predictions. One concern with the Ga experiments, that is the calibration of the detector, has been dealt with by GALLEX. In 1994 a manufactured ^{51}Cr neutrino source was used to calibrate

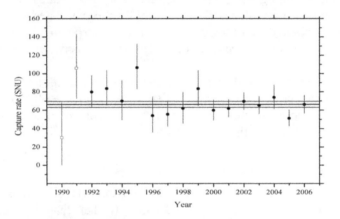

Figure 19.3. Results of the SAGE experiment from 1990 to 2006. From Hahn (2008). Copyright IOP Publishing. Reproduced with permission. All rights reserved.

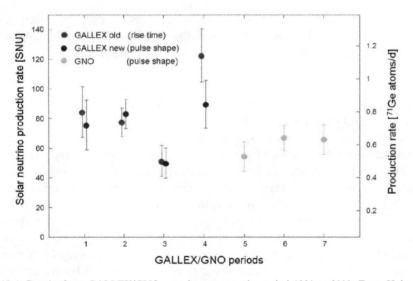

Figure 19.4. Results from GALLEX/GNO experiments over the period 1991 to 2003. From Hahn (2008). Copyright IOP Publishing. Reproduced with permission. All rights reserved.

the sensitivity of the detector. The ^{51}Cr is prepared by neutron irradiation of enriched ^{50}Cr,

$$^{50}\text{Cr} + \text{n} \rightarrow {}^{51}\text{Cr} + \gamma.$$

Neutrinos are produced by the electron capture decay of ^{51}Cr (lifetime 40 days),

$$\text{e}^- + {}^{51}\text{Cr} \rightarrow {}^{51}\text{V} + \nu_e.$$

The strength of the source (which is about 1.7 MCi) is determined by monitoring γ-ray intensity during neutron irradiation and monitoring γ-rays from the ^{51}V excited state de-excitation during the electron capture decay. Experimental results indicate that the measured neutrino flux is 97% ± 11% of the flux calculated on the basis of the measured source strength. This suggests that for the Ga experiments, at least, errors in detector calibration cannot account for the differences between measurement and theory.

Overall, the chlorine and gallium experiments to date have not provided satisfactory agreement with the expected solar neutrino flux. It would, therefore, be reasonable to look in more detail at the properties of neutrinos in order to understand the reasons for this discrepancy.

19.2 Neutrino flavor states

The experiments described above all have two important features in common. Firstly, they are designed to detect neutrinos (rather than antineutrinos) and secondly, they are designed to detect electron neutrinos (rather than muon neutrinos or tau neutrinos). Further to the discussion in chapter 14, it is important to look at the properties of neutrinos, particularly with regard to their flavor states, in more detail.

Experimental studies conducted at particle accelerators which can produce neutrinos of different flavors, have looked for processes that would indicate non-conservation of lepton generation number. Such processes include

$$\nu_\mu + \text{n} \rightarrow \text{p} + \text{e}^-. \tag{19.3}$$

Perhaps the most notable example of such an experiment was conducted in 1962 by Leon Lederman (1922–2018), Melvin Schwartz (1932–2006) and Jack Steinberger (1921–2020) using the 30 GeV Alternating Gradient Synchrotron at Brookhaven National Laboratory. Details of the operation of synchrotrons and other accelerators are given in Appendix C. In the case of the Lederman–Schwartz–Steinberger experiment, a high energy proton beam incident on a beryllium target produced pions which decayed to produce muons and muon neutrinos,

$$\pi^+ \rightarrow \mu^+ + \nu_\mu.$$

This, and other similar experiments have not detected the existence of processes such as in equation (19.3) and uphold the validity of lepton generation conservation.

Although different flavors of neutrinos appear to represent distinct particles, the possibility that oscillations between flavors can occur has been proposed. In 1957

Bruno Pontecorvo (1913–93) speculated that, if neutrinos had mass, then they would oscillate between flavors, analogous to the situation that had been observed for neutral K-mesons (see section 16.8). The basic principles of this phenomenon can be described in the context of Cabibbo mixing, which has been discussed in section 16.3 with regard to quark generation mixing.

Changes of neutrino flavor may be described by the time dependent evolution of the neutrino wavefunction, $\psi(t)$, in terms of the Schrödinger equation,

$$i\hbar\frac{\partial\psi(t)}{\partial t} = H\psi(t) \tag{19.4}$$

where H is the system Hamiltonian. For oscillations between (say) electron and muon neutrinos, equation (19.4) can be written in the form,

$$i\hbar\frac{d}{dt}\begin{pmatrix}\nu_e\\ \nu_\mu\end{pmatrix} = H\begin{pmatrix}\nu_e\\ \nu_\mu\end{pmatrix}.$$

Following the development in section 16.3 and equations (16.14) and (16.15), we express electron and muon neutrinos as admixtures of neutrino mass states ν_1 and ν_2 in terms of the neutrino mixing angle, θ_{12} as

$$\nu_e = \nu_1 \cdot \cos\theta_{12} + \nu_2 \cdot \sin\theta_{12} \tag{19.5}$$

and

$$\nu_\mu = -\nu_1 \cdot \sin\theta_{12} + \nu_2 \cdot \cos\theta_{12}. \tag{19.6}$$

In matrix form, the Hamiltonian is given in terms of the neutrino energy, E, as

$$H = \left(\frac{|m_2^2 - m_1^2|c^4}{4E}\right)\begin{pmatrix}-\cos\theta_{12} & \sin\theta_{12}\\ \sin\theta_{12} & \cos\theta_{12}\end{pmatrix}. \tag{19.7}$$

The wavelength of the neutrino flavor oscillations can then be found to be

$$\lambda = \frac{4\pi\hbar E}{|m_2^2 - m_1^2|c^3} = 2.48 \times \frac{E}{|m_2^2 - m_1^2|}. \tag{19.8}$$

In the form on the right-hand side of the equation, energy is given in MeV, mass is given in MeV c^{-2} and wavelength is in meters.

The Mikheyev–Smirnov–Wolfenstein (MSW) effect proposed by Stanislav Mikheyev (1940–2011), Alexei Smirnov (b. 1951) and Lincoln Wolfenstein (1923–2015) considers the effect of interactions with matter on the flavor oscillations of neutrinos. This effect describes the way in which the presence of charges can influence the neutrino wavefunction. An analogous situation occurs when photons propagate though a material with an index of refraction greater than unity. The interaction of the photon with charges in the medium will alter the wavelength of the electromagnetic radiation.

Based on the hypothesis that neutrinos may undergo flavor oscillations and that this effect is enhanced when the neutrinos pass through matter, it is possible to consider that electron neutrinos that are produced in reactions in the Sun's core and pass through nearly 700 000 km of matter before reaching the surface, may not be of a pure flavor state. This concept may be extended to include neutrinos that pass through the Earth before being detected.

An analysis of the consistency of results from radiological neutrino observation. i.e., chlorine and gallium experiments, must take into account several factors including the history of the neutrino between the time it was created and the time it was detected, as well as the energy spectrum of the neutrinos. Certainly, a different approach to neutrino detection is necessary to fully understand the behavior of neutrinos.

19.3 Real-time neutrino experiments

In order to understand the discrepancy between predicted and previously measured solar neutrino fluxes, it is important to undertake experiments that can, ideally, provide additional information about the neutrinos. This information would include:

- the time at which the neutrino was detected;
- the direction of travel of the neutrino;
- the energy of the neutrino;
- flavor of the neutrino.

The ability to obtain more detailed information about the neutrinos that are detected, also provides the opportunity for useful studies of neutrinos from non-solar sources, such as neutrinos produced artificially in accelerators or reactors, neutrinos that are by-products of radioactive decay in the Earth's interior and neutrinos that result from cosmic ray interactions in the Earth's atmosphere. Real-time neutrino detectors fall into three categories, water Cherenkov detectors, heavy water detectors and scintillation detectors. These three categories are overviewed below and notable examples of each are discussed.

19.3.1 Water Cherenkov detectors

The most significant facilities that utilize water Cherenkov detectors to study neutrinos are at the Kamioka Observatory operated by the Institute for Cosmic Ray Research, in the Gifu Prefecture, Japan. These are the Kamiokande and Super-Kamiokande experiments, as described in some detail in the previous chapter. As previously noted, Kamiokande was operational from 1983 to 1995. It was replaced by Super-Kamiokande, which became operational in 1996 and continues to collect data as of 2022. A larger water Cherenkov detector, Hyper-Kamiokande, is under construction at the observatory and is expected to become operational around 2027.

Kamiokande and Super-Kamiokande detect neutrinos through neutrino-electron scattering,

$$\nu_e + e^- \rightarrow \nu_e + e^-. \tag{19.9}$$

Electrons, liberated by their interaction with neutrinos, are observed by the Cherenkov radiation that they produce in the water-filled detector. The threshold energy for neutrino detection is about 5 MeV. Thus, only the high energy ^{8}B solar neutrinos can be detected. Neutrinos are observed in real time and since the direction of the electron's trajectory is the same as that of the incident neutrino, the origin of the neutrino flux can be investigated. Also, because the characteristics of the Cherenkov radiation are related to the electron velocity, the neutrino energy can be determined.

19.3.2 Heavy water detectors

The Sudbury Neutrino Observatory (SNO) in Ontario, Canada utilizes heavy water (D$_2$O) rather than light water as the detecting medium. It began taking data in the spring of 1999. This detector (see figure 19.5) consists of an acrylic sphere 12 m in

Figure 19.5. The detector at the Sudbury Neutrino Observatory. This [outside view of the Sudbury Neutrino Observatory detector] image has been obtained by the author from the Wikimedia website where it was made available by SNO (2018) under a CC BY-SA 4.0 licence. It is included within this chapter on that basis. It is attributed to SNO.

diameter that contains 10^6 kg of D_2O. This sphere is surrounded by a 17.8 m diameter structure containing 9456 inward-pointing 20 cm diameter photomultiplier tubes.

Three neutrino detection processes are possible,

$$\nu_e + d \rightarrow 2p + e^-, \qquad (19.10)$$

$$\nu_x + d \rightarrow \nu_x + p + n \qquad (19.11)$$

and

$$\nu_x + e^- \rightarrow \nu_x + e^-. \qquad (19.12)$$

The first reaction, equation (19.10), is the weak charged current (CC) process. It is only sensitive to electron neutrinos and has a threshold of about 7 MeV. The second reaction, equation (19.11), is the weak neutral current (NC) process and is neutrino flavor insensitive. The final reaction, equation (19.12), is elastic scattering (ES) and, again, is not sensitive to neutrino flavor. The charged current and elastic scattering processes are observed by the Cherenkov radiation that is produced by the electrons on the right-hand sides of the equations. The proton on the right-hand side of the equation for the neutral current process (equation (19.11)) is not sufficiently energetic to produce Cherenkov radiation in heavy water. This process is observed by detecting the neutron that is emitted. Over the course of operation of the SNO detector, three approaches to the observation of neutrons have been utilized, as described below.

Pure D_2O—Neutrons react with deuterons in the heavy water by the reaction

$$n + d \rightarrow t + \gamma. \qquad (19.13)$$

The γ-ray produced in this reaction transfers energy to electrons by means of Compton scattering. These electrons acquire sufficient energy to produce Cherenkov radiation, which is then detected by the photomultiplier tubes.

NaCl—2×10^3 kg of NaCl was dissolved in the D_2O. Neutrons interact with ^{35}Cl nuclei (76% abundant in natural chlorine) by the process

$$n + {}^{35}Cl \rightarrow {}^{36}Cl + \gamma. \qquad (19.14)$$

The γ-ray transfers energy to electrons by Compton scattering, which are then detected by their Cherenkov radiation, as above. As the cross section for the reaction in equation (19.14) is about two orders of magnitude greater than that for the reaction in equation (19.13), neutron detection efficiency was greatly improved.

Proportional counter—^3He gas filled proportional counters were suspended within the D_2O and used to detect neutrons by the reaction

$$n + {}^3He \rightarrow t + p.$$

The positively charged tritons and protons are detected by the proportional counter.

19.3.3 Scintillation detectors

A scintillator is a material that emits optical photons when it absorbs higher energy photons, such as γ-rays. These flashes of optical photons can then be detected by photomultiplier tubes. Neutrino detectors typically consist of a large volume of liquid scintillator in a vessel surrounded by photomultiplier tubes. Two major scintillator type detectors for neutrinos have been constructed, KamLAND at the Kamioka facility in the Gifu Prefecture in Japan and the Borexino facility located at the Laboratori Nazionali del Gran Sasso at the site of the previous GALLEX/GNO experiments.

A diagram of the KamLAND (KAMioka Liquid scintillator AntiNeutrino Detector) facility is shown in figure 19.6. The detector consists of an 18 m diameter stainless steel vessel lined with 1879 (43 cm or 50 cm diameter) photomultiplier tubes. An inner chamber 13 m in diameter and made of nylon contains 10^6 kg of liquid scintillator comprised of a mixture of mineral oil, benzene and fluorescent chemicals.

The Borexino detector, as shown in figure 19.7, consists of 2.78×10^5 kg of a mixture of 1,2,4-Trimethylbenzene and 2,5-Diphenyloxazole liquid scintillator, surrounded by 2200 photomultiplier tubes.

Scintillation detectors can detect solar neutrinos through the flavor neutral elastic scattering process

$$\nu_x + e^- \rightarrow \nu_x + e^-.$$

Since photon production is through scintillation, there is no Cherenkov threshold, and the detector is sensitive to low energy neutrinos. However, scintillation detectors have been most beneficial for the study of neutrinos from sources other than the Sun. These include reactors and accelerators, as well as radioactive nuclides within the Earth. For reactors and geological sources, the neutrinos are electron antineutrinos produced by β^- decays,

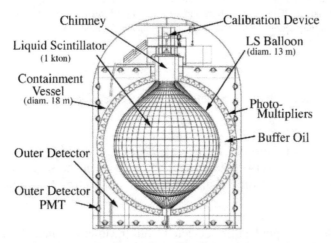

Figure 19.6. Diagram of the KamLAND detector. From Suzuki A *et al* (2005) © IOP Publishing. Reproduced with permission. All rights reserved.

Figure 19.7. Photograph of the Borexino detector. This [Borexino detector in LNGS in September 2015] image has been obtained by the author from the Wikimedia website https://commons.wikimedia.org/wiki/File: Laboratori_Nazionali_del_Gran_Sasso,_Borexino_detector.png (Borexino Collaboration 2015), where it is stated to have been released into the public domain. It is included within this chapter on that basis.

$$n \rightarrow p + e^- + \bar{\nu}_e. \tag{19.15}$$

These electron antineutrinos are detected by inverse β-decay,

$$\bar{\nu}_e + p \rightarrow n + e^+. \tag{19.16}$$

This process has a threshold energy of 1.8 MeV and the positron and neutron that are produced in the inverse β-decay can be detected in the scintillator detector as follows. Firstly, the positron will annihilate with an electron to produce two γ-rays,

$$e^+ + e^- \rightarrow 2\gamma.$$

Secondly, a delayed signal will be detected from the interaction of the neutron with a proton in the detector medium to form a deuteron,

$$n + p \rightarrow d + \gamma.$$

The detection of the electron antineutrino is, therefore, confirmed by the observation of these two signals with the proper time relationship.

19.4 More solar neutrino results

Super-Kamiokande and the SNO have provided very significant measurements of solar neutrino fluxes. The importance of these results comes from the real-time nature of the neutrino observations, as well as the ability to determine neutrino direction, energy and flavor.

Figure 19.8 shows the results of a directional study from Super-Kamiokande. These results were the first to show that the neutrinos that are being detected are actually produced by the Sun as they are travelling to the detector from the directional vector to the Sun.

Results from Super-Kamiokande showing the measured energy spectrum of observed solar neutrinos are shown in figure 19.9. These measured electron neutrinos are above the Cherenkov threshold of about 5 MeV, as noted previously, and are primarily the neutrinos produced in the ^{8}B β-decay, as illustrated in figure 19.1. This is the first observation that shows that the ratio of observed electron neutrino flux and the predicted solar neutrino flux is roughly constant as a function of energy.

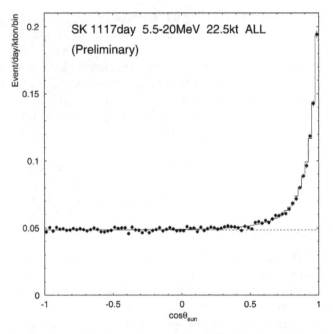

Figure 19.8. Neutrino flux measured by Super Kamiokande as a function of the angle to the Sun. Reprinted from Nishijima (2001). Copyright (2001). With permission from Elsevier.

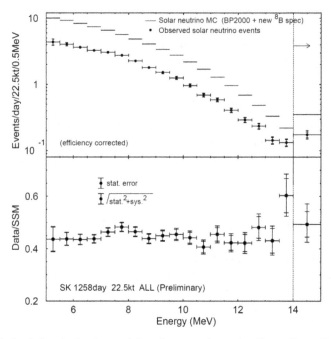

Figure 19.9. Calculated (standard solar model) and measured neutrino fluxes (Super-Kamiokande) as a function of energy (top panel) and ratio of the measured to the calculated values (bottom panel). Reprinted from Fukuda (2003). Copyright (2003). With permission from Elsevier.

Results from the SNO are of particular importance, as this detector is sensitive to all flavors of neutrinos through the weak neutral current and elastic scattering process but only to electron neutrinos through the weak charged current processes. This provides the unique ability to determine the ratio of electron neutrinos to total neutrinos and also to directly compare total neutrino flux to that predicted by the solar model. Some results from the SNO are summarized in figure 19.10. These results show a ratio of observed electron neutrinos to predicted solar neutrinos that is consistent with results obtained previously using radiochemical techniques, i.e., Homestake and gallium experiments. They also show, for the first time, the agreement between measured neutrino flux and predicted solar neutrino flux when all flavors of neutrino are included. This provides very convincing evidence for neutrino flavor oscillations, as discussed in further detail in the next section.

19.5 Atmospheric neutrino studies

In addition to experiments dealing with solar neutrinos, the measurement of the properties of neutrinos that are created in the atmosphere can provide important information. Atmospheric neutrinos are produced at altitudes of a few 10s of kilometers when cosmic rays interact with various particles in the Earth's atmosphere. They result primarily from the formation and decay of pions, e.g.,

$$\pi^+ \rightarrow \mu^+ + \nu_\mu.$$

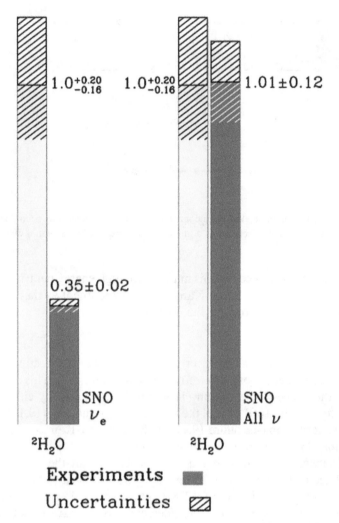

Figure 19.10. Solar neutrino results from the Sudbury Neutrino Observatory showing solar model predictions (yellow) and measured neutrino fluxes (blue). Electron neutrinos only (left) and all neutrinos (right). This [result of 30 years of research] image has been obtained by the author from the Wikimedia website where it was made available by Szdori (2006) under a CC BY-SA 3.0 licence. It is included within this chapter on that basis. It is attributed to Szdori. (Only results from SNO shown.)

This is followed by decay of the muons,

$$\mu^+ \to e^+ + \nu_e + \bar{\nu}_\mu.$$

A similar process can involve negative pions leading to the ratio,

$$\frac{n(\nu_\mu) + n(\bar{\nu}_\mu)}{n(\nu_e) + n(\bar{\nu}_e)} = 2.$$

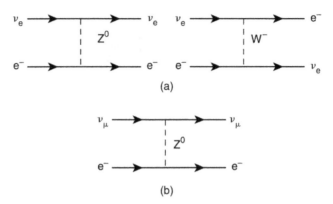

Figure 19.11. (a) Interaction of an electron and an electron neutrino mediated by neutral and charged weak bosons and (b) interaction of an electron and a muon neutrino mediated by a neutral weak boson.

The muon neutrinos produced in this manner are high energy neutrinos (>100 MeV) and can be observed by the Super-Kamiokande experiment as they interact via the weak interaction with electrons in the detector,

$$\nu_\mu + e^- \rightarrow \nu_\mu + e^-. \tag{19.17}$$

This interaction has a much smaller cross section (by about a factor of six) than the reaction of electron neutrinos and electrons in equation (19.9). This can be seen to be the case, as the process in equation (19.9) can be mediated by either a Z^0 or $W^\pm$ boson (see figure 19.11(a)) while the reaction in equation (19.17) can only be mediated by a Z^0 boson (see figure 19.11(b)). Since Super-Kamiokande is capable of measuring the direction from which neutrinos have entered the detector, it can distinguish between neutrinos that have been created in the atmosphere directly above the detector (within about 20 km) and those that have been created on the opposite side of the Earth (about 12 800 km away). Muon neutrinos that have traveled only about 20 km before being detected do not have time to oscillate into other flavors while those that travel from the other side of the Earth may no longer be muon neutrinos when they are incident on the detector. Thus, a measure of the muon neutrino flux as a function of angle (relative to the normal to the surface of the Earth) can be used to observe neutrino oscillations. Some experimental results are shown in figure 19.12. The angular dependence of the neutrino flux ratio, as seen in the figure, is a manifestation of neutrino flavor oscillations. At present, evidence suggests that the muon neutrinos oscillate to become primarily tau neutrinos.

19.6 Reactor neutrino studies

Probably the most extensive investigations of reactor neutrinos have been undertaken at the KamLAND facility. KamLAND is located in a region of Japan where there is extensive use of nuclear fission power. There are 55 nuclear reactors in the region at an average distance of 180 km from the KamLAND detector. Electron antineutrinos from these reactors come from actinides in the reactor core, primarily

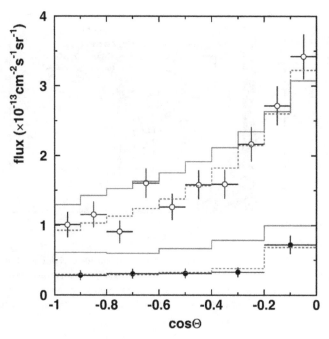

Figure 19.12. Muon neutrino flux as a measure of the muon neutrino flux as a function of angle (relative to the normal to the surface of the Earth) obtained by Super-Kamiokande. Upper data points are through-going and the lower data points are stopping neutrino induced upward-going muons. Solid histograms are the expected flux, and the dashed histograms are a fit including neutrino oscillations. Reprinted from Kajita *et al* (2018) CC BY 4.0.

^{235}U, ^{238}U, ^{239}Pu and ^{241}Pu. Smaller quantities come from radioactive fission products. These antineutrinos travel through varying depths of bedrock, depending on the reactor's distance from KamLAND, in a straight-line path to the detector, and are affected to different degrees by flavor oscillations due to the MSW effect.

Figure 19.13 shows some reactor neutrino results from KamLAND. The data show a deficiency of electron antineutrinos as a result of electron to muon antineutrino oscillations. As seen in the figure, data are well fit to predictions based on such a model.

19.7 Geoneutrino measurements

Geothermal energy is derived from several sources. The most important of these are believed to be:

- Residual primordial heat from the time the Earth was formed;
- Tidal friction in the liquid outer core;
- Latent heat of fusion associated with the liquid–solid interface between the inner and outer cores;
- Radioactive decay.

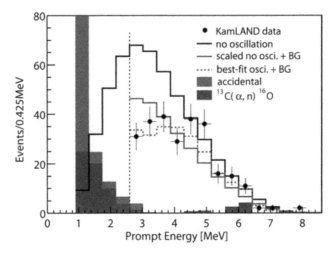

Figure 19.13. Neutrino energy spectrum for reactor electron antineutrinos, as discussed in the text, from the KamLAND detector. Experimental data are shown by the data points, the expected spectrum without neutrino oscillations is shown by the solid line and the expected spectrum with neutrino oscillations is shown by the broken line. From Suzuki *et al* (2005). Copyright IOP Publishing. Reproduced with permission. All rights reserved.

The presence of radioactive nuclides in the Earth's core can only be observed experimentally by the detection of neutrinos that result from β-decay processes. These are electron antineutrinos, as shown in equation (19.15). Scintillation detectors are the most appropriate means of detecting these antineutrinos as described in section 19.3, above. Of the radioactive nuclides in the Earth's core, it is only the antineutrinos from the ^{232}Th and ^{238}U decay chains that meet the threshold energy for inverse β-decay, as shown in equation (19.16).

The Borexino detector provided the first direct evidence of geoneutrinos in 2010. This represented the first real confirmation that the Earth's interior contained radioactive nuclides. Figure 19.14 shows some results from Borexino. In addition to showing the presence of radionuclides in the Earth's interior, these results also allow for the identification of the radioisotopes that are present, along with a measure of their relative concentrations.

19.8 Neutrino oscillations and masses

The neutrino experiments, as described above, as well as numerous other results, all provide overwhelming evidence supporting the concept of neutrino flavor oscillations.

The most significant contributions to this field have been provided by SNO and the Kamiokande experiments. Arthur B McDonald (b. 1943) from SNO and Takaaki Kajita (b. 1959) from the Kamiokande collaboration won the Nobel Prize in Physics in 2015 for their work on understanding the properties of neutrinos.

The presence of flavor oscillations implies that neutrinos are not massless particles, as shown by equation (19.8). The exact details of the mechanism by which

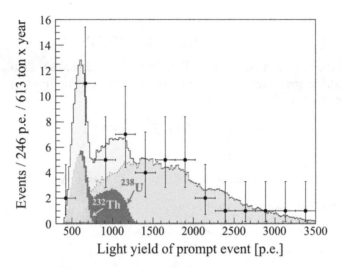

Figure 19.14. Electron antineutrino spectrum obtained by the Borexino detector shown by the data points in the figure. The reactor contribution is shown by the orange area and the blue and cyan areas show the contributions from geoneutrinos from ^{238}U and ^{232}Th, respectively. Calibration of the horizontal axis is approximately 500 p.e. = 1 MeV. CC BY 3.0 Reprinted from Bellini *et al* (2013). Copyright (2013). With permission from Elsevier.

neutrinos acquire mass are not known at present. This is in contrast to the situation for the weak bosons and the charged leptons, where the Higgs mechanism endows mass.

The simple two state approach as illustrated in equations (19.5) to (19.7), can be extended to three neutrino states. In this case, the relationship of flavor (ν_e, ν_μ, ν_τ) and mass (ν_1, ν_2, ν_3) states may be expressed in terms of the 3×3 mixing matrix, V, as

$$\begin{pmatrix} \nu_e \\ \nu_\mu \\ \nu_\tau \end{pmatrix} = V \begin{pmatrix} \nu_1 \\ \nu_2 \\ \nu_3 \end{pmatrix} = \begin{pmatrix} c_{13}c_{12} & c_{13}c_{12} & s_{13}e^{-i\delta} \\ -c_{23}s_{12} - s_{13}s_{23}c_{12}e^{i\delta} & c_{23}c_{12} - s_{13}s_{23}s_{12}e^{i\delta} & c_{13}s_{23} \\ s_{23}s_{12} - s_{13}c_{23}c_{12}e^{i\delta} & -s_{23}c_{12} - s_{13}c_{23}s_{12}e^{i\delta} & c_{13}s_{23} \end{pmatrix} \begin{pmatrix} \nu_1 \\ \nu_2 \\ \nu_3 \end{pmatrix} \quad (19.18)$$

where

$$c_{ij} = \cos \theta_{ij}$$

and

$$s_{ij} = \sin \theta_{ij}.$$

Here θ_{ij} is the mixing angle between states i and j and δ is a phase factor (still not yet determined). Following the development in section 19.2, the overall results from experimental observations to date provide mass-squared differences

$$\Delta m_{12}^2 = 7.59 \times 10^{-5}(\text{eV}c^{-2})^2$$

and

$$\Delta m_{23}^2 = 2.43 \times 10^{-3}(eVc^{-2})^2$$

Although the above values do not allow us to determine absolute mass values for the three neutrino mass states, they do allow us to put certain limits on the absolute values. We can write

$$m_2 = [m_1^2 + \Delta m_{12}^2]^{1/2} > [\Delta m_{12}^2]^{1/2}$$

and

$$m_3 = [m_1^2 + \Delta m_{12}^2 + \Delta m_{23}^2]^{1/2} > [\Delta m_{12}^2 + \Delta m_{23}^2]^{1/2}$$

From the numerical values given above it is found that

$$m_2 > 8.7 \times 10^{-3} eVc^{-2}$$

and

$$m_3 > 5.0 \times 10^{-2} eVc^{-2}$$

where the sum of the three neutrino masses is

$$\sum m_i > 0.059 \; eVc^{-2}. \tag{19.19}$$

This result will be discussed further in the next section.

The experimental data also allows for a determination of the mixing angles as

$$\sin^2(2\,\theta_{12}) = 0.846$$

$$\sin^2(2\,\theta_{23}) = 0.92$$

and

$$\sin^2(2\,\theta_{13}) = 0.093.$$

These results, in the context of the mixing matrix in equation (19.18), provide a relationship between the three flavor states and the three mass states, as shown in figure 19.15.

19.9 Other approaches to measuring neutrino masses

19.9.1 Cosmological observations

The neutrino masses as determined in the previous section are very small compared to the masses of other subatomic particles. As such, they would have very little influence on the processes that have been discussed previously in the present text. These processes would include, for example, β-decay. However, even a very small neutrino mass could have profound implications for the behavior of the Universe as a whole.

The ultimate future of the Universe depends on the relative contributions of kinetic and gravitational potential energy to the total energy. If the kinetic energy of the Universe is greater than the magnitude of its potential energy, then the theory of

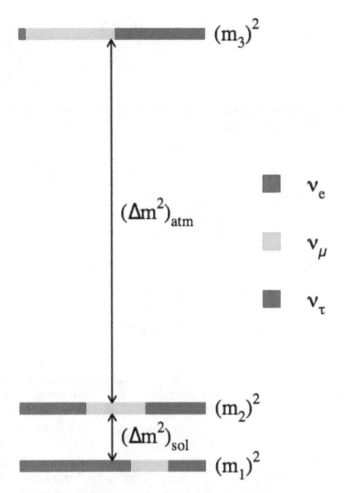

Figure 19.15. The relationship of neutrino flavor and mass states as determined from experimental measurements. The diagram assumes a normal mass hierarchy (i.e. $m(\nu_e) < m(\nu_\mu) < m(\nu_\tau)$). This [White paper: measuring the neutrino mass hierarchy image] has been obtained by the author from the Wikimedia website https://en.wikipedia.org/wiki/File:Hierfig.pdf (Cahn *et al* 2013) where it is stated to have been released into the public domain. It is included within this chapter on that basis.

general relativity indicates that space has a negative curvature, and the Universe will continue to expand indefinitely. If the magnitude of the gravitational potential energy is greater than the kinetic energy, then space is positively curved, and the Universe will ultimately collapse. If the magnitudes of the kinetic and gravitational potential energies are equal, then space is flat, and the Universe is said to be critical. We can view this problem quantitatively as described below.

The differential contribution to the potential energy, dU, from a mass element, dm, on the surface of a sphere of radius, r, is

$$dU = -\frac{GMdm}{r} = -\frac{4\pi Gr^3\rho}{3r}dm \qquad (19.20)$$

where G is the gravitational constant and ρ is the average density inside the sphere. The differential contribution to the kinetic energy, dT, from this mass element is

$$dT = \frac{v^2}{2} dm \tag{19.21}$$

where v is the velocity of the mass element. The velocity of matter in the Universe is governed by Hubble's law, which states

$$v = H_0 r \tag{19.22}$$

where H_0 is the Hubble constant. Combining equations (19.20) to (19.22) gives the condition for a critical Universe as

$$\frac{4\pi G r^3 \rho}{3r} dm = \frac{H_0^2 r^2}{2} dm.$$

This equality gives the critical density of the Universe as

$$\rho_{\text{crit}} = \frac{3H_0^2}{8\pi G}.$$

The Hubble constant has been measured by astronomical measurements to be 73.00 ± 1.75 km·s^{-1} per MPc, where 1 megaparsec (MPc) is 3.086×10^{22} m. It is interesting to note that the units of the Hubble constant are inverse time, so it may be written in SI units as 2.37×10^{-18} s^{-1}. The inverse of the Hubble constant gives the age of the Universe (since the big bang) as 4.23×10^{17} s = 13.4 billion years. This value of H_0 gives a critical density of $\rho_{\text{crit}} = 1.00 \times 10^{-26}$ kg·m^3. Recent studies of the large-scale structure in the Universe by the Wilkinson Microwave Anisotropy Probe (WMAP) satellite, have shown that space is flat within 15%. This means that the actual average density of the Universe does not vary substantially from the critical value and that the ultimate future of the Universe is not definitively known.

In order to understand the actual average density of the Universe, it is necessary to consider all contributions to its mass. These contributions include:
- photons;
- baryonic matter;
- dark matter;
- dark energy.

These four contributions are discussed below.

The equilibrium number density of photons at a temperature T_γ, is given by

$$n_\gamma = S \frac{g_\gamma \zeta(3)}{\pi^2} (k_B T_\gamma)^3$$

where k_B is the Boltzmann constant and $\zeta(3) = 1.202\,06$ is the third order Riemann zeta function. S is a statistical factor, $S = 1$ for bosons, and g_γ is the number of degrees of freedom, $g_\gamma = 2$ for photons. Using the temperature $T_\gamma = 2.275$ K corresponding to the cosmic radiation background gives $n_\gamma = 4.08 \times 10^8$ m^{-3}. Using the mass equivalent of the energy per photon and the number density as given, means that the contribution of photons to the density of the Universe is 5×10^{-31} kg·m^{-3}, an insignificant contribution.

According to the details of big bang theory, the relative number density of baryons to photons will be 6×10^{-10}. This gives the contribution of baryons to the density of the Universe as 4.1×10^{-28} kg·m^{-3}, only about 4% of the critical density.

Therefore, the question of whether the Universe will continue to expand or whether it will collapse depends on the contributions to the density of dark matter and dark energy. The exact nature of dark matter is not known; however, it is clear from an analysis of galactic dynamics that it does, in fact, contribute to the mass of the Universe in a significant way. As well, the nature and contribution of dark energy is not known. Since it has now been confirmed that neutrinos carry mass, they are obviously a component of dark matter that will contribute to the overall density of the Universe. In the absence of concrete information about other possible components of dark matter or dark energy, we can consider the possibility that massive neutrinos could play a role in the future of the Universe.

The equilibrium density of neutrinos in the Universe follows from the discussion above concerning photons. Thus, we can write

$$n_\nu = S \frac{g_\nu \zeta(3)}{\pi^2} (k_B T_\nu)^3 \tag{19.23}$$

where $S = 3/4$ for fermions and $g_\nu = 6$. The extra factor of 3 in the degrees of freedom comes from the three flavors of neutrinos. The equilibrium neutrino temperature can be evaluated through an analysis of the very early history of the Universe. When the big bang occurred, the very high temperature meant that there was electroweak symmetry, as discussed in detail in chapter 17. In this case, electron–positron annihilation could proceed via a neutral weak boson to produce neutrinos of any flavor,

$$e^+ + e^- \leftrightarrow \nu_x + \bar{\nu}_x$$

where the reaction can proceed in both directions. As the Universe cooled, spontaneous symmetry breaking occurred and electron–positron had to occur though a mediating photon to γ-rays,

$$e^+ + e^- \leftrightarrow \gamma + \gamma$$

Calculations show that the spontaneous breaking of electroweak symmetry occurred a few seconds after the Big Bang, after which electron–positron annihilations could not increase the number of neutrinos. Based on the conservation of entropy it can be shown that the relative equilibrium temperatures of neutrinos and photons are

$$\frac{T_\nu}{T_\gamma} = \left[\frac{4}{11} \right]^{1/3}.$$

Based on the measured cosmic radiation background value of 2.725 K, the equilibrium temperature of neutrinos is found to be 1.945 K. From equation (19.23), the equilibrium density of neutrinos of all flavors is 3.36×10^8 m^{-3}.

Assuming that massive neutrinos, as a component of dark matter, make up the majority of the mass in the Universe that would be necessary to achieve critical density, then the sum of the masses of the three neutrino generations can be calculated. This analysis gives

$$\sum m_i \approx 10 \text{ eV}c^{-2}.$$ (19.24)

This estimated neutrino mass is substantially greater than the value suggested by equation (19.19) and may suggest that other components of dark matter along with dark energy represents a more significant contribution to the total mass of the Universe than the mass of neutrinos.

While a direct approach to relating neutrino masses to the structure of the Universe provides minimal information, it is possible to gather more detailed information from somewhat different cosmological data. While the above discussion dealt with the average density of the Universe, it is obvious that there are local fluctuations in density. These fluctuations occur on a wide variety of scales, from denser regions representing planets and stars, to galaxies and galaxy clusters. The so-called matter power spectrum, as shown in figure 19.16, describes fluctuations in the density of the Universe as a function of wavenumber. If neutrinos are massless, then they will not be affected by the gravitational interaction and will be uniformly distributed in space. If, on the other hand, neutrinos carry mass, then they will be

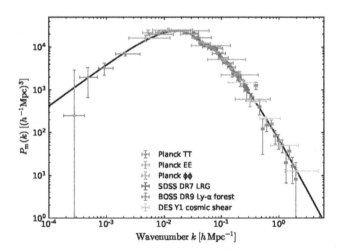

Figure 19.16. Matter power spectrum obtained from various cosmological probes. Note h is the reduced Hubble constant where $H_0 = (100 \text{ km·s}^{-1}\text{·Mpc}^{-1}) \times h$. This [linear matter power spectrum] image has been obtained by the author from the Wikimedia website https://en.wikipedia.org/wiki/File:Planck_2018_Linear_Matter_Power_Spectrum. pdf (ESA and the Planck Collaboration 2018), where it is stated to have been released into the public domain. It is included within this chapter on that basis.

influenced by gravitational interactions with other matter in the Universe and will contribute to the power spectrum as a result of their tendency to cluster in regions of increased density. Specifically, massive neutrinos will lower the power spectrum in the large wavenumber region of the spectrum. Analyses based on recent galaxy surveys allow for an upper limit on combined neutrino masses of

$$\sum m_i < 0.32 \mathrm{eV} c^{-2}. \tag{19.25}$$

This is, certainly, more consistent with results obtained from neutrino observatories than the average Universe density approach in equation (19.24).

19.9.2 Beta decay spectra

The most direct method of determining the mass of the neutrino is through β^- decay experiments. In this case, it is only the mass of the electron antineutrino that is obtained. Certain details of the energy spectrum of β^- decay electrons are affected by the presence or absence of neutrino mass. In the case of a massive neutrino, their rest mass must be considered in the determination of the electron energy spectrum. The main difference in the β^- decay electron energy spectrum for massless and massive neutrinos occurs near the end point energy and the observation of deviations from the massless neutrino curve, is evidence for neutrino mass. Specifically, when the neutrino is massless, electron energy spectrum would have zero slope where it intercepts the energy axis and when the neutrino carries mass, the spectrum would intercept the axis with infinite slope. Typical results can provide an estimate of the upper limit of neutrino mass.

In order to choose the most appropriate β^- decay to investigate these effects, we must consider several criteria. When all factors are taken into account, the β^- decay of ^{3}H,

$$^3\mathrm{H} \rightarrow {}^3\mathrm{He} + \mathrm{e}^- + \bar{\nu}_e$$

is the most suitable candidate for the study of neutrino masses. This is largely a result of the very small end point energy, 18.6 keV, and spectral features near this energy are illustrated in figure 19.17.

- It is a ground state to ground state transition.
- The nuclear eigenstates for the parent and daughter nuclei are well known.
- The end point energy is small.

Beginning around 1990, a number of experiments have studied the electron energy spectrum from the β^- decay of ^{3}H to look for spectral features that would be indicative of massive neutrinos. Figure 19.18 shows a summary of major experiments along these lines and the limits they have placed on $m(\bar{\nu}_e)^2$. The most recent experiments shown in the figure are from the Karlsruhe Tritium Neutrino (KATRIN) experiment at the Karlsruhe Institut für Technologie, Karlsruhe, Baden-Württemberg, Germany. Figure 19.19 shows a diagram of this experiment. Molecular tritium is used as a source of β^- decay electrons according to the decay process

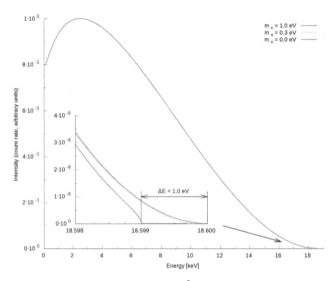

Figure 19.17. Electron energy spectrum of the β^- decay of ^{3}H for massless neutrinos (blue line) and neutrinos with masses of 0.3 eV (green line) and 1.0 eV (red line). This [energy spectrum of tritium beta decay used by the KATRIN neutrino experiment] image has been obtained by the author from the Wikimedia website where it was made available by Zykure (2012) under a CC BY-SA 3.0 licence. It is included within this chapter on that basis. It is attributed to Zykure.

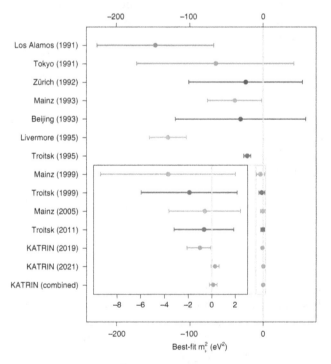

Figure 19.18. Results for $m(\bar{\nu}_e)^2$ from major ^{3}H β^- decay experiments. The original sources of data for the points shown on the diagram are given in KATRIN Collaboration (2022) CC BY 4.0.

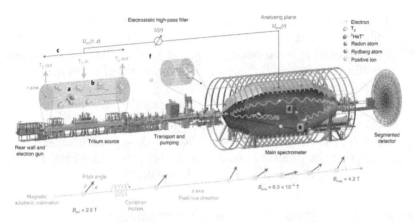

Figure 19.19. Diagram of the KATRIN ^{3}H β^- decay neutrino mass experiment. The total length of the beamline is 70 m. Reprinted from KATRIN Collaboration (2022) CC BY 4.0.

$$^3\text{H}_2 \rightarrow {}^3\text{He}^3\text{H}^+ + e^- + \bar{\nu}_e.$$

The combined results from KATRIN experiments in 2019 and 2020, give an upper limit to the electron antineutrino mass of $m(\bar{\nu}_e)^2 < 0.8$ eV c^{-2}. This value may be compared with the lower limit placed on the combined neutrino masses from neutrino observatories, as given in equation (19.19), and from cosmological observations, as shown in equation (19.25).

19.10 Summary

Recent experiments, particularly those conducted at Super-Kamiokande and the SNO, have provided definitive evidence for neutrino flavor oscillations. These oscillations clearly show the relevance of the MSW effect for cosmic ray and reactor neutrinos travelling through the Earth and solar neutrinos created in the Sun's interior. Consistent results have been obtained for mass-squared differences, although further work to define some mixing angles is needed. Although Super-Kamiokande and SNO results cannot provide absolute mass values for individual neutrino flavors, they can put some limits on the sum of masses. Cosmological studies, as well as direct measurements of ^{3}H β^- decay electron spectra can provide additional information concerning neutrino masses. Based on a combination of all recent measurements that provide limits to neutrino masses, probable masses of individual neutrino flavors likely fall within the range of 0.10 eV c^{-2} to 0.26 eV c^{-2}.

Problems

19.1. Solar neutrinos that are incident on the Earth have an average cross section of 10^{-20} b for interaction with a nucleus. Calculate the probability that a neutrino incident on the Earth along a line passing through the center of the Earth will interact with a nucleus.

19.2. Calculate the threshold energy for electron neutrino absorption by the following nuclei: (a) ^{55}Mn, (b) ^{40}Ca, (c) ^{125}Te, (d) ^{146}Nd and (e) ^{136}Xe.

19.3. Because of its low end point energy, ^{3}H is by far the most common nuclide used for experiments that attempt to observe the effect of neutrino mass on the β^- decay spectrum. However, the ground state to ground state β^- decay of ^{187}Re has a significantly lower end point energy.

1. (a) Calculate the end point energy for ^{187}Re.
2. (b) Why is ^{3}H preferred over ^{187}Re for these experiments.

19.4. Show that the process in equation (19.14) produces γ-rays with sufficient energy to be observed in a Cherenkov detector.

References and suggestions for further reading

Abe S *et al* 2008 Precision measurement of neutrino oscillation parameters with KamLAND *Phys. Rev. Lett.* **100** 221803

Bahcall J N 1996 Solar neutrinos: Where we are, where we are going *Ap. J.* **467** 475–84

Barger V, Marfita D and Whisnant K L 2012 *The Physics of Neutrinos* (Princeton: Princeton University Press)

Belinky S 2010 *Introduction to the Physics of Massive and Mixed Neutrinos* (New York: Springer) (Lecture Notes in Physics vol 817)

Bellini G *et al* 2013 Measurement of geo-neutrinos from 1353 days of Borexino *Phys. Lett.* B **722** 295–300

Borexino Collaboration 2015 Borexino detector in LNGS in September 2015 https://commons. wikimedia.org/wiki/File:Laboratori_Nazionali_del_Gran_Sasso,_Borexino_detector.png

Cahn R N *et al* 2013 White paper: Measuring the neutrino mass hierarchy https://en.wikipedia. org/wiki/File:Hierfig.pdf

Close F 2010 *Neutrino* (Oxford: Oxford)

Dunlap R A 2018 *Particle Physics* (San Rafael, CA: Morgan & Claypool)

ESA and the Planck Collaboration 2018 https://en.wikipedia.org/wiki/File:Planck_2018_ Linear_Matter_Power_Spectrum.pdf

Fukuda Y 2003 Status of solar neutrino observation at Super-Kamiokande *Nucl. Instrum. Meth.* A **503** 114–7

Gando A *et al* 2015 ^{7}Be solar neutrino measurement with KamLAND *Phys. Rev.* C **92** 055808

Gonzalvez-Garcia M C and Nir Y 2002 Neutrino masses and mixing: evidence and implications *Rev. Mod. Phys.* **75** 345–402

Hahn R L 2008 Radiochemical solar neutrino experiments—successful and otherwise *J. Phys. Conf. Ser.* **136** 022003

Hetherington N S 2014 *Encyclopedia of Cosmology (Routledge Revivals): Historical, Philosophical, and Scientific Foundations of Modern Cosmology* (Milton Park, Oxfordshire: Routledge)

Jelly N, McDonald A B and Robertson R G H 2009 The Sudbury Neutrino Observatory *Ann. Rev. Nucl. Part. Phys.* **59** 431–65

Kajita T *et al* 2018 Establishing atmospheric neutrino oscillations with Super-Kamiokande *Nucl. Phys.* B **908** 14–29

KATRIN Collaboration 2022 Direct neutrino-mass measurement with sub-electronvolt sensitivity *Nat. Phys.* **18** 160–6

Lyth D H 1993 Introduction to Cosmology: Lectures given at the Summer School in High Energy Physics and Cosmology *ICTP (Trieste)* **1993** http:/arxiv.org/pdf/astro-ph/9312022v1

Nishijima K 2001 The Super-Kamiokande experiment *Radiat. Phys. Chem.* **61** 247–53

SNO 2018 Outside view of the Sudbury Neutrino Observatory detector https://en.wikipedia.org/wiki/File:Sudbury_Neutrino_Observatory.detector_outside.jpg

Szdori 2006 The result of 30 years of research https://commons.wikimedia.org/wiki/File:K%C3%A9p1.png

Suzuki A *et al* 2005 Results from KamLAND Reactor Neutrino Detection *Phys. Scr.* **T121** 33–8

U.S. Department of Energy 2014 The first chemical detection of neutrinos produced by the Sun have been recorded at the Atomic Energy Commission's Brookhaven National Laboratory solar neutrino detector. c. 1972 www.flickr.com/photos/departmentofenergy/11967548426/

Zykure 2012 Energy spectrum of tritium beta decay used by the KATRIN neutrino experiment https://commons.wikimedia.org/wiki/File:KATRIN_Spectrum.svg

IOP Publishing

An Introduction to the Physics of Nuclei and Particles
(Second Edition)

Richard A Dunlap

Appendix A

Physical constants and conversion factors

Quantity	Symbol	Value	Units
Alpha particle binding energy	B_α	28.296	MeV
Alpha particle mass	m_α	4.001 506 18	u
		3727.409	MeV c^{-2}
Atomic mass unit	u	$1.660\ 5402 \times 10^{-27}$	kg
		931.494	MeV c^{-2}
Avogadro's number	N_A	$6.022\ 1367 \times 10^{23}$	mole^{-1}
Barn	b	10^{-28}	m^2
		10^2	fm^2
Bohr magneton	μ_B	$9.274\ 0154 \times 10^{-24}$	J·T^{-1}
		$5.788\ 3826 \times 10^{-5}$	eV·T^{-1}
Bohr radius	a_0	$5.291\ 772\ 49 \times 10^{-11}$	m
Boltzmann's constant	k_B	$1.380\ 658 \times 10^{-23}$	J·K^{-1}
		$8.617\ 38 \times 10^{-5}$	eV·K^{-1}
Coulomb constant	$1/4\pi\varepsilon_0$	$8.987\ 551 \times 10^{9}$	N·m^2·C^2
	$e^2/4\pi\varepsilon_0$	1.439 976	MeV·fm
Curie	Ci	3.7×10^{10}	decays·s^{-1}
Deuteron binding energy	B_d	2.224	MeV

(Continued)

doi:10.1088/978-0-7503-6094-4ch20

(Continued)

Quantity	Symbol	Value	Units
Deuteron mass	m_d	2.013 553 214	u
		1875.628	MeV c^{-2}
Electron mass	m_e	5.485 799 03 × 10^{-4}	u
		0.510 9988	MeV c^{-2}
Electron volt	eV	1.602 177 33 × 10^{-19}	J
Electronic charge	e	1.602 177 33 × 10^{-19}	C
Fine structure constant	α	0.007 297 353 08	
Gas constant	R	8.314 510	J·K^{-1}·mol^{-1}
Gravitational constant	G	6.672 59 × 10^{-11}	N·m^2·kg^{-2}
Neutron magnetic moment	μ_n	−1.913	μ_N
Neutron mass	m_n	1.008 664 904	u
		939.565 31	MeV c^{-2}
Nuclear magneton	μ_N	5.050 7866 × 10^{-27}	J·T^{-1}
		3.152 4517 × 10^{-8}	eV·T^{-1}
Permeability of free space	μ_0	4π × 10^{-7}	T·m·A^{-1}
Permittivity of free space	ε_0	8.854 187 817 × 10^{-12}	C^2·N^{-1}·m^{-2}
Planck's constant	h	6.626 0755 × 10^{-34}	J·s
	$\hbar$	4.135 70 × 10^{-15}	eV·s
	hc	1.054 572 66 × 10^{-34}	J·s
	$\hbar c$	6.582 17 × 10^{-16}	eV·s
		1.240 × 10^3	MeV·fm
		1.973 × 10^2	MeV·fm
Proton magnetic moment	μ_p	2.793	μ_N
Proton mass	m_p	1.007 276 470	u
		938.2723	MeV c^{-2}
Rydberg constant	R_∞	1.097 373 1534 × 10^7	m^{-1}
Speed of light	c	2.997 924 58 × 10^8	m·s^{-1}
Stefan–Boltzmann constant	σ	5.670 51 × 10^{-8}	W·m^{-2}·K^{-4}

A-2

IOP Publishing

An Introduction to the Physics of Nuclei and Particles
(Second Edition)

Richard A Dunlap

Appendix B

Properties of nuclides

The following table gives ground state properties of selected nuclides. Spin and parity states that are not known with certainty are indicated in parentheses. Natural abundances are shown for stable and naturally occurring nuclides. Masses are atomic masses of neutral atoms.

Z		A	Mass (u)	I^π	Abundance or half-life
1	H	1	1.007 825 032	$\frac{1}{2}^+$	99.985%
		2	2.014 101 778	1^+	0.015%
		3	3.016 049 268	$\frac{1}{2}^+$	12.33 y
2	He	3	3.016 029 31	$\frac{1}{2}^+$	0.000 14%
		4	4.002 603 25	0^+	99.999 86%
		5	5.012 223 628	$\frac{3}{2}^+$	
		6	6.018 888 072	0^+	806.7 ms
3	Li	6	6.015 122 281	1^+	7.5%
		7	7.016 004 049	$\frac{3}{2}^-$	92.5%
		8	8.022 486 67	2^+	838 ms
		9	9.026 789 122	$\frac{3}{2}^-$	178.3 ms
4	Be	6	6.019 725 804	0^+	
		7	7.016 929 246	$\frac{3}{2}^-$	53.29 d
		8	8.005 305 94	0^+	0.07 fs
		9	9.012 182 135	$\frac{3}{2}^-$	100%

(*Continued*)

doi:10.1088/978-0-7503-6094-4ch21

(*Continued*)

Z		A	Mass (*u*)	I^{π}	Abundance or half-life
		10	10.013 533 72	0^+	1.51 My
		11	11.021 657 65	$\frac{1}{2}^+$	13.81 s
		12	12.026 920 63	0^+	23.6 ms
5	B	8	8.024 606 713	2^+	770 ms
		9	9.013 329	$\frac{3}{2}^-$	0.85 as
		10	10.012 937 03	3^+	19.9%
		11	11.009 305 47	$\frac{3}{2}^-$	80.1%
		12	12.014 352 11	1^+	20.20 ms
		13	13.017 780 27	$\frac{3}{2}^-$	17.36 ms
		14	14.025 404 06	2^-	13.8 ms
6	C	9	9.031 040 087	$(\frac{3}{2}^-)$	126.5 ms
		10	10.016 853 11	0^+	19.255 s
		11	11.011 433 82	$\frac{3}{2}^-$	20.39 m
		12	12.000 000 00	0^+	98.9%
		13	13.003 354 84	$\frac{1}{2}^-$	1.1%
		14	14.003 241 99	0^+	5730 y
		15	15.010 599 26	$\frac{1}{2}^+$	2.449 s
		16	16.014 701 24	0^+	0.747 s
7	N	12	12.018 6132	1^+	11.000 ms
		13	13.005 738 58	$\frac{1}{2}^-$	9.965 m
		14	14.003 074 01	1^+	99.63%
		15	15.000 1089	$\frac{1}{2}^-$	0.37%
		16	16.006 101 42	2^-	7.13 s
		17	17.008 449 67	$\frac{1}{2}^-$	4.173 s
		18	18.014 081 83	1^-	624 ms
8	O	13	13.024 8104	$(\frac{3}{2}^-)$	8.58 ms
		14	14.008 595 29	0^+	70.606 s
		15	15.003 065 39	$\frac{1}{2}^-$	122.24 s
		16	15.994 914 62	0^+	99.76%
		17	16.999 1315	$\frac{5}{2}^+$	0.038%
		18	17.999 160 42	0^+	0.2%
		19	19.003 578 73	$\frac{5}{2}^+$	26.91 s
		20	20.004 076 15	0^+	13.51 s
		21	21.008 654 63	$(\frac{1}{2}, \frac{3}{2}, \frac{5}{2})^+$	3.42 s
		22	22.009 967 16	0^+	2.25 s
9	F	17	17.002 095 24	$\frac{5}{2}^+$	64.49 s
		18	18.000 937 67	1^+	109.77 m

(*Continued*)

(*Continued*)

Z		A	Mass (*u*)	I^π	Abundance or half-life
		19	18.998 403 21	$\frac{1}{2}^+$	100%
		20	19.999 981 32	2^+	11.00 s
		21	20.999 948 92	$\frac{5}{2}^+$	4.158 s
		22	22.002 999 25	$4^+,(3^+)$	4.23 s
		23	23.003 574 39	$(\frac{3}{2},\frac{5}{2})^+$	2.23 s
		24	24.008 099 37	$(1,2,3)^+$	0.34 s
10	Ne	17	17.017 697 57	$\frac{1}{2}^-$	109.2 ms
		18	18.005 697 07	0^+	1672 ms
		19	19.001 879 84	$\frac{1}{2}^+$	17.34 s
		20	19.992 440 18	0^+	90.48%
		21	20.993 846 74	$\frac{3}{2}^+$	0.27%
		22	21.991 385 51	0^+	9.25%
		23	22.994 467 34	$\frac{5}{2}^+$	37.24 s
		24	23.993 615 07	0^+	3.38 m
		25	24.997 7899	$(\frac{1}{2},\frac{3}{2})^+$	602 ms
		26	26.000 4615	0^+	0.23 s
11	Na	20	20.007 348 26	2^+	447.9 ms
		21	20.997 6551	$\frac{3}{2}^+$	22.49 s
		22	21.994 436 78	3^+	2.6019 y
		23	22.989 769 68	$\frac{3}{2}^+$	100%
		24	23.990 963 33	4^+	14.9590 h
		25	24.989 954 35	$\frac{5}{2}^+$	59.1 s
		26	25.992 5899	3^+	1.072 s
		27	26.994 0087	$\frac{5}{2}^+$	301 ms
		28	27.998 890 41	1^+	30.5 ms
12	Mg	20	20.018 862 74	0^+	95 ms
		21	21.011 714 17	$(\frac{3}{2},\frac{5}{2})^+$	122 ms
		22	21.999 574 06	0^+	3.857 s
		23	22.994 124 85	$\frac{3}{2}^+$	11.317 s
		24	23.985 0419	0^+	78.89%
		25	24.985 837 02	$\frac{5}{2}^+$	10%
		26	25.982 593 04	0^+	11.01%
		27	26.984 340 74	$\frac{1}{2}^+$	9.458 m
		28	27.983 8767	0^+	20.91 h
		29	28.988 554 74	$\frac{3}{2}^+$	1.30 s
		30	29.990 464 53	0^+	335 ms
13	Al	24	23.999 940 91	4^+	2.053 s
		25	24.990 428 56	$\frac{5}{2}^+$	7.183 s

(*Continued*)

(*Continued*)

Z		A	Mass (u)	I^π	Abundance or half-life
		26	25.986 891 66	5^+	0.74 My
		27	26.981 538 44	$\frac{5}{2}^+$	100%
		28	27.981 910 18	3^+	2.2414 m
		29	28.980 444 85	$\frac{5}{2}^+$	6.56 m
		30	29.982 9603	3^+	3.60 s
		31	30.983 946 02	$(\frac{3}{2}, \frac{5}{2})^+$	644 ms
		32	31.988 124 38	1^+	33 ms
14	Si	24	24.011 545 71	0^+	102 ms
		25	25.004 106 64	$\frac{5}{2}^+$	220 ms
		26	25.992 329 94	0^+	2.234 s
		27	26.986 704 76	$\frac{5}{2}^+$	4.16 s
		28	27.976 926 53	0^+	92.23%
		29	28.976 494 72	$\frac{1}{2}^+$	4.67%
		30	29.973 770 22	0^+	3.1%
		31	30.975 363 28	$\frac{3}{2}^+$	157.3 m
		32	31.974 148 13	0^+	172 y
		33	32.978 000 52	$(\frac{3}{2}^+)$	6.18 s
		34	33.978 575 75	0^+	2.77 s
15	P	28	27.992 312 33	3^+	270.3 ms
		29	28.981 801 38	$\frac{1}{2}^+$	4.142 s
		30	29.978 313 81	1^+	2.498 m
		31	30.973 761 51	$\frac{1}{2}^+$	100%
		32	31.973 907 16	1^+	14.262 d
		33	32.971 725 28	$\frac{1}{2}^+$	25.34 d
		34	33.973 636 38	1^+	12.43 s
		35	34.973 314 25	$\frac{1}{2}^+$	47.3 s
16	S	30	29.984 902 95	0^+	1.178 s
		31	30.979 554 42	$\frac{1}{2}^+$	2.572 s
		32	31.972 070 69	0^+	95.02%
		33	32.971 4585	$\frac{3}{2}^+$	0.75%
		34	33.967 866 83	0^+	4.21%
		35	34.969 032 14	$\frac{3}{2}^+$	87.51 d
		36	35.967 080 88	0^+	0.02%
		37	36.971 125 72	$\frac{7}{2}^-$	5.05 m
		38	37.971 163 44	0^+	170.3 m
		39	38.975 135 28	$(\frac{3}{2}, \frac{5}{2}, \frac{7}{2})^-$	11.5 s
		40	39.975 47	0^+	8.8 s
17	Cl	32	31.985 688 91	1^+	298 ms

(*Continued*)

(*Continued*)

Z		A	Mass (*u*)	I^π	Abundance or half-life
		33	32.977 4518	$\frac{3}{2}^+$	2.511 s
		34	33.973 761 97	0^+	1.5264 s
		35	34.968 852 71	$\frac{3}{2}^+$	75.77%
		36	35.968 306 95	2^+	0.301 My
		37	36.965 9026	$\frac{3}{2}^+$	24.23%
		38	37.968 010 55	2^-	37.24 m
		39	38.968 007 68	$\frac{3}{2}^+$	55.6 m
		40	39.970 415 56	2^-	1.35 m
		41	40.970 650 21	$(\frac{1}{2},\frac{3}{2})^+$	38.4 s
18	Ar	32	31.997 660 66	0^+	98 ms
		33	32.989 928 72	$\frac{1}{2}^+$	173.0 ms
		34	33.980 270 12	0^+	844.5 ms
		35	34.975 256 73	$\frac{3}{2}^+$	1.775 s
		36	35.967 546 28	0^+	0.337%
		37	36.966 775 91	$\frac{3}{2}^+$	35.04 d
		38	37.962 732 16	0^+	0.063%
		39	38.964 313 41	$\frac{7}{2}^-$	269 y
		40	39.962 383 12	0^+	99.6%
		41	40.964 500 83	$\frac{7}{2}^-$	109.34 m
		42	41.963 046 39	0^+	32.9 y
		43	42.965 6707	$(\frac{3}{2},\frac{5}{2})$	5.37 m
		44	43.965 365 27	0^+	11.87 m
		46	45.968 093 47	0^+	8.4 s
19	K	35	34.988 011 62	$\frac{3}{2}^+$	190 ms
		36	35.981 293 41	2^+	342 ms
		37	36.973 376 92	$\frac{3}{2}^+$	1.226 s
		38	37.969 080 11	3^+	7.636 m
		39	38.963 706 86	$\frac{3}{2}^+$	93.2581%
		40	39.963 998 67	4^-	0.0117%
		41	40.961 825 97	$\frac{3}{2}^+$	6.7302%
		42	41.962 403 06	2^-	12.360 h
		43	42.960 715 75	$\frac{3}{2}^+$	22.3 h
		44	43.961 556 15	2^-	22.13 m
		45	44.960 699 66	$\frac{3}{2}^+$	17.3 m
		46	45.961 9762	(2^-)	105 s
		47	46.961 677	$\frac{1}{2}^+$	17.5 s
		48	47.965 512 95	(2^-)	6.8 s
		51	50.976 38	$(\frac{1}{2}^+, \frac{3}{2}^+)$	365 ms

(*Continued*)

(*Continued*)

Z		A	Mass (*u*)	I^π	Abundance or half-life
20	Ca	36	35.993 087 23	0^+	100 ms
		37	36.985 871 51	$\frac{3}{2}+$	175 ms
		38	37.976 318 64	0^+	440 ms
		39	38.970 717 73	$\frac{3}{2}+$	859.6 ms
		40	39.962 591 16	0^+	96.941%
		41	40.962 278 35	$\frac{7}{2}-$	0.103 My
		42	41.958 618 34	0^+	0.647%
		43	42.958 766 83	$\frac{7}{2}-$	0.135%
		44	43.955 481 09	0^+	2.086%
		45	44.956 185 94	$\frac{7}{2}-$	163.8 d
		46	45.953 692 76	0^+	0.004%
		47	46.954 546 46	$\frac{7}{2}-$	4.536 d
		48	47.952 533 51	0^+	0.187%
		49	48.955 6733	$\frac{3}{2}-$	8.715 m
		50	49.957 518 29	0^+	13.9 s
21	Sc	40	39.977 964 01	4^-	182.3 ms
		41	40.969 251 32	$\frac{7}{2}-$	596.3 ms
		42	41.965 516 76	0^+	681.3 ms
		43	42.961 150 98	$\frac{7}{2}-$	3.891 h
		44	43.959 403 05	2^+	3.927 h
		45	44.955 910 24	$\frac{7}{2}-$	100%
		46	45.955 170 25	4^+	83.79 d
		47	46.952 408 03	$\frac{7}{2}-$	3.345 d
		48	47.952 234 99	6^+	43.67 h
		49	48.950 024 07	$\frac{7}{2}-$	57.2 m
		50	49.952 187 01	5^+	102.5 s
		51	50.953 6027	$(\frac{7}{2})^-$	12.4 s
		52	51.956 65	3^+	8.2 s
22	Ti	40	39.990 498 91	0^+	50 ms
		41	40.983 131	$\frac{3}{2}+$	80 ms
		42	41.973 031 62	0^+	199 ms
		43	42.968 523 34	$\frac{7}{2}-$	509 ms
		44	43.959 690 24	0^+	49 y
		45	44.958 124 35	$\frac{7}{2}-$	184.8 m
		46	45.952 629 49	0^+	8%
		47	46.951 763 79	$\frac{5}{2}-$	7.3%
		48	47.947 947 05	0^+	73.8%

(*Continued*)

(*Continued*)

Z		A	Mass (u)	I^π	Abundance or half-life
		49	48.947 870 79	$\frac{7}{2}-$	5.5%
		50	49.944 792 07	0^+	5.4%
		51	50.946 616 02	$\frac{3}{2}-$	5.76 m
		52	51.946 898 18	0^+	1.7 m
		53	52.949 731 71	$(\frac{3}{2})-$	32.7 s
23	V	45	44.965 782 29	$\frac{7}{2}-$	547 ms
		46	45.960 199 49	0^+	422.37 ms
		47	46.954 906 92	$\frac{3}{2}-$	32.6 m
		48	47.952 254 48	4^+	15.9735 d
		49	48.948 516 91	$\frac{7}{2}-$	330 d
		50	49.947 162 79	6^+	0.25%
		51	50.943 963 68	$\frac{7}{2}-$	99.75%
		52	51.944 779 66	3^+	3.743 m
		53	52.944 342 52	$\frac{7}{2}-$	1.61 m
		54	53.946 444 38	3^+	49.8 s
		55	54.947 238 19	$(\frac{7}{2}-)$	6.54 s
		55	54.947 238 19	$(\frac{7}{2}-)$	6.54 s
24	Cr	46	45.968 361 65	0^+	0.26 s
		47	46.962 906 51	$\frac{3}{2}-$	508 ms
		48	47.954 035 86	0^+	21.56 h
		49	48.951 341 14	$\frac{5}{2}-$	42.3 m
		50	49.946 049 61	0^+	4.345%
		51	50.944 771 77	$\frac{7}{2}-$	27.702 d
		52	51.940 511 90	0^+	83.790%
		53	52.940 653 78	$\frac{3}{2}-$	9.500%
		54	53.938 884 92	0^+	2.365%
		55	54.940 844 16	$\frac{3}{2}-$	3.497 m
		56	55.940 645 24	0^+	5.94 m
25	Mn	49	48.959 623 42	$\frac{5}{2}-$	384 ms
		50	49.954 243 96	0^+	283.07 ms
		51	50.948 215 49	$\frac{5}{2}-$	46.2 m
		52	51.945 570 08	6^+	5.591 d
		53	52.941 2947	$\frac{7}{2}-$	3.74 My
		54	53.940 363 25	3^+	312.3 d
		55	54.938 049 64	$\frac{5}{2}-$	100%
		56	55.938 909 37	3^+	2.5785 h
		57	56.938 287 46	$\frac{5}{2}-$	85.4 s
		58	57.939 986 45	3^+	65.3 s

(*Continued*)

(Continued)

Z		A	Mass (*u*)	I^{π}	Abundance or half-life
		59	58.940 447 17	$\frac{3-}{2},\frac{5-}{2}$	4.6 s
26	Fe	51	50.956 824 94	$(\frac{5-}{2})$	305 ms
		52	51.948 116 53	0^+	8.275 h
		53	52.945 312 28	$\frac{7-}{2}$	8.51 m
		54	53.939 614 84	0^+	5.9%
		55	54.938 298 03	$\frac{3-}{2}$	2.73 y
		56	55.934 942 13	0^+	91.72%
		57	56.935 398 71	$\frac{1-}{2}$	2.1%
		58	57.933 280 46	0^+	0.28%
		59	58.934 880 49	$\frac{3-}{2}$	44.503 d
		60	59.934 076 94	0^+	1.5 My
		61	60.936 749 46	$\frac{3-}{2},\frac{5-}{2}$	5.98 m
		62	61.936 7705	0^+	68 s
27	Co	53	52.954 224 99	$(\frac{7-}{2})$	240 ms
		54	53.948 464 15	0^+	193.23 ms
		55	54.942 003 15	$\frac{7-}{2}$	17.53 h
		56	55.939 843 94	4^+	77.27 d
		57	56.936 296 24	$\frac{7-}{2}$	271.79 d
		58	57.935 757 57	2^+	70.82 d
		59	58.933 200 19	$\frac{7-}{2}$	100%
		60	59.933 8222	5^+	5.2714 y
		61	60.932 479 38	$\frac{7-}{2}$	1.650 h
		62	61.934 054 21	2^+	1.50 m
		63	62.933 615 22	$(\frac{7-}{2})$	27.4 s
		64	63.935 813 52	1^+	0.30 s
		65	64.936 484 58	$(\frac{7-}{2})$	1.20 s
28	Ni	53	52.968 46	$(\frac{7-}{2})$	45 ms
		54	53.957 910 51	0^+	140 ms
		55	54.951 336 33	$\frac{7-}{2}$	212.1 ms
		56	55.942 136 34	0^+	5.9 d
		57	56.939 800 49	$\frac{3-}{2}$	35.60 h
		58	57.935 347 92	0^+	68.077%
		59	58.934 351 55	$\frac{3-}{2}$	0.076 My
		60	59.930 790 63	0^+	26.223%
		61	60.931 060 44	$\frac{3-}{2}$	1.14%
		62	61.928 348 76	0^+	3.634%
		63	62.929 672 95	$\frac{1-}{2}$	100.1 y

(Continued)

(*Continued*)

Z		A	Mass (u)	I^π	Abundance or half-life
		64	63.927 969 57	0^+	0.926%
		65	64.930 088 01	$\frac{5}{2}^-$	2.5172 h
		66	65.929 115 23	0^+	54.6 h
		67	66.931 569 64	$(\frac{1}{2}^-)$	21 s
29	Cu	57	56.949 2157	$\frac{3}{2}^-$	199.4 ms
		58	57.944 540 73	1^+	3.204 s
		59	58.939 504 11	$\frac{3}{2}^-$	81.5 s
		60	59.937 368 12	2^+	23.7 m
		61	60.933 462 18	$\frac{3}{2}^-$	3.333 h
		62	61.932 5873	1^+	9.74 m
		63	62.929 601 08	$\frac{3}{2}^-$	69.17%
		64	63.929 767 87	1^+	12.700 h
		65	64.927 793 71	$\frac{3}{2}^-$	30.83%
		66	65.928 873 04	1^+	5.088 m
		67	66.927 750 29	$\frac{3}{2}^-$	61.83 h
		68	67.929 620	1^+	31 s
		69	68.929 425 28	$\frac{3}{2}^-$	2.85 m
		70	69.932 409 29	1^+	4.5 s
30	Zn	59	58.949 267 07	$\frac{3}{2}^-$	182.0 ms
		60	59.941 832 03	0^+	2.38 m
		61	60.939 513 91	$\frac{3}{2}^-$	89.1 s
		62	61.934 334 13	0^+	9.186 h
		63	62.933 215 56	$\frac{3}{2}^-$	38.47 m
		64	63.929 146 58	0^+	48.6%
		65	64.929 245 08	$\frac{5}{2}^-$	244.26 d
		66	65.926 036 76	0^+	27.9%
		67	66.927 130 86	$\frac{5}{2}^-$	4.1%
		68	67.924 847 57	0^+	18.8%
		69	68.926 553 54	$\frac{1}{2}^-$	56.4 m
		70	69.925 324 87	0^+	0.6%
		71	70.927 7272	$\frac{1}{2}^-$	2.45 m
		72	71.926 861 12	0^+	46.5 h
		73	72.929 779 47	$(\frac{1}{2}^-)$	23.5 s
		74	73.929 458 26	0^+	96 s
		75	74.932 937 38	$(\frac{7}{2}^+)$	10.2 s
31	Ga	62	61.944 179 61	0^+	116.12 ms
		63	62.939 141 53	$\frac{3}{2}^-,\frac{5}{2}^-$	32.4 s
		64	63.936 838 31	0^+	2.630 m

(*Continued*)

(*Continued*)

Z		A	Mass (*u*)	I^π	Abundance or half-life
		65	64.932 739 32	$\frac{3}{2}-$	15.2 m
		66	65.931 592 36	0^+	9.49 h
		67	66.928 204 92	$\frac{3}{2}-$	3.2612 d
		68	67.927 9835	1^+	67.629 m
		69	68.925 580 91	$\frac{3}{2}-$	60.108%
		70	69.926 027 74	1^+	21.14 m
		71	70.924 705 01	$\frac{3}{2}-$	39.892%
		72	71.926 369 35	3^-	14.10 h
		73	72.925 169 83	$\frac{3}{2}-$	4.86 h
		74	73.926 941	(3^-)	8.12 m
		75	74.926 500 65	$\frac{3}{2}-$	126 s
		76	75.928 928 26	(3^-)	32.6 s
32	Ge	61	60.963 79	$(\frac{3}{2}-)$	40 ms
		64	63.941 572 64	0^+	63.7 s
		66	65.933 8468	0^+	2.26 h
		67	66.932 738 42	$\frac{1}{2}-$	18.9 m
		68	67.928 097 27	0^+	270.82 d
		69	68.927 972	$\frac{5}{2}-$	39.05 h
		70	69.924 250 37	0^+	21.23%
		71	70.924 953 99	$\frac{1}{2}-$	11.43 d
		72	71.922 076 18	0^+	27.66%
		73	72.923 459 36	$\frac{9}{2}+$	7.73%
		74	73.921 178 21	0^+	35.94%
		75	74.922 859 49	$\frac{1}{2}-$	82.78 m
		76	75.921 402 72	0^+	7.44%
		77	76.923 548 46	$\frac{7}{2}+$	11.30 h
		78	77.922 852 89	0^+	88.0 m
		79	78.925 401 56	$(\frac{1}{2})^-$	18.98 s
		80	79.925 444 76	0^+	29.5 s
33	As	67	66.939 190 42	$(\frac{5}{2}-)$	42.5 s
		69	68.932 280 15	$\frac{5}{2}-$	15.2 m
		70	69.930 927 81	$4(^+)$	52.6 m
		71	70.927 114 72	$\frac{5}{2}-$	65.28 h
		72	71.926 752 65	2^-	26.0 h
		73	72.923 825 29	$\frac{3}{2}-$	80.30 d
		74	73.923 929 08	2^-	17.77 d
		75	74.921 596 42	$\frac{3}{2}-$	100%

(*Continued*)

(*Continued*)

Z		A	Mass (u)	I^π	Abundance or half-life
		76	75.922 393 93	2^-	26.32 h
		77	76.920 6477	$\frac{3}{2}^-$	38.83 h
		78	77.921 828 58	2^-	90.7 m
		79	78.920 9485	$\frac{3}{2}^-$	9.01 m
		80	79.922 578 16	1^+	15.2 s
		81	80.922 132 88	$\frac{3}{2}^-$	33.3 s
34	Se	70	69.933 504	0^+	41.1 m
		71	70.932 268	$\frac{5}{2}^-$	4.74 m
		72	71.927 112 31	0^+	8.40 d
		73	72.926 7668	$\frac{9}{2}^+$	7.15 h
		74	73.922 476 56	0^+	0.89%
		75	74.922 523 57	$\frac{5}{2}^+$	119.779 d
		76	75.919 214 11	0^+	9.36%
		77	76.919 914 61	$\frac{1}{2}^-$	7.63%
		78	77.917 309 52	0^+	23.78%
		79	78.918 4998	$\frac{7}{2}^+$	0.65 My
		80	79.916 521 83	0^+	49.61%
		81	80.917 992 93	$\frac{1}{2}^-$	18.45 m
		82	81.9167	0^+	8.73%
		83	82.919 119 07	$\frac{9}{2}^+$	22.3 m
		84	83.918 464 52	0^+	3.1 m
		87	86.928 520 75	$(\frac{5}{2}^+)$	5.85 s
35	Br	75	74.925 776 41	$\frac{3}{2}^-$	96.7 m
		76	75.924 541 97	1^-	16.2 h
		77	76.921 380 12	$\frac{3}{2}^-$	57.036 h
		78	77.921 146 13	1^+	6.46 m
		79	78.918 337 65	$\frac{3}{2}^-$	50.69%
		80	79.918 529 95	1^+	17.68 m
		81	80.916 291 06	$\frac{3}{2}^-$	49.31%
		82	81.916 804 67	5^-	35.30 h
		83	82.915 180 22	$\frac{3}{2}^-$	2.40 h
		84	83.916 503	2^-	31.8 m
		85	84.915 608 03	$\frac{3}{2}^-$	2.90 m
		87	86.920 710 71	$\frac{3}{2}^-$	55.60 s
36	Kr	73	72.938 931 12	$\frac{5}{2}^-$	27.0 s
		74	73.933 258 23	0^+	11.50 m
		75	74.931 033 79	$(\frac{5}{2})^+$	4.3 m
		76	75.925 9483	0^+	14.8 h

(*Continued*)

(Continued)

Z		A	Mass (u)	I^π	Abundance or half-life
		77	76.924 667 88	$\frac{5}{2}+$	74.4 m
		78	77.920 386 27	0^+	0.35%
		79	78.920 082 99	$\frac{1}{2}-$	35.04 h
		80	79.916 378 04	0^+	2.25%
		81	80.916 592 42	$\frac{7}{2}+$	0.229 My
		82	81.913 4846	0^+	11.6%
		83	82.914 135 95	$\frac{9}{2}+$	11.5%
		84	83.911 506 63	0^+	57%
		85	84.912 526 95	$\frac{9}{2}+$	10.756 y
		86	85.910 610 31	0^+	17.3%
		87	86.913 354 25	$\frac{5}{2}+$	76.3 m
		88	87.914 446 95	0^+	2.84 h
		89	88.917 632 51	$(\frac{3}{2}+, \frac{5}{2}+)$	3.15 m
		90	89.919 5238	0^+	32.32 s
37	Rb	79	78.923 996 72	$\frac{5}{2}+$	22.9 m
		80	79.922 519 32	1^+	34 s
		81	80.918 994 17	$\frac{3}{2}-$	4.576 h
		82	81.918 207 69	1^+	1.273 m
		83	82.915 111 95	$\frac{5}{2}-$	86.2 d
		84	83.914 384 68	2^-	32.77 d
		85	84.911 789 34	$\frac{5}{2}-$	72.17%
		86	85.911 167 08	2^-	18.631 d
		87	86.909 183 47	$\frac{3}{2}-$	27.83%
		88	87.911 318 56	2^-	17.78 m
		90	89.914 808 94	0^-	158 s
		91	90.916 534 16	$\frac{3}{2}(^-)$	58.4 s
		92	91.919 725 44	0^-	4.492 s
		93	92.922 032 77	$\frac{5}{2}-$	5.84 s
38	Sr	79	78.929 707 08	$\frac{3}{2}(^-)$	2.25 m
		80	79.924 524 59	0^+	106.3 m
		81	80.923 2131	$\frac{1}{2}-$	22.3 m
		82	81.918 401 26	0^+	25.55 d
		83	82.917 555 03	$\frac{7}{2}+$	32.41 h
		84	83.913 424 78	0^+	0.56%
		85	84.912 932 69	$\frac{9}{2}+$	64.84 d
		86	85.909 262 35	0^+	9.86%
		87	86.908 879 32	$\frac{9}{2}+$	7%

(Continued)

(*Continued*)

Z		A	Mass (*u*)	I^π	Abundance or half-life
		88	87.905 614 34	0^+	82.58%
		89	88.907 452 91	$\frac{5}{2}^+$	50.53 d
		90	89.907 7376	0^+	28.78 y
		91	90.910 209 85	$\frac{5}{2}^+$	9.63 h
		92	91.911 0299	0^+	2.71 h
		93	92.914 022 41	$\frac{5}{2}^+$	7.423 m
		94	93.915 359 86	0^+	75.3 s
		95	94.919 358 21	$\frac{1}{2}^+$	23.90 s
		96	95.921 680 47	0^+	1.07 s
39	Y	83	82.922 352 57	$(\frac{9}{2}^+)$	7.08 m
		84	83.920 387 77	1^+	4.6 s
		85	84.916 427 08	$(\frac{1}{2})^-$	2.68 h
		86	85.914 887 72	4^-	14.74 h
		87	86.910 877 83	$\frac{1}{2}^-$	79.8 h
		88	87.909 503 36	4^-	106.65 d
		89	88.905 8479	$\frac{1}{2}^-$	100%
		90	89.907 151 44	2^-	64.10 h
		91	90.907 303 42	$\frac{1}{2}^-$	58.51 d
		92	91.908 946 83	2^-	3.54 h
		93	92.909 581 58	$\frac{1}{2}^-$	10.18 h
		94	93.911 594 01	2^-	18.7 m
		95	94.912 823 71	$\frac{1}{2}^-$	10.3 m
		96	95.915 897 79	0^-	5.34 s
		97	96.918 131 02	$(\frac{1}{2}^-)^-$	3.75 s
40	Zr	84	83.923 25	0^+	25.9 m
		85	84.921 465 22	$\frac{7}{2}^+$	7.86 m
		86	85.916 472 85	0^+	16.5 h
		87	86.914 816 58	$(\frac{9}{2})^+$	1.68 h
		88	87.910 226 18	0^+	83.4 d
		89	88.908 888 92	$\frac{9}{2}^+$	78.41 h
		90	89.904 703 68	0^+	51.45%
		91	90.905 644 97	$\frac{5}{2}^+$	11.22%
		92	91.905 040 11	0^+	17.15%
		93	92.906 475 63	$\frac{5}{2}^+$	1.53 My
		94	93.906 315 77	0^+	17.38%
		95	94.908 042 74	$\frac{5}{2}^+$	64.02 d
		96	95.908 275 68	0^+	2.8%
		97	96.910 950 72	$\frac{1}{2}^+$	16.90 h

(*Continued*)

(*Continued*)

Z		A	Mass (*u*)	I^π	Abundance or half-life
		98	97.912 746 37	0^+	30.7 s
		99	98.916 511 08	$(\frac{1}{2}^+)$	2.1 s
		100	99.917 7617	0^+	7.1 s
41	Nb	87	86.920 361 44	$(\frac{9}{2}^+$	2.6 m
		88	87.917 956	(8^+)	14.5 m
		89	88.913 4955	$(\frac{9}{2}^+)$	1.9 h
		90	89.911 264 11	8^+	14.60 h
		91	90.906 990 54	$\frac{9}{2}^+$	680 y
		92	91.907 193 21	$(7)^+$	34.7 My
		93	92.906 377 54	$\frac{9}{2}^+$	100%
		94	93.907 283 46	$(6)^+$	0.0203 My
		95	94.906 835 18	$\frac{9}{2}^+$	34.975 d
		96	95.908 100 08	6^+	23.35 h
		97	96.908 097 14	$\frac{9}{2}^+$	72.1 m
		98	97.910 330 69	1^+	2.86 s
		99	98.911 617 86	$\frac{9}{2}^+$	15.0 s
42	Mo	88	87.921 952 73	0^+	8.0 m
		89	88.919 480 56	$(\frac{9}{2}^+)$	2.04 m
		90	89.913 936 16	0^+	5.67 h
		91	90.911 750 75	$\frac{9}{2}^+$	15.49 m
		92	91.906 810 48	0^+	14.84%
		93	92.906 812 21	$\frac{5}{2}^+$	0.0040 My
		94	93.905 087 58	0^+	9.25%
		95	94.905 841 49	$\frac{5}{2}^+$	15.92%
		96	95.904 6789	0^+	16.68%
		97	96.906 021 03	$\frac{5}{2}^+$	9.55%
		98	97.905 407 85	0^+	24.13%
		99	98.907 7116	$\frac{1}{2}^+$	65.94 h
		100	99.907 477 15	0^+	9.63%
		101	100.910 3465	$\frac{1}{2}^+$	14.61 m
		102	101.910 2972	0^+	11.3 m
		103	102.913 2046	$(\frac{3}{2}^+)$	67.5 s
		104	103.913 7584	0^+	60 s
43	Tc	93	92.910 248 47	$\frac{9}{2}^+$	2.75 h
		94	93.909 656 31	7^+	293 m
		95	94.907 656 45	$\frac{9}{2}^+$	20.0 h
		96	95.907 8708	7^+	4.28 d

(*Continued*)

(*Continued*)

Z		A	Mass (*u*)	I^π	Abundance or half-life
		97	96.906 364 84	$\frac{9}{2}+$	2.6 My
		98	97.907 215 69	$(6)^+$	4.2 My
		99	98.906 254 55	$\frac{9}{2}+$	0.2111 My
		100	99.907 657 59	1^+	15.8 s
		101	100.907 3144	$(\frac{9}{2})^+$	14.22 m
		102	101.909 2129	1^+	5.28 s
		103	102.909 1788	$\frac{5}{2}+$	54.2 s
44	Ru	92	91.920 12	0^+	3.65 m
		93	92.917 051 52	$(\frac{9}{2})^+$	59.7 s
		94	93.911 359 57	0^+	51.8 m
		95	94.910 412 73	$\frac{5}{2}+$	1.643 h
		96	95.907 597 68	0^+	5.54%
		97	96.907 554 55	$\frac{5}{2}+$	2.9 d
		98	97.905 287 11	0^+	1.86%
		99	98.905 939 31	$\frac{5}{2}+$	12.7%
		100	99.904 219 66	0^+	12.6%
		101	100.905 5822	$\frac{5}{2}+$	17.1%
		102	101.904 3495	0^+	31.6%
		103	102.906 3237	$\frac{3}{2}+$	39.26 d
		104	103.905 4301	0^+	18.6%
		105	104.907 7503	$\frac{3}{2}+$	4.44 h
		106	105.907 3269	0^+	373.59 d
		107	106.909 9072	$(\frac{5}{2})^+$	3.75 m
		108	107.910 1922	0^+	4.55 m
		109	108.913 2016	$(\frac{5}{2}^+)$	34.5 s
45	Rh	97	96.911 336 64	$\frac{9}{2}+$	30.7 m
		98	97.910 716 43	$(2)^+$	8.7 m
		99	98.908 1321	$\frac{1}{2}-$	16.1 d
		100	99.908 116 63	1^-	20.8 h
		101	100.906 1635	$\frac{1}{2}-$	3.3 y
		102	101.906 8428	$(1^-,2^-)$	207 d
		103	102.905 5042	$\frac{1}{2}-$	100%
		104	103.906 6553	1^+	42.3 s
		105	104.905 6924	$\frac{7}{2}+$	35.36 h
		106	105.907 2846	1^+	29.80 s
		107	106.906 7505	$\frac{7}{2}+$	21.7 m
		108	107.908 7308	(5^+)	6.0 m
		109	108.908 7356	$\frac{7}{2}+$	80 s

(*Continued*)

(*Continued*)

Z		A	Mass (u)	I^π	Abundance or half-life
46	Pd	97	96.916 478 92	$(\frac{5}{2}^+)$	3.10 m
		98	97.912 720 75	0^+	17.7 m
		99	98.911 767 76	$(\frac{5}{2}^+)^+$	21.4 m
		100	99.908 5046	0^+	3.63 d
		101	100.908 2891	$(\frac{5}{2}^+)$	8.47 h
		102	101.905 6077	0^+	1.02%
		103	102.906 0872	$\frac{5}{2}^+$	16.991 d
		104	103.904 0349	0^+	11.14%
		105	104.905 084	$\frac{5}{2}^+$	22.33%
		106	105.903 4831	0^+	27.33%
		107	106.905 1285	$\frac{5}{2}^+$	6.5 My
		108	107.903 8945	0^+	26.46%
		109	108.905 9535	$\frac{5}{2}^+$	13.7012 h
		110	109.905 1524	0^+	11.72%
		111	110.907 644	$\frac{5}{2}^+$	23.4 m
		112	111.907 3133	0^+	21.03 h
		113	112.910 1513	$(\frac{5}{2})^+$	93 s
		114	113.910 3653	0^+	2.42 m
47	Ag	101	100.912 8021	$\frac{9}{2}^+$	11.1 m
		102	101.912	5^+	12.9 m
		103	102.908 9725	$\frac{7}{2}^+$	65.7 m
		104	103.908 6282	5^+	69.2 m
		105	104.906 5282	$\frac{1}{2}^-$	41.29 d
		106	105.906 6664	1^+	23.96 m
		107	106.905 093	$\frac{1}{2}^-$	51.839%
		108	107.905 9537	1^+	2.37 m
		109	108.904 7555	$\frac{1}{2}^-$	48.161%
		110	109.906 1105	1^+	24.6 s
		111	110.905 2947	$\frac{1}{2}^-$	7.45 d
		112	111.907 0041	$2(^-)$	3.130 h
		113	112.906 5657	$\frac{1}{2}^-$	5.37 h
		114	113.908 8079	1^+	4.6 s
		115	114.908 7623	$\frac{1}{2}^-$	20.0 m
48	Cd	102	101.914 7773	0^+	5.5 m
		103	102.913 419	$(\frac{5}{2})^+$	7.3 m
		104	103.909 8481	0^+	57.7 m
		105	104.909 4678	$\frac{5}{2}^+$	55.5 m

(*Continued*)

(*Continued*)

Z		A	Mass (*u*)	I^π	Abundance or half-life
		106	105.906 458	0^+	1.25%
		107	106.906 6142	$\frac{5}{2}^+$	6.50 h
		108	107.904 1834	0^+	0.89%
		109	108.904 9856	$\frac{5}{2}^+$	462.6 d
		110	109.903 0056	0^+	12.49%
		111	110.904 1816	$\frac{1}{2}^+$	12.8%
		112	111.902 7572	0^+	24.13%
		113	112.904 4009	$\frac{1}{2}^+$	12.22%
		114	113.903 3581	0^+	28.73%
		115	114.905 4306	$\frac{1}{2}^+$	53.46 h
		116	115.904 7554	0^+	7.49%
		117	116.907 2182	$\frac{1}{2}^+$	2.49 h
		118	117.906 9141	0^+	50.3 m
		119	118.909 9226	$\frac{3}{2}^+$	2.69 m
		120	119.909 8514	0^+	50.80 s
		121	120.912 9804	$(\frac{3}{2}^+)$	13.5 s
49	In	106	105.913 4611	7^+	6.2 m
		107	106.910 2922	$\frac{9}{2}^+$	32.4 m
		108	107.909 7197	7^+	58.0 m
		109	108.907 1541	$\frac{9}{2}^+$	4.2 h
		110	109.907 1688	7^+	4.9 h
		111	110.905 1107	$\frac{9}{2}^+$	2.8049 d
		112	111.905 5333	1^+	14.97 m
		113	112.904 0612	$\frac{9}{2}^+$	4.3%
		114	113.904 9168	1^+	71.9 s
		115	114.903 8783	$\frac{9}{2}^+$	95.7%
		116	115.905 26	1^+	14.10 s
		117	116.904 5157	$\frac{9}{2}^+$	43.2 m
		118	117.906 3546	1^+	5.0 s
		119	118.905 8463	$\frac{9}{2}^+$	2.4 m
		120	119.907 9615	1^+	3.08 s
		121	120.907 8488	$\frac{9}{2}^+$	23.1 s
		122	121.910 2771	1^+	1.5 s
50	Sn	107	106.915 6667	$(\frac{5}{2}^+)^+)$	2.90 m
		108	107.911 9653	0^+	10.30 m
		109	108.911 2869	$\frac{5}{2}^{(+)}$	18.0 m
		110	109.907 8527	0^+	4.11 h
		111	110.907 7354	$\frac{7}{2}^+$	35.3 m

(*Continued*)

(*Continued*)

Z		A	Mass (*u*)	I^{π}	Abundance or half-life
		112	111.904 8208	0^+	0.97%
		113	112.905 1734	$\frac{1}{2}^+$	115.09 d
		114	113.902 7818	0^+	0.65%
		115	114.903 346	$\frac{1}{2}^+$	0.36%
		116	115.901 7441	0^+	14.53%
		117	116.902 9538	$\frac{1}{2}^+$	7.68%
		118	117.901 6063	0^+	24.22%
		119	118.903 3089	$\frac{1}{2}^+$	8.58%
		120	119.902 1966	0^+	32.59%
		121	120.904 2369	$\frac{3}{2}^+$	27.06 h
		122	121.903 4401	0^+	4.63%
		123	122.905 7219	$\frac{11}{2}^-$	129.2 d
		124	123.905 2746	0^+	5.79%
		125	124.907 7849	$\frac{11}{2}^-$	9.64 d
		126	125.907 654	0^+	0.1 My
		127	126.910 351	$(\frac{11}{2}^-)$	2.10 h
		128	127.910 535	0^+	59.1 m
		129	128.913 44	$(\frac{3}{2}^+)$	2.16 m
51	Sb	115	114.906 5988	$\frac{5}{2}^+$	32.1 m
		116	115.906 7972	3^+	15.8 m
		117	116.904 8396	$\frac{5}{2}^+$	2.80 h
		118	117.905 5319	1^+	3.6 m
		119	118.903 9465	$\frac{5}{2}^+$	38.19 h
		120	119.905 0743	1^+	15.89 m
		121	120.903 818	$\frac{5}{2}^+$	57.36%
		122	121.905 1754	2^-	2.7238 d
		123	122.904 2157	$\frac{7}{2}^+$	42.64%
		124	123.905 9375	3^-	60.20 d
		125	124.905 2478	$\frac{7}{2}^+$	2.7582 y
		126	125.907 2482	$(8)^-$	12.46 d
		127	126.906 9146	$\frac{7}{2}^+$	3.85 d
		128	127.909 1673	8^-	9.01 h
		129	128.909 1501	$\frac{7}{2}^+$	4.40 h
52	Te	114	113.912 057	0^+	15.2 m
		115	114.911 5786	$\frac{7}{2}^+$	5.8 m
		116	115.908 4203	0^+	2.49 h
		117	116.908 6342	$\frac{1}{2}^+$	62 m

(*Continued*)

(*Continued*)

Z		A	Mass (*u*)	I^π	Abundance or half-life
		118	117.905 8252	0^+	6.00 d
		119	118.906 4081	$\frac{1}{2}+$	16.03 h
		120	119.904 0199	0^+	0.095%
		121	120.904 9298	$\frac{1}{2}+$	16.78 d
		122	121.903 0471	0^+	2.59%
		123	122.904 273	$\frac{1}{2}+$	0.905%
		124	123.902 8195	0^+	4.79%
		125	124.904 4247	$\frac{1}{2}+$	7.12%
		126	125.903 3055	0^+	18.93%
		127	126.905 2173	$\frac{3}{2}+$	9.35 h
		128	127.904 4614	0^+	31.7%
		129	128.906 5956	$\frac{3}{2}+$	69.6 m
		130	129.906 2228	0^+	33.87%
		131	130.908 5219	$\frac{3}{2}+$	25.0 m
		132	131.908 5238	0^+	3.204 d
		133	132.910 9391	$(\frac{3}{2}+)$	12.5 m
		134	133.911 5405	0^+	41.8 m
		135	134.916 4508	$(\frac{7}{2}-)$	19.0 s
53	I	119	118.910 1808	$\frac{5}{2}+$	19.1 m
		120	119.910 0478	2^-	81.0 m
		121	120.907 3661	$\frac{5}{2}+$	2.12 h
		122	121.907 5925	1^+	3.63 m
		123	122.905 5979	$\frac{5}{2}+$	13.27 h
		124	123.906 2114	2^-	4.18 d
		125	124.904 6242	$\frac{5}{2}+$	59.408 d
		126	125.905 6194	2^-	13.11 d
		127	126.904 4684	$\frac{5}{2}+$	100%
		128	127.905 8053	1^+	24.99 m
		129	128.904 9875	$\frac{7}{2}+$	15.7 My
		130	129.906 674	5^+	12.36 h
		131	130.906 1242	$\frac{7}{2}+$	8.020 70 d
		132	131.907 9945	4^+	2.295 h
		133	132.907 8065	$\frac{7}{2}+$	20.8 h
		134	133.909 8766	$(4)^+$	52.5 m
		135	134.910 0503	$\frac{7}{2}+$	6.57 h
54	Xe	118	117.916 5709	0^+	6 m
		119	118.915 5543	$(\frac{5}{2}+)$	5.8 m
		120	119.912 152	0^+	40 m

(*Continued*)

(*Continued*)

Z		A	Mass (*u*)	I^π	Abundance or half-life
		121	120.911 3865	$\frac{5}{2}(^+)$	40.1 m
		122	121.908 5484	0^+	20.1 h
		123	122.908 4707	$(\frac{1}{2})^+$	2.08 h
		124	123.905 8958	0^+	0.1%
		125	124.906 3982	$(\frac{1}{2})^+$	16.9 h
		126	125.904 2689	0^+	0.09%
		127	126.905 1796	$(\frac{1}{2}^+)$	36.4 d
		128	127.903 5304	0^+	1.91%
		129	128.904 7795	$\frac{1}{2}+$	26.4%
		130	129.903 5079	0^+	4.1%
		131	130.905 0819	$\frac{3}{2}+$	21.2%
		132	131.904 1545	0^+	26.9%
		133	132.905 9057	$\frac{3}{2}+$	5.243 d
		134	133.905 3945	0^+	10.4%
		135	134.907 2075	$\frac{3}{2}+$	9.14 h
		136	135.907 2195	0^+	8.9%
		137	136.911 5629	$\frac{7}{2}-$	3.818 m
		138	137.913 9885	0^+	14.08 m
55	Cs	126	125.909 448	1^+	1.64 m
		127	126.907 4176	$\frac{1}{2}+$	6.25 h
		128	127.907 7479	1^+	3.62 m
		129	128.906 0634	$\frac{1}{2}+$	32.06 h
		130	129.906 7062	1^+	29.21 m
		131	130.905 4602	$\frac{5}{2}+$	9.689 d
		132	131.906 4298	2^+	6.479 d
		133	132.905 4469	$\frac{7}{2}+$	100%
		134	133.906 7134	4^+	2.0648 y
		135	134.905 9719	$\frac{7}{2}+$	2.3 My
		136	135.907 3057	5^+	13.16 d
		137	136.907 0835	$\frac{7}{2}+$	30.07 y
		138	137.911 0105	3^-	33.41 m
		139	138.913 3579	$\frac{7}{2}+$	9.27 m
		140	139.917 2771	1^-	63.7 s
		141	140.920 044	$\frac{7}{2}+$	24.94 s
		142	141.924 2923	0^-	1.70 s
		143	142.927 3303	$\frac{3}{2}+$	1.78 s
		144	143.932 0274	1	1.01 s

(*Continued*)

(*Continued*)

Z		A	Mass (*u*)	I^π	Abundance or half-life
		145	144.935 3882	$\frac{3}{2}+$	0.594 s
56	Ba	125	124.914 6202	$\frac{1}{2}(+)$	3.5 m
		126	125.911 2441	0^+	100 m
		127	126.911 1213	$(\frac{1}{2}+)$	12.7 m
		128	127.908 3089	0^+	2.43 d
		129	128.908 6737	$\frac{1}{2}+$	2.23 h
		130	129.906 3105	0^+	0.106%
		131	130.906 9308	$\frac{1}{2}+$	11.50 d
		132	131.905 0562	0^+	0.101%
		133	132.906 0024	$\frac{1}{2}+$	10.52 y
		134	133.904 5033	0^+	2.42%
		135	134.905 6827	$\frac{3}{2}+$	6.593%
		136	135.904 5701	0^+	7.85%
		137	136.905 8214	$\frac{3}{2}+$	11.23%
		138	137.905 2413	0^+	71.7%
		139	138.908 8354	$\frac{7}{2}-$	83.06 m
		140	139.910 5995	0^+	12.752 d
		141	140.914 4064	$\frac{3}{2}-$	18.27 m
		142	141.916 4482	0^+	10.6 m
		143	142.920 6172	$\frac{5}{2}-$	14.33 s
		144	143.922 9405	0^+	11.5 s
57	La	131	130.910 1085	$\frac{3}{2}+$	59 m
		132	131.910 1104	2^-	4.8 h
		133	132.908 3964	$\frac{5}{2}+$	3.912 h
		134	133.908 4896	1^+	6.45 m
		135	134.906 971	$\frac{5}{2}+$	19.5 h
		136	135.907 6512	1^+	9.87 m
		137	136.906 4657	$\frac{7}{2}+$	0.06 My
		138	137.907 1068	5^+	0.0902%
		139	138.906 3482	$\frac{7}{2}+$	99.9098%
		140	139.909 4726	3^-	1.6781 d
		141	140.910 957	$(\frac{7}{2}+)$	3.92 h
		142	141.914 0745	2^-	91.1 m
		143	142.916 0586	$(\frac{7}{2})+$	14.2 m
		144	143.919 5917	(3^-)	40.8 s
58	Ce	132	131.911 49	0^+	3.51 h
		133	132.911 55	$\frac{9}{2}-$	4.9 h
		134	133.909 0264	0^+	3.16 d

(*Continued*)

(*Continued*)

Z		A	Mass (u)	I^{π}	Abundance or half-life
		135	134.909 1456	$\frac{1}{2}(^+)$	17.7 h
		136	135.907 1436	0^+	0.19%
		137	136.907 7776	$\frac{3}{2}+$	9.0 h
		138	137.905 9856	0^+	0.25%
		139	138.906 6466	$\frac{3}{2}+$	137.640 d
		140	139.905 434	0^+	88.43%
		141	140.908 2711	$\frac{7}{2}-$	32.501 d
		142	141.909 2397	0^+	11.13%
		143	142.912 3812	$\frac{3}{2}-$	33.039 h
		144	143.913 6427	0^+	284.893 d
		145	144.917 2279	$(\frac{3}{2})^-$	3.01 m
		146	145.918 6897	0^+	13.52 m
		147	146.922 511	$(\frac{5}{2}^-)$	56.4 s
59	Pr	136	135.912 6469	2^+	13.1 m
		137	136.910 6784	$\frac{5}{2}+$	1.28 h
		138	137.910 7489	1^+	1.45 m
		139	138.908 9322	$\frac{5}{2}+$	4.41 h
		140	139.909 0712	1^+	3.39 m
		141	140.907 6477	$\frac{5}{2}+$	100%
		142	141.910 0399	2^-	19.12 h
		143	142.910 8122	$\frac{7}{2}+$	13.57 d
		144	143.913 3006	0^-	17.28 m
		145	144.914 5069	$\frac{7}{2}+$	5.984 h
		146	145.917 588	$(2)^-$	24.15 m
		147	146.918 979	$(\frac{3}{2}^+)$	13.4 m
		148	147.922 1832	1^-	2.27 m
60	Nd	136	135.915 0205	0^+	50.65 m
		137	136.914 6397	$\frac{1}{2}+$	38.5 m
		138	137.911 93	0^+	5.04 h
		139	138.911 9242	$\frac{3}{2}+$	29.7 m
		140	139.909 3098	0^+	3.37 d
		141	140.909 6048	$\frac{3}{2}+$	2.49 h
		142	141.907 7186	0^+	27.13%
		143	142.909 8096	$\frac{7}{2}-$	12.18%
		144	143.910 0826	0^+	23.8%
		145	144.912 5688	$\frac{7}{2}-$	8.3%
		146	145.913 1121	0^+	17.19%

(*Continued*)

(*Continued*)

Z		A	Mass (u)	I^π	Abundance or half-life
		147	146.916 0958	$\frac{5}{2}-$	10.98 d
		148	147.916 8885	0^+	5.76%
		149	148.920 1442	$\frac{5}{2}-$	1.728 h
		150	149.920 8866	0^+	5.64%
		151	150.923 8247	$(\frac{3}{2})^+$	12.44 m
		152	151.924 6824	0^+	11.4 m
61	Pm	137	136.920 713	$\frac{11}{2}-$	2.4 m
		138	137.919 445	1^+	10 s
		139	138.916 7598	$(\frac{5}{2})^+$	4.15 m
		140	139.915 8016	1^+	9.2 s
		141	140.913 6066	$\frac{5}{2}+$	20.90 m
		142	141.912 9507	1^+	40.5 s
		143	142.910 9276	$\frac{5}{2}+$	265 d
		144	143.912 5858	5^-	363 d
		145	144.912 7439	$\frac{5}{2}+$	17.7 y
		146	145.914 6922	3^-	5.53 y
		147	146.915 1339	$\frac{7}{2}+$	2.6234 y
		148	147.917 4678	1^-	5.370 d
		149	148.918 3292	$\frac{7}{2}+$	53.08 h
		150	149.920 9795	(1^-)	2.68 h
		151	150.921 2027	$\frac{5}{2}+$	28.40 h
		152	151.923 4906	1^+	4.1 m
		153	152.924 1132	$\frac{5}{2}-$	5.4 m
62	Sm	138	137.923 54	0^+	3.1 m
		139	138.922 302	$(\frac{1}{2})^+$	2.57 m
		140	139.918 991	0^+	14.82 m
		141	140.918 4685	$\frac{1}{2}+$	10.2 m
		142	141.915 1933	0^+	72.49 m
		143	142.914 6236	$\frac{3}{2}+$	8.83 m
		144	143.911 9947	0^+	3.1%
		145	144.913 4056	$\frac{7}{2}-$	340 d
		146	145.913 0368	0^+	103 My
		147	146.914 8933	$\frac{7}{2}-$	15%
		148	147.914 8179	0^+	11.3%
		149	148.917 1795	$\frac{7}{2}-$	13.8%
		150	149.917 2715	0^+	7.4%
		151	150.919 9284	$\frac{5}{2}-$	90 y
		152	151.919 7282	0^+	26.7%

(*Continued*)

(*Continued*)

Z		A	Mass (u)	I^π	Abundance or half-life
		153	152.922 0939	$\frac{3}{2}+$	46.27 h
		154	153.922 2053	0^+	22.7%
		155	154.924 6359	$\frac{3}{2}-$	22.3 m
		156	155.925 5262	0^+	9.4 h
63	Eu	145	144.916 2613	$\frac{5}{2}+$	5.93 d
		146	145.917 1997	4^-	4.59 d
		147	146.916 7412	$\frac{5}{2}+$	24.1 d
		148	147.918 1538	5^-	54.5 d
		149	148.917 9259	$\frac{5}{2}+$	93.1 d
		150	149.919 6983	$5(^-)$	35.8 y
		151	150.919 846	$\frac{5}{2}+$	47.8%
		152	151.921 7404	3^-	13.542 y
		153	152.921 2262	$\frac{5}{2}+$	52.2%
		154	153.922 9754	3^-	8.593 y
		155	154.922 8894	$\frac{5}{2}+$	4.7611 y
		156	155.924 7509	0^+	15.19 d
		157	156.925 4194	$\frac{5}{2}+$	15.18 h
		158	157.927 8419	(1^-)	45.9 m
		159	158.929 0845	$\frac{5}{2}+$	18.1 m
64	Gd	145	144.921 6875	$\frac{1}{2}+$	23.0 m
		146	145.918 3053	0^+	48.27 d
		147	146.919 0894	$\frac{7}{2}-$	38.06 h
		148	147.918 1098	0^+	74.6 y
		149	148.919 3364	$\frac{7}{2}-$	9.28 d
		150	149.918 6555	0^+	1.79 My
		151	150.920 3443	$\frac{7}{2}-$	124 d
		152	151.919 7879	0^+	0.2%
		153	152.921 7463	$\frac{3}{2}-$	241.6 d
		154	153.920 8623	0^+	2.18%
		155	154.922 6188	$\frac{3}{2}-$	14.8%
		156	155.922 1196	0^+	20.47%
		157	156.923 9567	$\frac{3}{2}-$	15.65%
		158	157.924 1005	0^+	24.84%
		159	158.926 3851	$\frac{3}{2}-$	18.479 h
		160	159.927 0506	0^+	21.86%
		161	160.929 6657	$\frac{5}{2}-$	3.66 m
		162	161.930 9812	0^+	8.4 m

(*Continued*)

(*Continued*)

Z		A	Mass (*u*)	I^π	Abundance or half-life
		163	162.933 99	$(\frac{5}{2}^-)$	68 s
65	Tb	149	148.923 2416	$\frac{1}{2}^+$	4.118 h
		150	149.923 6542	(2^-)	3.48 h
		151	150.923 0982	$\frac{1}{2}(^+)$	17.609 h
		152	151.924 0713	2^-	17.5 h
		153	152.923 4309	$\frac{5}{2}^+$	2.34 d
		154	153.924 6862	0	21.5 h
		155	154.923 5004	$\frac{3}{2}^+$	5.32 d
		156	155.924 7437	3^-	5.35 d
		157	156.924 0212	$\frac{3}{2}^+$	99 y
		158	157.925 4103	3^-	180 y
		159	158.925 3431	$\frac{3}{2}^+$	100%
		160	159.927 164	3^-	72.3 d
		161	160.927 5663	$\frac{3}{2}^+$	6.88 d
		162	161.929 4848	1^-	7.60 m
		163	162.930 6439	$\frac{3}{2}^+$	19.5 m
66	Dy	150	149.925 5797	0^+	7.17 m
		151	150.926 1796	$\frac{7}{2}(^-)$	17.9 m
		152	151.924 7139	0^+	2.38 h
		153	152.925 7609	$\frac{7}{2}(^-)$	6.4 h
		154	153.924 422	0^+	3.0 My
		155	154.925 749	$\frac{3}{2}^-$	9.9 h
		156	155.924 2783	0^+	0.06%
		157	156.925 4613	$\frac{3}{2}^-$	8.14 h
		158	157.924 4046	0^+	0.1%
		159	158.925 7357	$\frac{3}{2}^-$	144.4 d
		160	159.925 1937	0^+	2.34%
		161	160.926 9296	$\frac{5}{2}^+$	18.9%
		162	161.926 7947	0^+	25.5%
		163	162.928 7275	$\frac{5}{2}^-$	24.9%
		164	163.929 1712	0^+	28.2%
		165	164.931 6998	$\frac{7}{2}^+$	2.334 h
		166	165.932 8032	0^+	81.6 h
		167	166.935 649	$(\frac{1}{2}^-)$	6.20 m
		168	167.937 23	0^+	8.7 m
67	Ho	157	156.928 1881	$\frac{7}{2}^-$	12.6 m
		158	157.928 9457	5^+	11.3 m
		159	158.927 7085	$\frac{7}{2}^-$	33.05 m

(*Continued*)

(*Continued*)

Z		A	Mass (u)	I^{π}	Abundance or half-life
		160	159.928 7257	5^+	25.6 m
		161	160.927 8517	$\frac{7}{2}-$	2.48 h
		162	161.929 0924	1^+	15.0 m
		163	162.928 7303	$\frac{7}{2}-$	4570 y
		164	163.930 2306	1^+	29 m
		165	164.930 3192	$\frac{7}{2}-$	100%
		166	165.932 2813	0^-	26.83 h
		167	166.933 1262	$\frac{7}{2}-$	3.1 h
		168	167.935 4964	3^+	2.99 m
		169	168.936 8683	$\frac{7}{2}-$	4.7 m
68	Er	154	153.932 7773	0^+	3.73 m
		155	154.933 2043	$\frac{7}{2}-$	5.3 m
		156	155.931 015	0^+	19.5 m
		157	156.931 9455	$\frac{3}{2}-$	18.65 m
		158	157.929 912	0^+	2.24 h
		159	158.930 6807	$\frac{3}{2}-$	36 m
		160	159.929 0789	0^+	28.58 h
		161	160.930 0013	$\frac{3}{2}-$	3.21 h
		162	161.928 7749	0^+	0.14%
		163	162.930 0293	$\frac{5}{2}-$	75.0 m
		164	163.929 197	0^+	1.61%
		165	164.930 7228	$\frac{5}{2}-$	10.36 h
		166	165.930 29	0^+	33.6%
		167	166.932 0454	$\frac{7}{2}+$	22.95%
		168	167.932 3678	0^+	26.8%
		169	168.934 5881	$\frac{1}{2}-$	9.40 d
		170	169.935 4603	0^+	14.9%
		171	170.938 0259	$\frac{5}{2}-$	7.516 h
		172	171.939 3521	0^+	49.3 h
		173	172.9424	$(\frac{7}{2}-)$	1.4 m
69	Tm	161	160.933 398	$\frac{7}{2}+$	33 m
		162	161.933 9701	1^-	21.70 m
		163	162.932 6476	$\frac{1}{2}+$	1.810 h
		164	163.933 451	1^+	2.0 m
		165	164.932 4325	$\frac{1}{2}+$	30.06 h
		166	165.933 5531	2^+	7.70 h
		167	166.932 8488	$\frac{1}{2}+$	9.25 d

(*Continued*)

(*Continued*)

Z		A	Mass (*u*)	I^π	Abundance or half-life
		168	167.934 1704	3^+	93.1 d
		169	168.934 2111	$\frac{1}{2}+$	100%
		170	169.935 7979	1^-	128.6 d
		171	170.936 4258	$\frac{1}{2}+$	1.92 y
		172	171.938 3961	2^-	63.6 h
		173	172.939 6003	$(\frac{1}{2}+)$	8.24 h
		174	173.942 1646	$(4)^-$	5.4 m
70	Yb	160	159.937 56	0^+	4.8 m
		161	160.937 853	$\frac{3}{2}-$	4.2 m
		162	161.935 75	0^+	18.87 m
		163	162.936 2655	$\frac{3}{2}-$	11.05 m
		164	163.934 52	0^+	75.8 m
		165	164.935 3976	$\frac{5}{2}-$	9.9 m
		166	165.933 8796	0^+	56.7 h
		167	166.934 9469	$\frac{5}{2}-$	17.5 m
		168	167.933 8945	0^+	0.13%
		169	168.935 1871	$\frac{7}{2}+$	32.026 d
		170	169.934 7587	0^+	3.05%
		171	170.936 3223	$\frac{1}{2}-$	14.3%
		172	171.936 3777	0^+	21.9%
		173	172.938 2068	$\frac{5}{2}-$	16.12%
		174	173.938 8581	0^+	31.8%
		175	174.941 2725	$\frac{7}{2}-$	4.185 d
		176	175.942 5684	0^+	12.7%
		177	176.945 2571	$(\frac{9}{2}+)$	1.911 h
		178	177.946 6434	0^+	74 m
		179	178.950 17	$(\frac{1}{2}-)$	8.0 m
71	Lu	167	166.938 3071	$\frac{7}{2}+$	51.5 m
		168	167.938 6986	(6^-)	5.5 m
		169	168.937 6488	$\frac{7}{2}+$	34.06 h
		170	169.938 4722	0^+	2.00 d
		171	170.937 9099	$\frac{7}{2}+$	8.24 d
		172	171.939 0822	4^-	6.70 d
		173	172.938 9269	$\frac{7}{2}+$	1.37 y
		174	173.940 3335	$(1)^-$	3.31 y
		175	174.940 7679	$\frac{7}{2}+$	97.41%
		176	175.942 6824	7^-	2.59%
		177	176.943 755	$\frac{7}{2}+$	6.734 d

(*Continued*)

(Continued)

Z		A	Mass (*u*)	I^π	Abundance or half-life
		178	177.945 9514	$1^{(+)}$	28.4 m
		179	178.947 3242	$\frac{7}{2}^{(+)}$	4.59 h
		180	179.949 88	$(3)^+$	5.7 m
		181	180.951 97	$(\frac{7}{2}^+)$	3.5 m
72	Hf	167	166.9426	$(\frac{5}{2}^-)$	2.05 m
		168	167.940 63	0^+	25.95 m
		169	168.941 1586	$(\frac{5}{2})^-$	3.24 m
		170	169.939 65	0^+	16.01 h
		171	170.940 49	$(\frac{7}{2}^+)$	12.1 h
		172	171.939 458	0^+	1.87 y
		173	172.940 65	$\frac{1}{2}^-$	23.6 h
		174	173.940 0402	0^+	0.162%
		175	174.941 503	$\frac{5}{2}^-$	70 d
		176	175.941 4018	0^+	5.206%
		177	176.943 22	$\frac{7}{2}^-$	18.606%
		178	177.943 6977	0^+	27.297%
		179	178.945 8151	$\frac{9}{2}^+$	13.629%
		180	179.946 5488	0^+	35.1%
		181	180.949 0991	$\frac{1}{2}^-$	42.39 d
		182	181.950 5529	0^+	9 My
73	Ta	175	174.943 65	$\frac{7}{2}^+$	10.5 h
		176	175.944 7406	$(1)^-$	8.09 h
		177	176.944 4718	$\frac{7}{2}^+$	56.56 h
		178	177.945 7503	1^+	9.31 m
		179	178.945 9341	$\frac{7}{2}^+$	1.82 y
		180	179.947 4657	1^+	0.012%
		181	180.947 9963	$\frac{7}{2}^+$	99.988%
		182	181.950 1524	3^-	114.43 d
		183	182.951 3732	$\frac{7}{2}^+$	5.1 d
		184	183.954 0093	(5^-)	8.7 h
74	W	174	173.946 16	0^+	31 m
		175	174.946 77	$(\frac{1}{2}^-)$	35.2 m
		176	175.945 59	0^+	2.5 h
		177	176.946 62	$(\frac{1}{2}^-)$	135 m
		178	177.945 8484	0^+	21.6 d
		179	178.947 0717	$(\frac{7}{2})^-$	37.05 m
		180	179.946 7057	0^+	0.12%

(Continued)

(Continued)

Z		A	Mass (*u*)	I^{π}	Abundance or half-life
		181	180.948 1981	$\frac{9}{2}+$	121.2 d
		182	181.948 2055	0^+	26.3%
		183	182.950 2245	$\frac{1}{2}-$	14.28%
		184	183.950 9326	0^+	30.7%
		185	184.953 4206	$\frac{3}{2}-$	75.1 d
		186	185.954 3622	0^+	28.6%
		187	186.957 1584	$\frac{3}{2}-$	23.72 h
		188	187.958 487	0^+	69.4 d
		189	188.961 9122	$(\frac{3}{2}-)$	11.5 m
75	Re	179	178.949 981	$(\frac{5}{2})^+$	19.5 m
		180	179.950 7877	$(1)^-$	2.44 m
		181	180.950 0646	$\frac{5}{2}+$	19.9 h
		182	181.951 2114	7^+	64.0 h
		183	182.950 8213	$\frac{5}{2}+$	70.0 h
		184	183.952 5243	$3(^-)$	38.0 d
		185	184.952 9557	$\frac{5}{2}+$	37.4%
		186	185.954 9865	1^-	90.64 h
		187	186.955 7508	$\frac{5}{2}+$	62.6%
		188	187.958 1123	1^-	16.98 h
		189	188.959 2284	$\frac{5}{2}+$	24.3 h
		190	189.961 8161	$(2)^-$	3.1 m
76	Os	178	177.953 3482	0^+	5.0 m
		179	178.953 951	$(\frac{1}{2}-)$	6.5 m
		180	179.952 351	0^+	21.5 m
		181	180.953 2745	$\frac{1}{2}-$	105 m
		182	181.952 1862	0^+	22.10 h
		183	182.953 11	$\frac{9}{2}+$	13.0 h
		184	183.952 4908	0^+	0.02%
		185	184.954 043	$\frac{1}{2}-$	93.6 d
		186	185.953 8384	0^+	1.58%
		187	186.955 7479	$\frac{1}{2}-$	1.6%
		188	187.955 836	0^+	13.3%
		189	188.958 1449	$\frac{3}{2}-$	16.1%
		190	189.958 4452	0^+	26.4%
		191	190.960 928	$\frac{9}{2}-$	15.4 d
		192	191.961 479	0^+	41%
		193	192.964 1481	$\frac{3}{2}-$	30.5 h
		194	193.965 1793	0^+	6.0 y

(Continued)

(*Continued*)

Z		A	Mass (u)	I^π	Abundance or half-life
77	Ir	183	182.956 814	$\frac{5}{2}^-$	58 m
		184	183.957 3883	5^-	3.09 h
		185	184.956 59	$\frac{5}{2}^-$	14.4 h
		186	185.957 9511	5^+	16.64 h
		187	186.957 3608	$\frac{3}{2}^+$	10.5 h
		188	187.958 852	1^-	41.5 h
		189	188.958 7165	$\frac{3}{2}^+$	13.2 d
		190	189.960 5923	$(4)^+$	11.78 d
		191	190.960 5912	$\frac{3}{2}^+$	37.3%
		192	191.962 6022	$4(^+)$	73.831 d
		193	192.962 9237	$\frac{3}{2}^+$	62.7%
		194	193.965 0756	1^-	19.15 h
		195	194.965 9768	$\frac{3}{2}^+$	2.5 h
		196	195.968 3799	(0^-)	52 s
		197	196.969 6365	$\frac{3}{2}^+$	5.8 m
78	Pt	181	180.963 177	$\frac{1}{2}^-$	51 s
		182	181.961 2676	0^+	2.2 m
		183	182.961 729	$\frac{1}{2}^-$	6.5 m
		184	183.959 895	0^+	17.3 m
		185	184.960 7538	$(\frac{9}{2}^+)$	70.9 m
		186	185.959 4323	0^+	2.0 h
		187	186.960 558	$\frac{3}{2}^-$	2.35 h
		188	187.959 3957	0^+	10.2 d
		189	188.960 8319	$\frac{3}{2}^-$	10.87 h
		190	189.959 9301	0^+	0.01%
		191	190.961 6847	$\frac{3}{2}^-$	2.9 d
		192	191.961 0352	0^+	0.79%
		193	192.962 9845	$\frac{1}{2}^-$	50 y
		194	193.962 6636	0^+	32.9%
		195	194.964 7744	$\frac{1}{2}^-$	33.8%
		196	195.964 9349	0^+	25.3%
		197	196.967 3234	$\frac{1}{2}^-$	18.3 h
		198	197.967 876	0^+	7.2%
		199	198.970 5762	$\frac{5}{2}^-$	30.80 m
79	Au	191	190.963 6492	$\frac{3}{2}^+$	3.18 h
		192	191.964 8101	1^-	4.94 h
		193	192.964 1317	$\frac{3}{2}^+$	17.65 h

(*Continued*)

(*Continued*)

Z		A	Mass (*u*)	I^π	Abundance or half-life
		194	193.965 3389	1^-	38.02 h
		195	194.965 0179	$\frac{3}{2}+$	186.098 d
		196	195.966 5513	2^-	6.183 d
		197	196.966 5516	$\frac{3}{2}+$	100%
		198	197.968 2252	2^-	2.6935 d
		199	198.968 748	$\frac{3}{2}+$	3.139 d
		200	199.970 7179	$1(^-)$	48.4 m
		201	200.971 6408	$\frac{3}{2}+$	26 m
		202	201.973 7884	(1^-)	28.8 s
		203	202.975 1373	$\frac{3}{2}+$	53 s
80	Hg	188	187.967 555	0^+	3.25 m
		189	188.968 132	$\frac{3}{2}-$	7.6 m
		190	189.966 277	0^+	20.0 m
		191	190.967 0631	$(\frac{3}{2}-)$	49 m
		192	191.965 572	0^+	4.85 h
		193	192.966 6442	$\frac{3}{2}-$	3.80 h
		194	193.965 3818	0^+	520 y
		195	194.966 639	$\frac{1}{2}-$	9.9 h
		196	195.965 8148	0^+	0.15%
		197	196.967 1953	$\frac{1}{2}-$	64.14 h
		198	197.966 7518	0^+	9.97%
		199	198.968 2625	$\frac{1}{2}-$	16.87%
		200	199.968 3087	0^+	23.1%
		201	200.970 2853	$\frac{3}{2}-$	13.1%
		202	201.970 6256	0^+	29.86%
		203	202.972 8571	$\frac{5}{2}-$	46.612 d
		204	203.973 4756	0^+	6.87%
		205	204.976 0561	$\frac{1}{2}-$	5.2 m
		206	205.977 4987	0^+	8.15 m
81	Tl	195	194.969 65	$\frac{1}{2}+$	1.16 h
		196	195.970 515	2^-	1.84 h
		197	196.969 5362	$\frac{1}{2}+$	2.84 h
		198	197.970 4663	2^-	5.3 h
		199	198.969 8138	$\frac{1}{2}+$	7.42 h
		200	199.970 9454	2^-	26.1 h
		201	200.970 8038	$\frac{1}{2}+$	72.912 h
		202	201.972 0906	2^-	12.23 d
					29.524%

(*Continued*)

(*Continued*)

Z		A	Mass (*u*)	I^π	Abundance or half-life
		203	202.972 3291	$\frac{1}{2}+$	
		204	203.973 8486	2^-	3.78 y
		205	204.974 4123	$\frac{1}{2}+$	70.476%
		206	205.976 0953	0^-	4.199 m
		207	206.977 4079	$\frac{1}{2}+$	4.77 m
82	Pb	195	194.974 471	$\frac{3}{2}-$	15 m
		196	195.972 71	0^+	37 m
		197	196.973 38	$\frac{3}{2}-$	8 m
		198	197.971 98	0^+	2.40 h
		199	198.972 9094	$\frac{3}{2}-$	90 m
		200	199.971 8156	0^+	21.5 h
		201	200.972 8466	$\frac{5}{2}-$	9.33 h
		202	201.972 1438	0^+	0.0525 My
		203	202.973 3755	$\frac{5}{2}-$	51.873 h
		204	203.973 0288	0^+	1.4%
		205	204.974 4671	$\frac{5}{2}-$	15.3 My
		206	205.974 449	0^+	24.1%
		207	206.975 8806	$\frac{1}{2}-$	22.1%
		208	207.976 6359	0^+	52.4%
		209	208.981 0748	$\frac{9}{2}+$	3.253 h
		210	209.984 1731	0^+	22.3 y
		211	210.988 7315	$\frac{9}{2}+$	36.1 m
		212	211.991 8875	0^+	10.64 h
		213	212.9965	$(\frac{9}{2}+)$	10.2 m
83	Bi	200	199.978 142	7^+	36.4 m
		201	200.976 9707	$\frac{9}{2}-$	108 m
		202	201.977 6745	5^+	1.72 h
		203	202.976 8681	$\frac{9}{2}-$	11.76 h
		204	203.977 8052	6^+	11.22 h
		205	204.977 3747	$\frac{9}{2}-$	15.31 d
		206	205.978 4829	$6(^+)$	6.243 d
		207	206.978 4552	$\frac{9}{2}-$	31.55 y
		208	207.979 7267	$(5)^+$	0.368 My
		209	208.980 3832	$\frac{9}{2}-$	100%
		210	209.984 1049	1^-	5.013 d
		211	210.987 2581	$\frac{9}{2}-$	2.14 m
		212	211.991 2715	$1(^-)$	60.55 m
		213	212.994 3748	$\frac{9}{2}-$	45.59 m

(*Continued*)

(*Continued*)

Z		A	Mass (u)	I^π	Abundance or half-life
		214	213.998 6987	1^-	19.9 m
84	Po	201	200.982 209	$\frac{3}{2}-$	15.3 m
		202	201.980 704	0^+	44.7 m
		203	202.981 4129	$\frac{5}{2}-$	36.7 m
		204	203.980 3071	0^+	3.53 h
		205	204.981 1654	$\frac{5}{2}-$	1.66 h
		206	205.980 4652	0^+	8.8 d
		207	206.981 5782	$\frac{5}{2}-$	5.80 h
		208	207.981 2311	0^+	2.898 y
		209	208.982 4158	$\frac{1}{2}-$	102 y
		210	209.982 8574	0^+	138.376 d
		211	210.986 6369	$\frac{9}{2}+$	0.516 s
		212	211.988 8518	0^+	0.299 μs
		213	212.992 8425	$\frac{9}{2}+$	4.2 μs
		214	213.995 1859	0^+	164.3 μs
		215	214.999 4146	$\frac{9}{2}+$	1.781 ms
85	At	205	204.986 0364	$\frac{9}{2}-$	26.2 m
		206	205.986 5992	$(5)^+$	30.0 m
		207	206.985 7759	$\frac{9}{2}-$	1.80 h
		208	207.986 5825	6^+	1.63 h
		209	208.986 1587	$\frac{9}{2}-$	5.41 h
		210	209.987 1313	$(5)^+$	8.1 h
		211	210.987 4808	$\frac{9}{2}-$	7.214 h
		212	211.990 7347	(1^-)	0.314 s
		213	212.992 9212	$\frac{9}{2}-$	125 ns
		214	213.996 3564	1^-	558 ns
		215	214.998 6412	$\frac{9}{2}-$	0.10 ms
		216	216.002 4088	$1(^-)$	0.30 ms
		217	217.004 7096	$\frac{9}{2}-$	32.3 ms
86	Rn	204	203.991 365	0^+	1.24 m
		205	204.991 668	$\frac{5}{2}-$	2.8 m
		206	205.990 16	0^+	5.67 m
		207	206.990 7268	$\frac{5}{2}-$	9.25 m
		208	207.989 6312	0^+	24.35 m
		209	208.990 3766	$\frac{5}{2}-$	28.5 m
		210	209.989 6799	0^+	2.4 h
		211	210.990 5854	$\frac{1}{2}-$	14.6 h
		212	211.990 6889	0^+	23.9 m

(*Continued*)

(*Continued*)

Z		A	Mass (*u*)	I^π	Abundance or half-life
		213	212.993 8684	$(\frac{9}{2}^+)$	25.0 ms
		214	213.995 3463	0^+	0.27 μs
		215	214.998 7292	$\frac{9}{2}+$	2.30 μs
		217	217.003 9146	$\frac{9}{2}+$	0.54 ms
		218	218.005 5863	0^+	35 ms
		219	219.009 4748	$\frac{5}{2}+$	3.96 s
		221	221.015 455	$\frac{7}{2}(^+)$	25 m
		222	222.017 5705	0^+	3.8235 d
		223	223.021 79	$\frac{7}{2}$	23.2 m
		224	224.024 09	0^+	107 m
		225	225.028 44	$\frac{7}{2}-$	4.5 m
		226	226.030 89	0^+	6.0 m
87	Fr	207	206.996 8594	$\frac{9}{2}-$	14.8 s
		208	207.997 1338	7^+	59.1 s
		209	208.995 9154	$\frac{9}{2}-$	50.0 s
		210	209.996 3983	6^+	3.18 m
		211	210.995 5293	$\frac{9}{2}-$	3.10 m
		212	211.996 195	5^+	20.0 m
		213	212.996 1748	$\frac{9}{2}-$	34.6 s
		214	213.998 9547	(1^-)	5.0 ms
		215	215.000 326	$\frac{9}{2}-$	86 ns
		217	217.004 6165	$\frac{9}{2}-$	22 μs
		218	218.007 5633	(1^-)	1.0 ms
		219	219.009 2408	$\frac{9}{2}-$	20 ms
		221	221.014 2457	$\frac{5}{2}-$	4.9 m
		222	222.017 544	2^-	14.2 m
		223	223.019 7307	$\frac{3}{2}(^-)$	21.8 m
		224	224.023 2355	$1(^-)$	3.30 m
		225	225.025 6069	$\frac{3}{2}-$	4.0 m
88	Ra	218	218.007 1239	0^+	25.6 μs
		219	219.010 0688	$(\frac{7}{2})^+$	10 ms
		220	220.011 0147	0^+	25 ms
		221	221.013 9078	$\frac{5}{2}+$	28 s
		222	222.015 3618	0^+	38.0 s
		223	223.018 4971	$\frac{3}{2}+$	11.435 d
		224	224.020 202	0^+	3.66 d
		225	225.023 6045	$\frac{1}{2}+$	14.9 d

(*Continued*)

(*Continued*)

Z		A	Mass (*u*)	I^π	Abundance or half-life
		226	226.025 4026	0^+	1600 y
		227	227.029 1707	$\frac{3}{2}+$	42.2 m
		228	228.031 0641	0^+	5.75 y
		229	229.034 8203	$\frac{5}{2}(^+)$	4.0 m
		230	230.037 0848	0^+	93 m
89	Ac	223	223.019 126	$(\frac{5}{2}^-)$	2.10 m
		225	225.023 2206	$(\frac{3}{2}^-)$	10.0 d
		226	226.026 0898	1^-	29 h
		227	227.027 747	$\frac{3}{2}-$	21.773 y
		229	229.032 9309	$(\frac{3}{2}^+)$	62.7 m
		230	230.036 0251	(1^+)	122 s
		231	231.038 5515	$(\frac{1}{2}^+)$	7.5 m
90	Th	223	223.020 7952	$(\frac{5}{2})^+$	0.60 s
		224	224.021 4593	0^+	1.05 s
		225	225.023 9414	$(\frac{3}{2})^+$	8.72 m
		226	226.024 8907	0^+	30.9 m
		227	227.027 6989	$(\frac{1}{2}^+)$	18.72 d
		228	228.028 7313	0^+	1.9131 y
		229	229.031 7553	$\frac{5}{2}+$	7340 y
		230	230.033 1266	0^+	0.075 38 My
		231	231.036 2971	$\frac{5}{2}+$	25.52 h
		232	232.038 0504	0^+	100%
		233	233.041 5769	$\frac{1}{2}+$	22.3 m
		234	234.043 5955	0^+	24.10 d
		235	235.047 5044	$(\frac{1}{2}^+)$	7.1 m
91	Pa	227	227.028 7932	$(\frac{5}{2}^-)$	38.3 m
		228	228.031 0369	(3^+)	22 h
		229	229.032 0886	$(\frac{5}{2}^+)$	1.50 d
		230	230.034 5326	(2^-)	17.4 d
		231	231.035 8789	$\frac{3}{2}-$	32760 y
		232	232.038 5817	(2^-)	1.31 d
		233	233.040 2402	$\frac{3}{2}-$	26.967 d
		234	234.043 3023	4^+	6.70 h
		235	235.045 4368	$(\frac{3}{2}^-)$	24.5 m
92	U	226	226.029 3398	0^+	0.5 s
		227	227.031 1401	$(\frac{3}{2}^+)$	1.1 m
		229	229.033 4961	$(\frac{3}{2}^+)$	58 m
		230	230.033 9274	0^+	20.8 d

(*Continued*)

(*Continued*)

Z		A	Mass (*u*)	I^π	Abundance or half-life
		231	231.036 2892	$(\frac{5}{2}^-)$	4.2 d
		232	232.037 1463	0^+	68.9 y
		233	233.039 6282	$\frac{5}{2}^+$	0.1592 My
		234	234.040 9456	0^+	0.0055%
		235	235.043 9231	$\frac{7}{2}^-$	0.72%
		236	236.045 5619	0^+	23.42 My
		237	237.048 724	$\frac{1}{2}^+$	6.75 d
		238	238.050 7826	0^+	99.2745%
		239	239.054 2878	$\frac{5}{2}^+$	23.45 m
		240	240.056 5857	0^+	14.1 h
93	Np	233	233.040 7324	$(\frac{5}{2}^+)$	36.2 m
		234	234.042 8886	(0^+)	4.4 d
		235	235.044 0559	$\frac{5}{2}^+$	396.1 d
		236	236.046 5597	(6^-)	0.154 My
		237	237.048 1673	$\frac{5}{2}^+$	2.14 My
		238	238.050 9405	2^+	2.117 d
		239	239.052 9314	$\frac{5}{2}^+$	2.3565 d
		240	240.056 1688	(5^+)	61.9 m
		241	241.058 2463	$(\frac{5}{2}^+)$	13.9 m
		242	242.061 635	6^-	5.5 m
94	Pu	234	234.043 3047	0^+	8.8 h
		235	235.045 2815	$(\frac{5}{2}^+)$	25.3 m
		236	236.046 0481	0^+	2.858 y
		237	237.048 4038	$\frac{7}{2}^-$	45.2 d
		238	238.049 5534	0^+	87.7 y
		239	239.052 1565	$\frac{1}{2}^+$	24110 y
		240	240.053 8075	0^+	6564 y
		241	241.056 8453	$\frac{5}{2}^+$	14.35 y
		242	242.058 7368	0^+	0.3733 My
		243	243.061 997	$\frac{7}{2}^+$	4.956 h
		244	244.064 1977	0^+	80.8 My
		245	245.067 7387	$(\frac{9}{2}^-)$	10.5 h
		246	246.070 1984	0^+	10.84 d
95	Am	237	237.049 9707	$\frac{5}{2}(^-)$	73.0 m
		238	238.051 9778	1^+	98 m
		239	239.053 0185	$(\frac{5}{2})^-$	11.9 h
		240	240.055 2878	(3^-)	50.8 h

(*Continued*)

(*Continued*)

Z		A	Mass (u)	I^π	Abundance or half-life
		241	241.056 8229	$\frac{5}{2}-$	432.2 y
		242	242.059 543	1^-	16.02 h
		243	243.061 3727	$\frac{5}{2}-$	7370 y
		244	244.064 2794	(6^-)	10.1 h
		245	245.066 4454	$(\frac{5}{2})^+$	2.05 h
		246	246.069 7684	(7^-)	39 m
		247	247.072 086	$(\frac{5}{2})$	23.0 m
96	Cm	243	243.061 3822	$\frac{5}{2}+$	29.1 y
		244	244.062 7463	0^+	18.10 y
		245	245.065 4856	$\frac{7}{2}+$	8500 y
		246	246.067 2176	0^+	4730 y
		247	247.070 3468	$\frac{9}{2}-$	15.6 My
		248	248.072 3422	0^+	0.340 My
		249	249.075 9471	$\frac{1}{2}(^+)$	64.15 m
		250	250.078 3507	0^+	9000 y
		251	251.082 2779	$(\frac{1}{2}+)$	16.8 m
97	Bk	243	243.063 0016	$(\frac{3}{2}-)$	4.5 h
		244	244.065 1679	(1^-)	4.35 h
		245	245.066 3554	$\frac{3}{2}-$	4.94 d
		246	246.068 6668	$2(^-)$	1.80 d
		247	247.070 2985	$(\frac{3}{2}-)$	1380 y
		248	248.073 08	(6^+)	9 y
		249	249.074 9799	$\frac{7}{2}+$	320 d
		250	250.078 3105	2^-	3.217 h
98	Cf	246	246.068 7988	0^+	35.7 h
		247	247.070 992	$(\frac{7}{2}+)$	3.11 h
		248	248.072 1781	0^+	333.5 d
		249	249.074 8468	$\frac{9}{2}-$	351 y
		250	250.0764	0^+	13.08 y
		251	251.079 5801	$\frac{1}{2}+$	898 y
		252	252.081 6196	0^+	2.645 y
		253	253.085 1268	$(\frac{7}{2}+)$	17.81 d
		254	254.087 3162	0^+	60.5 d
99	Es	249	249.076 405	$\frac{7}{2}(^+)$	102.2 m
		250	250.078 654	(6^+)	8.6 h
		251	251.079 9836	$(\frac{3}{2}-)$	33 h
		252	252.082 9722	(5^-)	471.7 d
		253	253.084 818	$\frac{7}{2}+$	20.47 d

(*Continued*)

(*Continued*)

Z		A	Mass (*u*)	I^π	Abundance or half-life
		254	254.088 016	(7^+)	275.7 d
		255	255.090 2664	$(\frac{7}{2}^+)$	39.8 d
100	Fm	251	251.081 5665	$(\frac{9}{2}^-)$	5.30 h
		252	252.082 4601	0^+	25.39 h
		253	253.085 1763	$\frac{1}{2}^+$	3.00 d
		254	254.086 8478	0^+	3.240 h
		255	255.089 9555	$\frac{7}{2}^+$	20.07 h
		256	256.091 7665	0^+	157.6 m
		257	257.095 0986	$(\frac{9}{2}^+)$	100.5 d
		258	258.097 069	0^+	370 μs
101	Md	255	255.091 0752	$(\frac{7}{2}^-)$	27 m
		256	256.094 0528	$(0^-,1^-)$	76 m
		257	257.095 5346	$(\frac{7}{2}^-)$	5.3 h
		258	258.098 4253	(8^-)	55 d
		259	259.100 503	$(\frac{7}{2}^-)$	103 m
102	No	253	253.090 649	$(\frac{9}{2}^-)$	1.7 m
		254	254.090 9487	0^+	55 s
		255	255.093 2324	$(\frac{1}{2}^+)$	3.1 m
		256	256.094 2759	0^+	3.3 s
		257	257.096 8528	$(\frac{7}{2}^+)$	25 s
		258	258.0982	0^+	1.2 ms
		259	259.101 024	$(\frac{9}{2}^+)$	58 m
103	Lr	252	252.095 33		1 s
		253	253.095 258		1.3 s
		257	257.099 606	$(\frac{9}{2}^+)$	0.646 s
		258	258.101 883		4.3 s
		259	259.102 99		5.4 s
		260	260.105 320		180 s

IOP Publishing

An Introduction to the Physics of Nuclei and Particles
(Second Edition)

Richard A Dunlap

Appendix C

An overview of particle accelerators

Particle accelerators are a means of accelerating subatomic particles or ionized atoms to high energy in order to investigate collisions with other particles or nuclei. They are a necessary experimental tool for the study of both nuclear physics and particle physics. There are several different accelerator designs, and each has its applications to the study of particular phenomena. In this appendix the various types of accelerators are reviewed along with the physics of their operation and their particular applications.

Accelerators can be classified according to the energies of the particles that they can produce.

Low energy accelerators produce particle beams that are typically in the range of about 1–100 MeV. These energies are typical of the nucleon binding energies in nuclei and the energies associated with excited nuclear states. They are applicable to the investigation of reactions such as those described in chapter 11.

Medium energy accelerators typically produce particles with energies in the range of about 100 MeV to 1 GeV. These energies are comparable to the rest mass energies of light mesons, such as the π-mesons, and collisions involving nucleons in this energy range can be used for the production and study of mesons, as in figure 16.11.

High energy accelerators have energies in excess of about 1 GeV (approximately the rest mass energy of a nucleon) and current machines can produce particle beams with energies in excess of 1 TeV. These machines are designed for the production and study of new particles as described, for example, in chapter 15. Such accelerators have been essential for the discovery of real weak bosons, see section 16.2, and the Higgs boson, see chapter 17.

For some experiments, particularly those at relatively low energies that look at interactions of systems of nucleons, such as those described in section 11.3, the accelerated particles are ions. We begin here with a brief description of an ion source. We then overview the different types of accelerators.

doi:10.1088/978-0-7503-6094-4ch22

C.1 Ion sources

Figure C.1 shows the design of an ion source. Gas is introduced into the plasma chamber where it is ionized by 2.45 GHz microwaves which enter through the waveguide coupling. The plasma is confined by a magnetic field provided by rare earth permanent magnets around the plasma chamber. The ions are extracted by the extraction electrodes by a potential difference of 10–20 kV between the plasma chamber and the electrodes.

The process described above will produce a beam of positive ions, which can then be injected into a suitable accelerator. In some cases, it is desirable to have a beam of negative ions. This can be produced from the output of the ion source by passing the ions through a region of neutral gas atoms, comprised of atoms with a low ionization potential, such as alkali atoms. The velocity of the negative ions with an energy of about 10 keV, is comparable to the orbital velocity of the atomic electrons on the neutral gas atoms and there is, therefore, a probability that the positive ions will capture electrons to become negatively charged ions.

C.2 Electrostatic accelerators

The first particle accelerator was developed in 1932 by John Cockcroft (1897–1967) and Ernest Walton (1903–95). It utilizes the Cockcroft–Watton voltage multiplier shown in figure C.2. This circuit was developed in 1919 by Heinrich Greinacher (1880–1974) and is, therefore, sometimes known as a Greinacher multiplier. An early example of a Cockcroft–Walton generator is shown in figure C.3.

The Cockcroft–Walton multiplier circuit consists of an arrangement of capacitors and rectifiers, as shown in the figure, that is connected to an AC voltage source. The time constant of the circuit is long compared to the inverse frequency of the AC voltage, so the capacitors fully charge. Because of the placement of the diodes, the

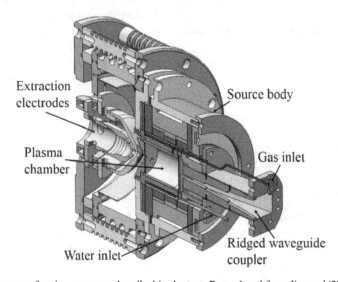

Figure C.1. Diagram of an ion source as described in the text. Reproduced from Jin *et al* (2021) CC BY 4.0.

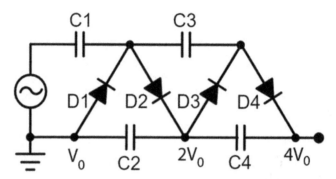

Figure C.2. Cockcroft–Walton voltage multiplier circuit. This [Cockcroft–Walton voltage multiplier circuit, also called the Greinacher multiplier] image has been obtained by the author from the Wikimedia website https://commons.wikimedia.org/wiki/File:Cockcroft_Walton_voltage_multiplier_circuit.svg (Chetvorno 2015) where it is stated to have been released into the public domain. It is included within this chapter on that basis.

voltage across the even numbered capacitors, C2, C4, etc, will be DC and will be doubled at each stage of the circuit, as shown in the figure.

Charged particles from an appropriate source are accelerated by the electric field resulting from the potential difference across the Cockcroft–Walton generator. Cockcroft and Walton used their accelerator to create the first artificial nuclear reaction using protons accelerated to 800 keV,

$$p + {}^{7}Li \rightarrow {}^{4}He + {}^{4}He.$$

Cockcroft–Walton generators are limited to a few MeV and are in fairly common use today as a source of particles to inject into more powerful accelerators.

The van de Graaff generator was developed by Robert J van de Graaff (1901–67) in the 1930s. A simplified diagram of a van de Graaff generator and accelerator is shown in figure C.4. Electric charge is distributed onto the bottom of an insulating rotating belt. The belt carries the charge to a conducting sphere at the top of the device. Once the charge is deposited onto the conducting sphere it accumulates on the outside of the sphere and this allows more charge to be extracted from the moving belt and transferred to the inside of the sphere. Charged particles can be accelerated by the high voltage collected on the sphere, as illustrated in the figure.

An important variation on the van de Graaf accelerator is the tandem van de Graaf accelerator. The principle of operation of a tandem accelerator is shown in figure C.5. Negatively charged ions, produced by an ion source as described above, are injected at low energy on the right side of the device shown in the figure. These negatively charged ions are accelerated towards a positive high voltage electrode at the center of the device where they pass through a gas or foil stripper which removes electrons from the ions, producing positively charged ions. The positively charged ions are accelerated towards the right end of the device (at zero potential) where they emerge at high energy.

Figure C.3. A Cockcroft–Walton generator constructed in 1937 by Philips in Eindhoven, Netherlands. It is currently located at the National Science Museum in London, UK. This [Cockcroft–Walton generator in the National Science Museum in London, England] image has been obtained by the author from the Wikimedia website where it was made available by Geni (2012) under a CC BY-SA 4.0 licence. It is included within this chapter on that basis. It is attributed to Geni.

Contemporary van de Graaff generators (see figure C.6) can produce up to about 20 MeV and are frequently used to inject charged particles into a cyclotron or synchrotron.

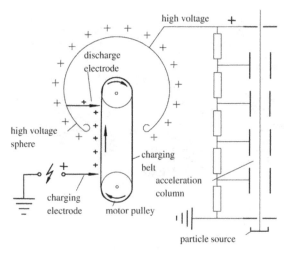

Figure C.4. Diagram of a van de Graaff generator. Reproduced from Wiedemann (2015) CC BY 4.0.

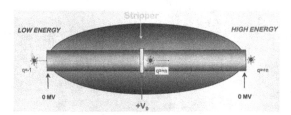

Figure C.5. Basic design of a tandem accelerator. This [acceleratore tandem] image been obtained by the author from the Wikimedia website where it was made available by Mauro (2018) under a CC BY-SA 4.0 licence. It is included within this chapter on that basis. It is attributed to Davide Mauro.

Figure C.6. Tandem van de Graaff accelerator at Hönggerberg, Laboratorium für Kernphysik ETH Zürich. Reproduced from ETH Zürich (1964). Photograph: Markwalder, Max / Ans_02683 / CC BY-SA 4.0.

C.3 Linear accelerators

The concept of the linear particle accelerator was first proposed in 1924 by Gustaf Ising (1883–1960) and the first operational prototype was constructed in 1928 by Rolf Widerøe (1902–96). Figure C.7 shows a basic diagram of a linear accelerator (sometimes called a 'linac'). Charged particles, provided by an ion source (or alternately electrons or positrons), pass through a series of hollow conductive tubes (referred to as drift tubes). The drift tubes are alternately connected to the output of an rf generator, as shown in the figure. As adjacent drift tubes have opposite polarities, there is an electric field in the gap between the drift tubes that accelerates the charged particles. Figure C.7 shows that the polarity of the tubes changes every half cycle of the rf voltage on the drift tubes and this frequency is synchronized to the time it takes for a particle to travel from one gap to the next. As the drift tubes are conducting, there is no electric field inside the tube and the particles travel through the tube at constant velocity. Since the particles increase velocity as they travel through the device, the drift tubes become progressively longer so that the time to traverse the length of each tube remains the same and the particle motion remains synchronized with the rf frequency.

We can look at the design of the linear accelerator in more detail for the case where the particles are non-relativistic. If the voltage difference across the gap between adjacent drift tubes is V_0, then after traversing n gaps the kinetic energy, T_n, of a particle of charge q and mass m will be,

$$T_n = nq\,V_0$$

and its non-relativistic velocity will be,

$$v_n = \left[\frac{2nqV_0}{m}\right]^{1/2}.$$

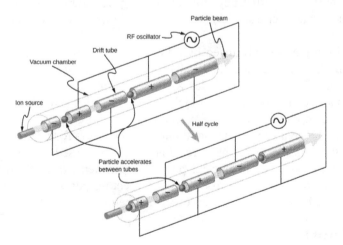

Figure C.7. Diagram of a linear accelerator showing particle locations and drift tube polarity at a half-cycle interval. Reproduced from LibreTexts Physics (2023) CC BY 4.0.

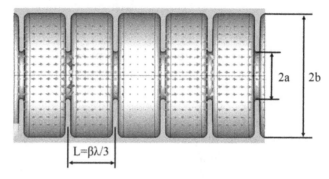

Figure C.8. Cross section of a disk loaded rf waveguide for use in a linear accelerator. Reprodced from Kutsaev (2021) CC BY 4.0.

If the frequency of the rf voltage is f, then the length of the nth drift tube, l_n, will be,

$$l_n = \left[\frac{nq V_0}{2mf^2} \right]^{1/2}.$$

This shows that for non-relativistic particles the drift tubes increase in length as $n^{1/2}$.

For particles that become relativistic then the velocity approaches c and does not increase proportionately with the increase in energy, so the drift tubes become nearly equal in length. This is particularly the case for electron or positron linear accelerators, due to the particle's low mass. At high energies traveling waves are more efficient for accelerating particles than the drift tube design described above. A typical disk loaded traveling waveguide designed for this purpose is shown in figure C.8.

Modern linear accelerators often use superconducting materials to reduce resistive losses in the rf cavities. Non-superconducting linear accelerators may have less than 25% efficiency for converting rf energy to particle energy. Superconducting accelerators, such as illustrated in figure C.9, can have efficiencies as high as 99%.

The maximum energy produced by a linear accelerator is largely limited by the practical length of the device. The longest linear accelerator constructed was the 3 km long Stanford Linear Collider at the Stanford Linear Accelerator Center (SLAC). This machine produces 50 GeV electrons and positrons. In the center of mass frame, the colliding beams have an energy of 100 GeV.

Because of the greater particle mass, linear accelerators for protons or ions, have lower energy limits. As such they are most commonly used for applications that require lower energy, such as nuclear physics experiments or medical radioisotope production, or are used as the front end for injecting particles into higher energy accelerators, such as the synchrotron described below.

C.4 Cyclotrons

The cyclotron was developed by Ernest Orlando Lawrence (1901–58) beginning in 1929. The first functional cyclotron was constructed by Lawrence in 1931. It had a

Figure C.9. Cryomodules in the SNS Superconducting Linac at Oak Ridge National Laboratory, Oak Ridge, TN. This linear accelerator is about 320 m long and accelerates protons for use in the Oak Ridge Spallation Neutron Source. Reproduced from Oak Ridge National Laboratory (2014) Flickr CC BY 2.0.

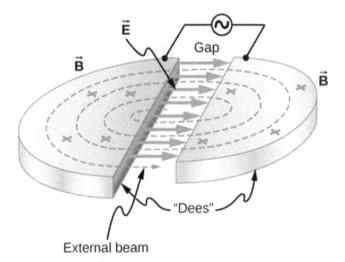

Figure C.10. Diagram of a cyclotron. Reproduced from LibreTexts Physics (2023) CC BY 4.0.

diameter of 22 cm and accelerated protons to 80 keV. The basic design of the cyclotron is shown in figure C.10. It consists of two 'dee' shaped metal chambers (called dees) separated by a gap. The chambers are placed between the poles of an electromagnet which provides a magnetic field perpendicular to the dees. Charged particles are injected into the center of the device and travel in circular paths as a result of the applied magnetic field. An alternating voltage is applied to the dees so that each time the charged particle traverses the gap it is subject to an electric potential that accelerates the particle. As the particle gains energy, it spirals outward and eventually leaves the cyclotron. An important aspect of the cyclotron is that the orbital period of the particle is independent of the radius of its orbit. In this way, the alternating voltage applied to the dees can be synchronized with the orbit of the particle so that the electric field changes direction every half period of the particle orbit.

The motion of the charged particle may be described by an analysis of the forces acting on it. Assuming a non-relativistic particle with charge q, the Lorentz force produced by the applied magnetic field, B, is,

$$\overrightarrow{F} = q\overrightarrow{v} \times \overrightarrow{B} \tag{C.1}$$

where $\overrightarrow{v}$ is the particle velocity. In the case illustrated in figure C.10, the applied magnetic field is perpendicular to the particle velocity and equation (C.1) may be written as

$$F = qvB = \frac{mv^2}{r}$$

where m is the particle mass and r is the radius of its orbit. Solving for v, gives

$$v = \frac{qrB}{m}. \tag{C.2}$$

We can calculate the time for the particle to complete one orbit as

$$t = \frac{2\pi r}{v} = \frac{2\pi m}{qB}.$$

The necessary AC voltage frequency, f, to synchronize the electric field with the particle motion is

$$f = \frac{1}{t} = \frac{qB}{2\pi m}.$$

This frequency is referred to as the cyclotron frequency, sometimes written as

$$\omega_c = 2\pi f = \frac{qB}{m}.$$

For a cyclotron with a radius R, the maximum energy will be achieved for $r = R$ so from equation (C.2) the maximum kinetic energy T_{max} can be found as

$$T_{max} = \frac{mv_{max}^2}{2} = \frac{q^2 R^2 B^2}{2m}. \tag{C.3}$$

A modern cyclotron is shown in figure C.11. In order to minimize resistive losses, many current machines, such as the one shown in the figure, utilize superconducting magnets.

Figure C.11. Superconducting Ring Cyclotron at the RIKEN Nishina Center for Accelerator-Based Science in Saitama, Japan. The outer diameter of the cyclotron is 18.4 m. This [Superconducting Ring Cyclotron (SRC) of Radioactive Isotope Beam Factory] image has been obtained by the author from the Wikimedia website where it was made available by Kestrel (2018) under a CC BY-SA 4.0 licence. It is included within this chapter on that basis. It is attributed to Kestrel.

Following along the discussion for the linear accelerator, it is clear that the above description of the cyclotron is accurate only for non-relativistic particles. This means that because of their low mass, cyclotrons are not suitable for accelerating electrons. For protons and heavier ions, cyclotrons are useful for relatively low energy applications, such as medical radioisotope production. At higher energies there are ways of dealing with the relativistic properties of the particles. The solution to this problem was fairly straight forward for the linear accelerator but is not so simple for the cyclotron. An analysis of the above equations shows that in order to maintain the proper resonance condition for relativistic particles the cyclotron field or frequency (or both) must be adjusted.

The synchrocyclotron uses a constant magnetic field and adjusts the frequency as the particles move outward from the center of the devices towards the edge. It is obvious that that this device cannot operate continuously. Rather, the particles travel in bunches from the center to the outside. As each bunch travels through the device, the frequency is swept from its maximum value (when the particles start in the center) to its minimum value (when the particles exit the synchrocyclotron).

The other approach, the isochronous cyclotron (or isocyclotron), utilizes a constant frequency but varies the strength of the magnetic field as a function of distance from the center of the device. In this case, the pole pieces of the magnet are shaped so that the field becomes stronger with increasing distance from the center. By itself, this approach tends to de-focus the particle beam. It is, therefore, typically used in conjunction with magnets that also vary the field strength azimuthally so that stable particle orbits are no longer circular. Such devices are referred to a AVF (or

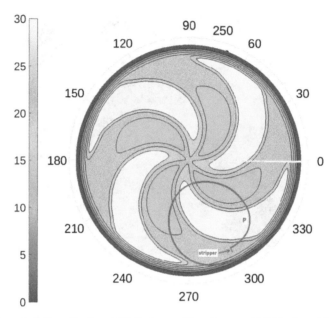

Figure C.12. Magnetic field profiles in a AVF Cyclotron. From Rao *et al* (2023). Copyright IOP Publishing. Reproduced with permission. All rights reserved.

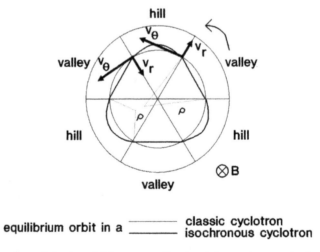

Figure C.13. Comparison of circular orbit in a conventional cyclotron and a non-circular orbit in an AVF cyclotron. Reprinted from Strijckmans (2001). Copyright (2001). With permission from Elsevier.

azimuthally varying field) cyclotrons. Figure C.12 shows a typical magnetic field profile in an AVF cyclotron and figure C.13 shows a typical non-circular particle orbit. AVF cyclotrons have the advantage over synchrocyclotrons that the particle beam is continuous.

As equation (C.3) shows that the maximum energy that can be obtained using a cyclotron is a function of the radius of the device, a major factor that limits the energy that can be produced by a cyclotron is the diameter of an electromagnet that is viable. For protons, maximum energies around 500 MeV have been achieved by cyclotrons. For higher energies, synchrotrons, as discussed below, are used.

C.5 Synchrotrons

The synchrocyclotron and isocyclotron described above take different approaches to dealing with relativistic effects in the design of the cyclotron by using variable frequencies and variable magnetic fields, respectively. The synchrotron uses a design that incorporates both varying frequencies, as well as varying magnetic fields. An important consequence of this approach is that the particle orbits can be constrained to be circular, rather than spiraling outward as the particle energy increases. This has the advantage that the magnetic field has to be applied over the region of the orbit rather than over the entire area of the spiraling orbit and this greatly reduces the cost of a large accelerator.

Figure C.14 shows the design of a synchrotron. Many synchrotrons use the so-called 'racetrack' geometry which consists of curved sections of beam path connected together with straight sections. Particles travel in bunches around the synchrotron and are accelerated during each orbit. Particles in a typical synchrotron might make 10^7 orbits before reaching maximum energy. Five of the important components in the design on the synchrotron are the injection magnet, the extraction magnet, bending magnets, focusing magnets and an accelerating cavity.

Since the synchrotron cannot accelerate particles from zero velocity, particles from another accelerator, frequently a linear accelerator, must be injected into the

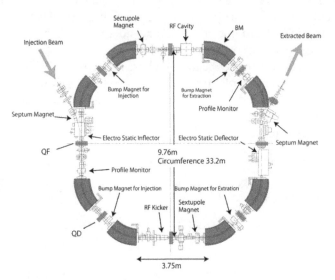

Figure C.14. Diagram of a synchrotron. Reprinted from Shiokawa *et al* (2021). Copyright (2021). With permission from Elsevier.

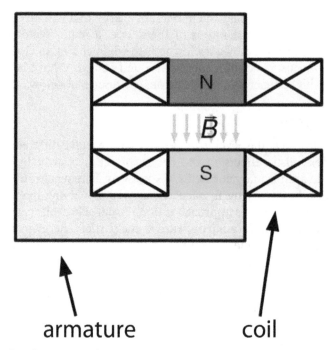

armature   coil

Figure C.15. Bending dipole magnet from a synchrotron. This [sketch of a dipole magnet] image has been obtained by the author from the Wikimedia website where it was made available by Florian (2005a) under a CC BY-SA 3.0 licence. It is included within this chapter on that basis. It is attributed to Florian D O. (Labels translated to English.)

synchrotron. This process utilizes an injection magnet to align the input particle beam with the trajectory of the particles in the synchrotron. The extraction magnet removes the particles from the accelerator and directs them to an appropriate experimental area.

Bending magnets direct the beam of charged particles by applying a dipole magnetic field. The geometry of a typical bending magnet is shown in figure C.15. As the energy of the accelerated particles increases, the magnetic field provided by the bending magnets is increased to maintain a constant radius orbit. Using the relativistic form of equation (C.2) gives the magnetic field as

$$B = \frac{\gamma m v}{q r}$$

where the Lorentz factor is

$$\gamma = \left[1 - \frac{v^2}{c^2} \right]^{-1/2}.$$

Thus, as the particles are accelerated, the magnetic field in the dipole bending magnets is ramped up, and if the particle energy is maintained, then the magnetic

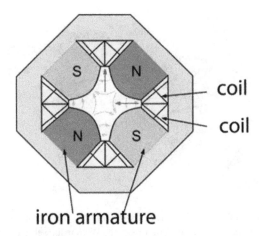

Figure C.16. Quadrupole focusing magnet from a synchrotron. Grey arrows: magnetic field lines, blue arrows: forces on particles. This [sketch of a quadrupole magnet] image has been obtained by the author from the Wikimedia website where it was made available by Florian (2005b) under a CC BY-SA 3.0 licence. It is included within this chapter on that basis. It is attributed to Florian D O. (Labels translated to English.)

field can be kept constant. Typically, modern synchrotrons utilize superconducting magnets in order to maximize the applied field from the bending magnets, and, therefore, to maximize the particle energy.

The particle beam tends to spread as it travels around the synchrotron and quadrupole focusing magnets are used to confine the beam. Figure C.16 shows the design of a focusing magnet. The magnet focuses the beam in one direction (the horizontal direction in the figure) and de-focuses it in an orthogonal direction (the vertical direction in the figure). Placing the focusing magnets in pairs of alternating orientation along the straight sections of the beam path, will focus the beam in both directions. This technique is known as alternating gradient focusing, or sometimes strong focusing.

The rf cavity, as seen in figure C.14, provides energy to the particles once every orbit around the synchrotron. As the energy increases, the frequency of the energy supplied to the cavity must increase as well. For a circular orbit the frequency is related to the Lorentz factor as

$$f = \frac{qB}{2\pi\gamma m}.$$

It is important to note that a charged particle in a synchrotron loses energy due to synchrotron radiation. This is because, even if the particle is traveling at a constant velocity, it is accelerating in its orbit. The power radiated by the particle can be expressed as

$$P = \frac{1}{6\pi\varepsilon_0} \frac{q^2 E^4}{c^7 m^4 r^2} \tag{C.4}$$

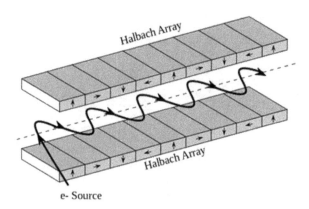

e- Source

Figure C.17. Basic geometry of an undulator used for producing synchrotron radiation. This [free electron laser design] image has been obtained by the author from the Wikimedia website https://commons.wikimedia. org/wiki/File:HalbachArrayFEL.png (Hiltonj 2006), where it is stated to have been released into the public domain. It is included within this chapter on that basis.

where ε_0 is the vacuum permittivity, E is the total energy and r is the radius of curvature of the orbit. It is important to note the mass dependence, m^{-4}, in equation (C.4). This means that under the same conditions, electrons will radiate about $(2000)^4 \approx 10^{13}$ times as much power as protons. This fact has significant consequences for the use of synchrotrons for the acceleration of electrons compared to protons or heavier ions. In fact, this feature has been put to practical use for the production of synchrotron radiation by electron synchrotrons. This radiation is in the form of very intense x-rays and is in common use for x-ray diffraction and x-ray imaging experiments. Typically, accelerated particles are injected into a storage ring, where the energy is kept constant. Additional magnets, i.e., undulators as shown in figure C.17, are included in the beam path in order to provide regions of alternating field where the particles undergo a trajectory with small radii of curvature. This increases the intensity of the synchrotron radiation that is produced.

For the study of particle physics, synchrotrons are typically used to accelerate protons or other ions and have been instrumental in recent discoveries, such as the Higgs boson discussed in chapter 17. Figure C.18 shows an example of a typical proton synchrotron used for particle studies.

The largest and highest energy synchrotron is the Large Hadron Collider (LHC) constructed by the European Organization for Nuclear Research (CERN) on the France-Switzerland border near Geneva (see Evans 2007). It first became operational in 2008. The collider is 26.7 km in circumference or 4.24 km radius and typically collides protons with protons. The maximum proton energy is 6.8 TeV and the center-of-mass energy for the p-p collisions is 13.6 TeV. Experiments of collisions of Pb–Pb ions or Pb–protons can also be conducted.

The basic layout of the LHC is shown in figure C.19. Protons are first accelerated by a linear accelerator to 160 MeV and are then injected into the Proton Synchrotron Booster, where the energy is increased to 2 GeV. The 2 GeV protons are then injected into the Proton Synchrotron, which produces an energy of 26 GeV.

Figure C.18. Main ring of the 50 GeV Synchrotron at the Japan Proton Accelerator Research Complex. This [Main Ring 50 GeV Synchrotron of J-PARC] image has been obtained by the author from the Wikimedia website where it was made available by Kestrel (2016) under a CC BY-SA 4.0 licence. It is included within this chapter on that basis. It is attributed to Kestrel.

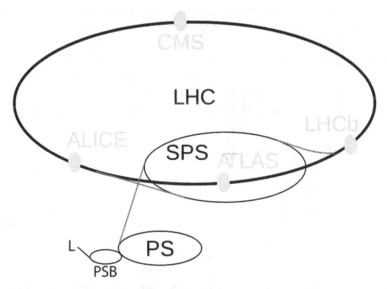

Figure C.19. Schematic diagram of the LHC at CERN L: Linear Accelerator (Linac), PSB: Proton Synchrotron Booster, PS: Proton Synchrotron, SPS: Super Proton Synchrotron, LHC: Large Hadron Collider. This [LHC experiments and the preaccelerators] image has been obtained by the author from the Wikimedia website where it was made available by Horvath (2006) under a CC BY-SA 2.5 licence. It is included within this chapter on that basis. It is attributed to Arpad Horvath. (PSB label added, two linacs replaced with linac L.)

Figure C.20. A portion of the LHC beamline. The blue cylinders contain the two evacuated beamlines for the two counter rotating beams and the superconducting dipole bending magnets along with their associated cryogenic systems. The curvature of the beamline can be seen at the far left in the image. Reproduced from CERN (2009) CC BY 4.0.

The Super Proton Synchrotron then yields protons at 450 GeV, and these are introduced into the LHC where the energy is increased to a maximum of 6.8 TeV. A portion of the main beamline of the LHC is shown in figure C.20.

The LHC uses two evacuated beam tubes to carry counter-circulating proton beams which intersect at four points around the circumference of the collider where experiments are performed. These are indicated by the yellow dots in figure C.19. The injection of the clockwise and counterclockwise rotating proton beams from the Super Proton Synchrotron into the LHC are shown by the red lines in the figure. Protons are accelerated in bunches, and it takes about 20 min to ramp-up their energy to 6.8 TeV. Once the desired energy is achieved, the proton bunches can be circulated for 5–24 h, during which time experiments can be conducted.

References and suggestions for further reading

CERN 2009 Views of the LHC tunnel in sector 3–4 (http://cds.cern.ch/record/1211045)

Chetvorno 2015 Cockcroft–Walton voltage multiplier circuit, also called the Greinacher multiplier (https://commons.wikimedia.org/wiki/File:Cockcroft_Walton_voltage_multiplier_circuit.svg)

Zürich E T H 1964 Zürich, ETH Zürich, Hönggerberg, Laboratorium für Kernphysik (HIK/HPK), Maschinentrakt, Versuchslabor Tandem-Van-de-Graaff-Beschleuniger (https://ba.e-pics.ethz.ch/catalog/ETHBIB.Bildarchiv/r/3107/viewmode=infoview)

Evans L 2007 The large Hadron collider *New J. Phys.* **9** 335

Florian D O 2005a Sketch of a dipole magnet (https://commons.wikimedia.org/wiki/File:Dipole_de.png)

Florian D O 2005b Sketch of a quadrupole magnet (https://commons.wikimedia.org/wiki/File:Quadrupole_de.svg)

Geni 2012 Cockcroft–Walton generator in the National Science Museum in London, England (https://commons.wikimedia.org/wiki/File:Cockcroft%E2%80%93Walton_generator_2012.JPG)

Hiltonj 2006 Free electron laser design (https://commons.wikimedia.org/wiki/File:HalbachArrayFEL.png)

Horvath A 2006 The LHC experiments and the preaccelerators (https://commons.wikimedia.org/wiki/File:LHC.svg)

Jin Q Y, Liu Y G, Zhou Y, Wu Q, Zhai Y J and Sun L T 2021 RF and microwave ion sources study at Institute of Modern Physics *Plasma* **4** 332–44

Kain V 2016 *Beam Dynamics and Beam Losses - Circular Machines* (https://researchgate.net/figure/Typical-layout-of-synchrotron_fig9_305995044)

Kestrel 2016 The main ring 50 GeV synchrotron of J-PARC (https://commons.wikimedia.org/wiki/File:J-PARC_Main_Ring_P7311328.jpg)

Kestrel 2018 Superconducting ring cyclotron (SRC) of Radioactive Isotope Beam Factory (https://commons.wikimedia.org/wiki/File:RIKEN_Supercomducting_Ring_Cyclotron_P4213471.jpg)

Kutsaev S V 2021 Electron bunchers for industrial RF linear accelerators: theory and design guide *Eur. Phys. J. Plus* **136** 446

LibreTexts Physics 2023 *Particle accelerators and detectors* 11 5 https://phys.libretexts.org/Bookshelves/University_Physics/Book%3A_University_Physics_(OpenStax)/University_Physics_III_-_Optics_and_Modern_Physics_(OpenStax)/11%3A_Particle_Physics_and_Cosmology/11.05%3A_Particle_Accelerators_and_Detectors

Mauro D 2018 Acceleratore tandem (https://commons.wikimedia.org/wiki/File:Tandem_Accelerator_Diagram.jpg)

Oak Ridge National Laboratory 2014 Cryomodules in the SNS superconducting linac (https://flickr.com/photos/oakridgelab/12508030203/)

Rao Y N, Zhang L G, Baartman R, Bylinskii Y, Koay H W and Planche T 2023 A high-intensity superconducting H^+_3 cyclotron for isotope production *JINST* **18** P03023

Shiokawa T, Okugawa Y, Kurita T and Nakanishi T 2021 Slow beam extraction method from synchrotron for uniform spill and fast beam switching using an RF knockout method of multi-band colored noise signal—POP experiment and simulation *Nucl. Instrum. Methods Phys. Res.* **1010** 165560

Strijckmans K 2001 The isochronous cyclotron: principles and recent developments *CMIG* **25** 69–78

Wiedemann H 2015 *Particle Accelerator Physics* (Cham: Springer) https://link.springer.com/book/10.1007/978-3-319-18317-6

IOP Publishing

An Introduction to the Physics of Nuclei and Particles
(Second Edition)

Richard A Dunlap

Appendix D

Solutions to even numbered problems

3.2. Data may be analyzed on the basis of equation (3.10). This may be written in terms of the incident energy, E, as

$$\frac{d\sigma}{d\Omega} = \left(\frac{Zze^2}{4\pi\varepsilon_0}\right)^2\left(\frac{1}{4E}\right)^2 \csc^4\left(\frac{\theta}{2}\right).$$

Measurements are made for a fixed angle, θ, as a function of energy. Since the detector will subtend a solid angle, Ω, the total cross section for scattering into the detector will be

$$\sigma = \int \frac{d\sigma}{d\Omega} d\Omega = 2\pi \left(\frac{Zze^2}{4\pi\varepsilon_0}\right)^2\left(\frac{1}{4E}\right)^2 \int \csc^4\left(\frac{\theta}{2}\right) \sin\theta \; d\theta.$$

The number of scattered particles observed will be proportional to σ and hence to E^{-2}.

3.4.

    (a) The scattering angle, θ, is related to the impact parameter, b, as

$$\tan\frac{\theta}{2} = \frac{Zze^2}{4\pi\varepsilon_0 m v_0^2 b}.$$

Solving for b and expressing the velocity in terms of the initial kinetic energy, E_0;

$$b = \frac{e^2}{4\pi\varepsilon_0}\frac{Zz}{2E_0 \tan\frac{\theta}{2}}.$$

doi:10.1088/978-0-7503-6094-4ch23

Using $Z = 79$, $z = 2$, $E_0 = 8$ MeV and $\theta_, = 90°$ gives

$$b = 14.2 \text{ fm}.$$

(b) Conservation of energy gives

$$E_0 = \frac{1}{2}mv_c^2 + \frac{e^2}{4\pi\varepsilon_0}\frac{Zz}{r_c}$$

where the subscript c denotes the point of closest approach. Conservation of angular momentum gives

$$mv_0b = mv_c r_c.$$

Solving for v_c and substituting into the expression for E_0 gives

$$E_0 = E_0\frac{b^2}{r_c^2} + \frac{e^2}{4\pi\varepsilon_0}\frac{Zz}{r_c}.$$

Substituting $E_0 = 8$ MeV and $b = 14.2$ fm gives a quadratic in r_c;

$$8r_c^2 - 228r_c - 1618 = 0.$$

Solving for r_c gives $r_c = 34.4$ fm.

(c) The kinetic energy will be

$$E_c = \frac{1}{2}mv_c^2 = E_0\frac{b^2}{r_c^2}.$$

Substituting numerical values gives $E_c = 1.36$ MeV.

3.6.

(a) All three nuclei have $\rho(0) \approx 0.16$ fm^{-3}. The values of r_{90} and r_{10} are found so that

$$\rho(r_{90}) = 0.9 \times 0.16 = 0.144 \text{ fm}^{-3}$$

$$\rho(r_{10}) = 0.1 \times 0.16 = 0.016 \text{ fm}^{-3}.$$

The width of the surface region is then given as $r_{10}-r_{90}$. Reading values from the graph gives the values table D.1.

(b) The values of r_{10} and r_{90} are defined in terms of the mean radius as

$$r_{90} = R_0 - 1.2 \text{ (fm)}$$

and

$$r_{10} = R_0 + 1.2 \text{ (fm)}$$

Table D.1. Calculated width of surface regions for ^{16}O, ^{118}Sn and ^{197}Au.

Nucleus	r_{10} (fm)	r_{90} (fm)	$(r_{10}-r_{90})$ (fm)
^{16}O	3.6	1.2	2.4
^{118}Sn	6.3	3.9	2.4
^{197}Au	7.7	5.3	2.4

Equation (3.13) may be rewritten as

$$\rho_{90} = \frac{0.16}{1 + \exp\left[-\frac{1.2}{a}\right]}$$

and

$$\rho_{10} = \frac{0.16}{1 + \exp\left[+\frac{1.2}{a}\right]}.$$

Either equation may be solved to give $a = 0.55$ fm.

3.8. The relativistic scattering cross section is given as

$$\left(\frac{d\sigma}{d\Omega}\right)_{rel} = \left(\frac{d\sigma}{d\Omega}\right)_{nonrel}\left[1 - \frac{v^2}{c^2}\sin^2\frac{\theta}{2}\right].$$

We define the relative size of the relativistic correction as

$$f = \frac{\left|\left(\frac{d\sigma}{d\Omega}\right)_{rel} - \left(\frac{d\sigma}{d\Omega}\right)_{nonrel}\right|}{\left(\frac{d\sigma}{d\Omega}\right)_{nonrel}} = \frac{v^2}{c^2}\sin^2\frac{\theta}{2}.$$

For $E = 0.1$ MeV then $mv^2/2 \ll mc^2$ and we calculate

$$\frac{v^2}{c^2} = \frac{2\left(\frac{1}{2}mv^2\right)}{c^2} = \frac{0.2}{0.511} = 0.39.$$

For $E = 1$ MeV or 100 MeV the $E > mc^2$ so $v \approx c$ giving $v^2/c^2 = 1$. Using these values in the above expression gives the results in table D.2.

3.10.

(a) Conservation of energy gives

$$m_a(v_{ai}^2 - v_{af}^2) = m_A v_{Af}^2.$$

This may be rearranged as

$$m_a(v_{ai} - v_{af})(v_{ai} + v_{af}) = m_A v_{Af}^2.$$

Table D.2. Relativistic corrections for scattering cross sections as a function of energy and scattering angle.

E (MeV)	v^2/c^2	θ (°)	f
0.1	0.39	20	0.011
0.1	0.39	90	0.195
1.0	1.0	20	0.12
1.0	1.0	90	0.5
100	1.0	20	0.12
100	1.0	90	0.5

Table D.3. Recoil energies for ^{118}Sn and ^{197}Au.

Nucleus	m_A (u)	R_A (MeV)
^{118}Sn	117.902	1.27
^{197}Au	196.967	0.78

Conservation of momentum gives

$$m_\alpha(v_{\alpha i} - v_{\alpha f}) = m_A v_{Af}.$$

Combining these expressions yields

$$v_{\alpha i} + v_{\alpha f} = v_{Af}.$$

Using this with the conservation of momentum equation gives

$$v_{Af} = 2\frac{m_\alpha v_{\alpha i}}{m_\alpha + m_A}.$$

The recoil energy (kinetic energy) is obtained from this as

$$R_A = \frac{2m_A m_\alpha^2 v_{\alpha i}^2}{(m_\alpha + m_A)^2}.$$

Using

$$E_{\alpha i} = \frac{1}{2}m_\alpha v_{\alpha i}^2$$

this may be rewritten as

$$R_A = \frac{4m_\alpha m_A E_{\alpha i}}{(m_\alpha + m_A)^2}.$$

Using $m_\alpha = 4.0015$ u, $m(^{16}O) = 15.995$ u and $E_{\alpha i} = 10$ MeV we obtain $R_A = 6.4$ MeV.

(b) Using the masses for ^{118}Sn and ^{197}Au we obtain the results in table D.3.

4.2.

(a) We use the equation for B/A with values of the constants $a_V = 15.5$, $a_s = 16.8$, $a_C = 0.72$, $a_{sym} = 23.2$ and $a_p = -34$, 0, +34 in MeV. For the nuclei considered here the pairing term is

$$^3\text{H (e} - \text{o)} \; a_p = 0$$

$$^4\text{He (e} - \text{e)} \; a_p = -34 \text{ MeV}$$

$$^{64}\text{Cu (o} - \text{o)} \; a_p = +34 \text{ MeV}$$

$$^{119}\text{Sn (o} - \text{e)} \; a_p = 0$$

A simple computer analysis gives the results in table D.4.

(b) Using

$$\frac{B}{A} = -\left[m - Nm_n - Z(m_p + m_e) \right] \frac{c^2}{A}$$

and measured particle masses and atomic masses gives the results in table D.5.

(c) A comparison of parts (a) and (b) shows that the liquid drop model is not successful for light nuclei but works very well for heavy nuclei.

Table D.4. Binding energy per nucleon from the semiempirical mass formula.

Nuclide	$(B/A)_{\text{SEM}}$
^3H	1.27
^4He	7.69
^{64}Cu	8.78
^{119}Sn	8.47

Table D.5. Calculated binding energy per nucleon values from measured masses.

Nuclide	$(B/A)_{\text{measured}}$
^3H	2.83
^4He	7.07
^{64}Cu	8.74
^{119}Sn	8.50

4.4.

(a) Ignoring the electronic binding we write

$$m = Nm_n + Z(m_p + m_e) - \frac{B}{c^2}$$

or

$$B = \left[Nm_n + Z(m_p + m_e) - m \right]c^2.$$

For ^{13}C ($N = 7$, $Z = 6$) and ^{13}N ($N = 6$, $Z = 7$) we substitute values of the particle and atomic masses from the tables to obtain $B(^{13}C) = 97.109$ MeV and $B(^{13}N) = 94.105$ MeV.

(b) For the semiempirical mass expression, for the binding energy all terms are the same for ^{13}C and ^{13}N except the Coulomb term.

(c) The Coulomb term in the binding energy is written as

$$B_c = a_c \frac{Z(Z-1)}{A^{1/3}}.$$

Using $R_0 = 1.2\, A^{1/3}$ (fm) gives $A^{1/3} = 0.83 R_0$. Then

$$B_c = a_c \frac{Z(Z-1)}{0.83 R_0}.$$

Taking the differential

$$\Delta B_c = \Delta Z \frac{\partial B_c}{\partial Z} = \Delta Z \frac{a_c}{0.83 R_0} [2Z - 1]$$

and solving for R_0

$$R_0 = \frac{a_c[2Z - 1]\Delta Z}{0.83 \Delta B_c}.$$

Using ΔB_c from part (a) for the difference between ^{13}C and ^{13}N, $Z = 6$ (for carbon), $a_C = 0.72$ MeV and $\Delta Z = 7 - 6 = 1$ gives $R_0 = 3.17$ fm. Compare this with $R_0 = 1.2\, A^{1/3} = 1.2 \times 15^{1/3} = 2.96$ fm.

4.6. From the equations for the separation energies

$$S_n = \left[m\left(^{A-1}_{Z}X^{N-1} \right) + m_n - m\left(^{A}_{Z}X^{N} \right) \right]c^2$$

$$S_p = \left[m\left(^{A-1}_{Z-1}Y^{N} \right) + m_p + m_e - m\left(^{A}_{Z}X^{N} \right) \right]c^2$$

where the masses are atomic masses. Using tabulated mass values, we find the results in table D.6.

Table D.6. Calculated neutron and proton separation energies.

$_Z^A X^N$	$_Z^{A-1} X^{N-1}$	$_{Z-1}^{A-1} Y^N$	S_n (MeV)	S_p (MeV)
^{235}U	^{234}U	^{234}Pa	5.298	6.711
^{236}U	^{235}U	^{235}Pa	6.545	7.173
^{235}Np	^{234}Np	^{234}U	6.984	4.392
^{236}Np	^{235}Np	^{235}U	5.739	4.833

Table D.7. Number of neutrons and protons in the nuclei from table D.6.

Nuclide	N	Z
^{235}U	143	92
^{236}U	144	92
^{235}Np	142	93
^{236}Np	143	93

Table D.8. Calculated Z_{min} for different values of A.

A	Z_{min}
3	1.48
19	9
99	43
201	80

Consider these in the context of the number of neutrons and protons in the parent nucleus given in table D.7

In all cases the odd nucleon is less tightly bound—as we would expect.

4.8. Using the equation for Z_{min} for various A as shown we obtain the results in table D.8.

Using $A = N + Z$ then $N = A - Z$ and

$$\frac{N}{Z} = \frac{A - Z}{Z}$$

giving the values in table D.9.

Note that there is a large round-off error for $A = 3$ and the semiempirical mass formula is not expected to work well for small A. Otherwise, this shows good agreement with observed stable nuclei.

Table D.9. N/Z ratios as a function of A.

A	N/Z
3	1.03
19	1.11
99	1.30
201	1.51

4.10.

(a) Beginning with equation (4.4) we convert the expression for A and Z to A and N by substituting $Z = A - N$. This gives

$$m(A, N) = Nm_n + (A - N)(m_p + m_e) - \frac{a_V A}{c^2} + \frac{a_S A^{2/3}}{c^2} + \frac{a_C\left(A^2 - A(2N + 1) + N^2 + N\right)}{A^{1/2}c^2}$$
$$+ \frac{a_{sym}\left(A^2 - 4NA + 4N^2\right)}{Ac^2} + \frac{a_p}{A^{3/4}c^2}.$$

(b) The binding energy may be determined from the above expression as

$$B(A, N) = a_V A - a_S A^{2/3} - \frac{a_C(A^2 - A(2N + 1) + N^2 + N)}{A^{1/3}} - \frac{a_{sym}(A^2 - 4NA + 4N^2)}{A} - \frac{a_p}{A^{3/4}}.$$

The two-proton separation energy is given by

$$\Delta B = \Delta A \frac{\partial B}{\partial A}\bigg|_{N=const}$$

where $\Delta A = -2$ (i.e., A decreases by 2 while N remains unchanged). We combine terms according to powers of A in order to calculate the derivative

$$B(A, N) = -a_C A^{5/3} + (a_V - a_{sym})A + (a_C(2N + 1) - a_S)A^{2/3}$$
$$+ 4a_{sym}N - a_C(N^2 + N)A^{-1/3} - a_p A^{-3/4} - 4a_{sym}NA^{-1}.$$

Taking the derivative gives

$$\Delta B = -2\left\{-\frac{5}{3}a_C A^{2/3} + (a_V - a_{sym}) + \frac{2}{3}(a_C(2N + 1) - a_S)A^{-1/3}\right.$$
$$\left. + \frac{1}{3}a_C(N^2 + N)A^{-1/3} + \frac{3}{4}a_p A^{-7/4} + 4a_{sym}N^2A^{-2}\right\}.$$

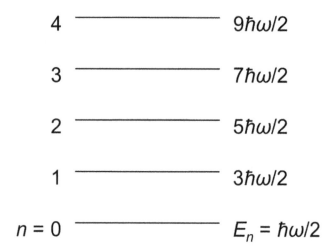

Figure D.1. Energy levels as calculated in part (a).

(c) By looking at the two-proton separation energy, the value of a_p does not change. However, the presence of the pairing term in the final expression cannot be eliminated because of the A dependence of this term.

5.2.

(a) A nice description of the quantum harmonic oscillator can be found in J Powell and B Craseman *Quantum Mechanics* Addison-Wesley (1961). The energy levels are given in terms of the principal quantum number, n, as

$$E_n = \hbar\omega\left(n + \frac{1}{2}\right)$$

where $n = 0,1,2,....$ Thus E_n represents the energy of the $(n+1)$th level. The energy levels are evenly spaced with a separation $\hbar\omega$. Each level, therefore, represents a filled shell and determines the relevant magic number.

(b) The energy level diagram as given by the above expression is shown in figure D.1.

The degeneracy of each level depends on the allowed values of the good quantum numbers. For each level as determined by n, the spin degeneracy gives a factor of 2 and the ℓ degeneracy gives a factor of $2\ell+1$. It is important to determine the allowed values of ℓ for a given value of n. The form of the solution to Schrödinger's equation requires that ℓ has positive values such that $\ell = n, n-2, n-4, ...$ The total degeneracy in n is the sum of the degeneracies in ℓ with an additional factor of 2 for the spin degeneracy, as given in table D.10.

Table D.10. Calculated degeneracies for different values of n.

n	ℓ	Degeneracy
0	0	2
1	1	6
2	0,2	12
3	1,3	20
4	0,2,4	30
5	1,3,5	42

Table D.11. Calculated magic numbers from part (b).

Energy level ($n+1$)	Magic number
1	2
2	8
3	20
4	40
5	70
6	112

(c) The magic numbers are given by the total degeneracy up to a given energy level as shown in table D.11.

5.4.

(a) From equation (4.2) we find

$$B = \left[Nm_n + Z(m_p + m_e) - m \right] c^2$$

where we have ignored the electronic binding energy. Thus,

$$\frac{B}{A} = \frac{1}{A}\left[Nm_n + Z(m_p + m_e) - m \right] c^2$$

and using tabulated values of the masses we find the values in table D.12.

(b) We see that ^{40}Ca ($N = 20$, $Z = 20$) has a higher B/A than ^{38}K ($N = 19$, $Z = 19$) and ^{42}Sc ($N = 21$, $Z = 21$). In terms of the semiempirical mass formula ^{40}Ca is an even–even nucleus while ^{38}K and ^{42}Sc are odd–odd nuclei. The former occurs on the e–e parabola and has an abnormally small mass compared with the later nuclei that occur on the o–o parabola. Hence

Table D.12. Calculated binding energies per nucleon.

Nuclide	B/A (MeV)
^{38}K	8.44
^{40}Ca	8.55
^{42}Sc	8.44

the difference in the value of B/A. In terms of the shell model ^{40}Ca is doubly magic and is characterized by a large binding energy per nucleon.

6.2.

(a) The energy associated with the (quantum) rotated levels is

$$E = \frac{I(I+1)\hbar^2}{2I}$$

giving

$$I = \frac{I(I+1)\hbar^2}{2E}.$$

Units are as follows:

$$\hbar = 1.055 \times 10^{-34} \text{ kg} \cdot \text{m}^2 \text{ s}^{-1}$$

$$E \text{ in kg} \cdot \text{m}^2 \text{ s}^{-1} \text{ is } 1.6 \times 10^{-13} E \text{ (MeV)}$$

$$I = \frac{I(I+1)}{2} \cdot \frac{(1.055 \times 10^{-34})^2}{(1.6 \times 10^{-13})} \cdot \frac{1}{E(\text{MeV})} \cdot \text{kg}^2 \cdot \text{m}^4 \text{ s}^{-2} \cdot \text{s}^2 \cdot \text{kg}^{-1} \text{m}^{-2}$$

$$= \frac{I(I+1)}{2} \cdot \frac{6.96 \times 10^{-56}}{E(\text{MeV})} \text{ kg m}^2$$

This is the correct SI unit for I. The information tabulated from figure 6.11 is given in table D.13.

The slight increase as a function of I (or E) results from the greater deformation of nuclei with higher rotational energy.

(b) Classically for a sphere

$$I = \frac{2}{5}mR^2$$

where

$$R = 1.2 \times 10^{-15}(174)^{1/3} = 6.69 \times 10^{-15} \text{ m}.$$

Table D.13. Calculated moments of inertia for excited rotational states of ^{174}W.

Level	I	E	I (kg· m^2)
Ground	0	–	–
1st	2	0.11	1.90×10^{-54}
2nd	4	0.36	1.93×10^{-54}
3rd	6	0.70	2.09×10^{-54}
4th	8	1.14	2.20×10^{-54}
5th	10	1.64	2.33×10^{-54}
6th	12	2.19	2.48×10^{-54}

Table D.14. Ground state neutron and proton configurations.

Nuclide	(N, Z)	Neutron configuration	Proton configuration
^{14}N	(7, 7)	$1s_{1/2}{}^2 1p_{3/2}{}^4 1p_{1/2}{}^1$	$1s_{1/2}{}^2 1p_{3/2}{}^4 1p_{1/2}{}^1$
^{20}F	(11, 9)	$1s_{1/2}{}^2 1p_{3/2}{}^4 1p_{1/2}{}^2 1d_{5/2}{}^3$	$1s_{1/2}{}^2 1p_{3/2}{}^4 1p_{1/2}{}^2 1d_{5/2}{}^1$
^{24}Na	(13, 11)	$1s_{1/2}{}^2 1p_{3/2}{}^4 1p_{1/2}{}^2 1d_{5/2}{}^5$	$1s_{1/2}{}^2 1p_{3/2}{}^4 1p_{1/2}{}^2 1d_{5/2}{}^3$
^{26}Al	(13, 13)	$1s_{1/2}{}^2 1p_{3/2}{}^4 1p_{1/2}{}^2 1d_{5/2}{}^5$	$1s_{1/2}{}^2 1p_{3/2}{}^4 1p_{1/2}{}^2 1d_{5/2}{}^5$

Table D.15. Unpaired neutron and proton spins and parities.

Nuclide	Neutron	Proton
^{14}N	$1/2^-$	$1/2^-$
^{20}F	$5/2^+$	$5/2^+$
^{24}Na	$5/2^+$	$5/2^+$
^{26}Al	$5/2^+$	$5/2^+$

$$m = 174 \times 1.66 \times 10^{-27} \text{ kg} = 2.89 \times 10^{-25} \text{ kg}.$$

Then we can calculate I as

$$I = 0.4 \times (2.89 \times 10^{-25} \text{ kg}) \times (6.69 \times 10^{-15} \text{ m})^2$$
$$= 5.15 \times 10^{-54} \text{ kg m}^2$$

about a factor of 2 too large, but correct order of magnitude.

6.4. The ground state neutron and proton states may be determined from the degeneracy of each of the shell model states and are given in table D.14.
The I^{π} of the unpaired nucleons are shown in table D.15.

The parities follow immediately by combining the neutron and proton parities. The range of I for the nucleus is given by

$$\left| j_n - j_p \right| \leqslant |I| \leqslant \left| j_n + j_p \right|.$$

These may be tabulated and compared with the actual values, as given in table D.16. In all cases, the actual values lie within the range of predicted values and the parity is in agreement.

6.6. The spin and parity of ^{59}Ni are due to the properties of the unpaired neutron ($N = 31$). The configuration for the neutrons is $1s_{1/2}{}^2\ 1p_{3/2}{}^4\ 1p_{1/2}{}^2\ 1d_{5/2}{}^6\ 1d_{3/2}{}^4$ $2s_{1/2}{}^2\ 1f_{7/2}{}^8\ 2p_{3/2}{}^3$. The ground state $I^{\pi} = 3/2^-$ is given by the unpaired $2p_{3/2}$ neutron. For the excited state I^{π} we have the possible configurations given in table D.17. We can see the occupancy of the levels (above $1f_{7/2}$) that correspond to the ground state and these excited states in figure D.2.

Table D.16. Comparison of calculated and actual spins and parities.

	Calculated			Actual	
Nuclide	I_{min}	I_{max}	π	I	π
^{14}N	0	1	+	1	+
^{20}F	0	5	+	2	+
^{24}Na	0	5	+	4	+
^{26}Al	0	5	+	5	+

Table D.17. Possible excited state configurations for ^{59}Ni.

I^{π}	State
$5/2^-$	$1f_{5/2}$
$1/2^-$	$2p_{1/2}$
$3/2^-$	$2p_{3/2}$

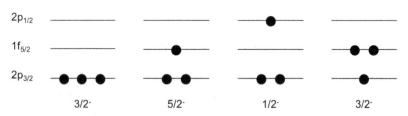

Figure D.2. Excited state neutron occupancies for ^{59}Ni.

7.2. One gram of natural carbon contains

$$\frac{6.02 \times 10^{23} \text{ atoms mol}^{-1}}{12 \text{ g}^{-1} \text{ mol}} = 5.02 \times 10^{22} \text{ atoms g}^{-1}$$

of which 1 in 7.69×10^{11} are ^{14}C, for a total of 6.53×10^{10} atoms of ^{14}C g^{-1} of natural carbon. Assuming an average human weighs about 70 kg then the number of ^{14}C atoms in the body is

$$N = (70 \times 10^3 \text{ g}) \times 0.2 \times (6.53 \times 10^{10} \text{ atoms g}^{-1}) = 9.14 \times 10^{14} \text{ atoms.}$$

The decay rate is given by equation (7.1) as

$$\left| \frac{dN}{dt} \right| = \lambda N.$$

From Appendix B the half-life of ^{14}C is 5730 years giving a decay constant of

$$\lambda = \frac{\ln 2}{\tau_{1/2}} = 3.83 \times 10^{-12} \text{ s}^{-1}.$$

Solving for the decay rate gives

$$\left| \frac{dN}{dt} \right| = 3.5 \times 10^3 \text{ decays s}^{-1}.$$

Since each decay gives off one electron, the activity in Curies will be

$$\frac{3.5 \times 10^3}{3.7 \times 10^{10}} = 9.4 \times 10^{-8} \text{ Ci.}$$

7.4. We know that $N_{235}(0)/N_{238}(0) = 1$ and that $N_{235}(t)/N_{238}(t) = 7.3 \times 10^{-3}$ where t is the age of the earth. The number of nuclei of each species may be written as

$$N_{235}(t) = N_{235}(0)e^{-\lambda_{235}t}$$

and similarly, for ^{238}U. From the above we may write

$$\frac{N_{235}(t)}{N_{238}(t)} = \frac{N_{235}(0)}{N_{238}(0)} \cdot \frac{e^{-\lambda_{235}t}}{e^{-\lambda_{238}t}} = e^{(\lambda_{238}-\lambda_{235})t} = 7.3 \times 10^{-3}.$$

Solving for t gives

$$t = \frac{\ln(7.3 \times 10^{-3})}{\lambda_{238} - \lambda_{235}}.$$

The lifetimes of ^{235}U and ^{238}U are 1.03×10^9 years and 6.49×10^9 years, respectively, giving decay constants of

$$\lambda_{235} = 9.71 \times 10^{-10} \text{ years}^{-1}.$$
$$\lambda_{238} = 1.54 \times 10^{-10} \text{ years}^{-1}$$

Solving for t gives $t = 6.0 \times 10^9$ years.

Table D.18. Calculated decay rates in decays per second for different lifetimes at different times.

τ	1 s	1 m	1 h	1 d
			t	
1 m	3.64×10^7	1.36×10^7	3.2×10^{-19}	≈ 0
1 h	3.7×10^7	3.64×10^7	1.36×10^7	1.4×10^{-1}
1 d	3.7×10^7	3.697×10^7	3.55×10^7	1.36×10^7

7.6. For a given species the activity at $t = 0$ will be given by

$$\left|\frac{dN}{dt}\right| = \lambda N(0).$$

We may calculate the number of radioactive nuclei as

$$N(0) = \frac{1}{\lambda}\left|\frac{dN}{dt}\right| = \tau\left|\frac{dN}{dt}\right|.$$

For an activity A (in Ci)

$$N(0) = 3.7 \times 10^{10}\ \tau A.$$

The decay rate may, therefore, be written as a function of time as

$$\left|\frac{dN(t)}{dt}\right| = \lambda N(0)e^{-\lambda t} = 3.7 \times 10^{10}\ Ae^{-t/\tau}.$$

Here $A = 1$ mCi and for the lifetimes given in the problem we obtain the results in table D.18.

7.8. The number of nuclei of type A will be

$$N_A(t) = N_A(0)e^{-t/\tau_A}.$$

For type B

$$N_B(t) = N_B(0)e^{-t/\tau_B}$$

The ratio of these two numbers will be

$$\frac{N_A(t)}{N_B(t)} = \frac{N_A(0)}{N_B(0)} \times \exp\left[\frac{t}{\tau_B} - \frac{t}{\tau_A}\right].$$

From the initial conditions stated

$$\frac{N_A(0)}{N_B(0)} = 2$$

and the lifetimes are $\tau_A = \tau$ and $\tau_B = 3\tau$ giving

$$\frac{N_A(t)}{N_B(t)} = 2\exp\left[\frac{t}{3\tau} - \frac{t}{\tau}\right].$$

Setting

$$\frac{N_A(t)}{N_B(t)} = 1$$

we can solve for t as

$$t = \frac{\tau \ln 2}{\left[1 - \frac{1}{3}\right]} \approx 1.04\tau.$$

8.2. The Q for α-decay is given by

$$Q = \left[m\left(^A X\right) - m\left(^{A-4} Y\right) - m\left(^4 He\right)\right]c^2$$

in terms of the atomic masses. Using values from the table of masses gives the results in table D.19.

These are in good agreement with the values given in the *Table of Isotopes*.

8.4. The decay

$$^{226}Ra \rightarrow {}^{222}Rn + \alpha$$

has a Q of

$$Q = \left[m\left(^{226}Ra\right) - m\left(^{222}Rn\right) - m\left(^4He\right)\right]c^2$$
$$= 4.844 \text{ MeV}.$$

The theoretical value of the lifetime is given by equations (8.10) and (8.11) where

$$\tau = \tau_0 e^G$$

and

$$G = \frac{4Z_D e^2}{4\pi\varepsilon_0}\sqrt{\frac{2m}{\hbar^2 Q}}\left[\cos^{-1}\sqrt{\frac{a}{b}} - \sqrt{\frac{a}{b}\left[1 - \frac{a}{b}\right]}\right]$$

Table D.19. Calculated α-decay energies.

$^A X$	$^{A-4} Y$	Q (MeV)
^{208}Po	^{204}Pb	5.216
^{222}Ra	^{218}Rn	6.681
^{240}Pu	^{236}U	5.256
^{252}Fm	^{248}Cf	7.153

Here $Z_D = 86$, $m = m_\alpha m_D/(m_\alpha + m_D) = 3.932$ u and $\tau_0 = 6.3 \times 10^{-23}$ s. Using the experimental value of the lifetime $\tau = 7.4 \times 10^{10}$ s gives

$$G = \ln(\tau/\tau_0) = 76.15$$

The expression for G above is rearranged to give

$$G = \frac{4Z_D}{\hbar c} \frac{e^2}{4\pi\varepsilon_0} \sqrt{\frac{2mc^2}{Q}} [\,]$$

where [ ] is the $\cos^{-1}$ term as shown above. Substituting values for the constants

$$G = \frac{4 \times 86}{197.3} \cdot 1.44 \sqrt{\frac{2 \times 3.932 \times 931.5}{4.844}} [\,] = 97.62[\,].$$

Note that the value of G is unitless. The value of b is given by the Coulomb relation

$$Q = \frac{2 \times 86 \times e^2}{4\pi\varepsilon_0 b}$$

or

$$b = \frac{2 \times 86 \times e^2}{4\pi\varepsilon_0 \times 4.844} = 51.1 \text{ fm.}$$

The value of a is the combined radii of the ^{222}Rn and the α-particle where

$$r\left(^{222}\text{Rn}\right) = a - r_\alpha = a - 1.2 \times 4^{1/3} = a - 1.9 \text{ fm.}$$

It is easiest to solve for a using an iterative method and

$$[\,] = \frac{76.15}{97.62} = 0.780.$$

table D.20 illustrates the process.

Interpolating gives $a \approx 8.55$ fm or $r(^{222}\text{Rn}) = 6.65$ fm. (This may be compared with the value $r(^{222}\text{Rn}) = 1.2 \times 222^{1/3} = 7.25$ fm).

Table D.20. Illustration of iterative process for finding the value of the term in [ ] in equation (8.17).

a	[ ]
7.00	0.851
7.50	0.826
8.00	0.803
8.50	0.782
9.00	0.760

8.6.

(a) The decay probability is e^{-G} where G is given by equation (8.11) as

$$G = \frac{4Z_De^2}{4\pi\varepsilon_0}\sqrt{\frac{2m}{\hbar^2Q}}\left[\cos^{-1}\sqrt{\frac{a}{b}} - \sqrt{\frac{a}{b}\left[1 - \frac{a}{b}\right]}\right].$$

Here a is the radius of the daughter nucleus plus the radius of the α-particle and b is defined as

$$b = \frac{2Z_De^2}{4\pi\varepsilon_0Q}.$$

Using typical values of $Q = 6$ MeV, $Z_D = 90$ and $A_D = 240$ gives values of

$$b = 43 \text{ fm}$$

and

$$\frac{4Z_De^2}{4\pi\varepsilon_0}\sqrt{\frac{2m}{\hbar^2Q}} = 92.6.$$

Here we have used the relation

$$\frac{m}{\hbar^2} = \frac{mc^2}{(\hbar c)^2} = \frac{3727 \text{ MeV c}^{-2}}{(197.3)^2 \text{ MeV}^2 \text{ fm}^2}$$

to simplify the calculation. The radius of the α-particle is

$$R_\alpha = 1.2 \times 4^{1/3} = 1.9 \text{ fm}$$

and the radius of the daughter nucleus is

$$R_D = 1.2 \times 240^{1/3} = 7.46 \text{ fm}.$$

If the nucleus is a prolate ellipsoid with a ratio of the major/minor axes of 1.5 then (see chapter 12)

$$R_D{}^3 = r_{maj}r_{min}{}^2 = \frac{r_{maj}{}^3}{1.5^2}.$$

Solving for the semimajor and semiminor axes gives

$$r_{maj} = 9.78 \text{ fm}$$

$$r_{min} = 6.52 \text{ fm}.$$

We now need to calculate the term in [ ] from equation (8.11) and thus G with the value of a taking on the values of the semimajor and semiminor axes. This gives the values in table D.21.

Table D.21. Values of G along the semimajor and semiminor axes of a nonspherical nucleus.

a (fm)	[ ]	G
9.78	0.655	60.65
6.52	0.812	75.19

Table D.22. ^{241}Cm α-decay properties to various ^{237}Pu states.

Daughter state	I^π	E (MeV)	f
Ground	7/2$^-$	6.184	0
1st	1/2$^+$	6.039	0.7
2nd	3/2$^+$	6.028	0.17
3rd	5/2$^+$	5.982	0.13

The ratios of the decay probabilities along the two axes is

$$\frac{P_{maj}}{P_{min}} = \frac{e^{-G_{maj}}}{e^{-G_{min}}} = e^{14.54} = 2.1 \times 10^6.$$

(b) Since it is much more probable that the α-particle will be emitted along the semimajor axis than along the semiminor axis the radiation pattern will be most intense along the major axis directions forming two beams along the 'polar' directions. Along similar lines we can see that for an oblate nucleus the radiation pattern will consist of a plane of radiation emitted along the 'equitorial' plane.

8.8.

(a) For the decay

$$^{241}\text{Cm} \rightarrow {}^{237}\text{Pu} + \alpha$$

the Q is given by

$$Q = \left[m\left(^{241}\text{Cm}\right) - m\left(^{237}\text{Pu}\right) - m\left(^4\text{He}\right) \right]c^2 = 6.184 \text{ MeV}.$$

(b) The ^{241}Cm ground state has $I^\pi = 1/2^+$. The transitions to the various ^{237}Pu states are described in table D.22.

Table D.23. Minimum angular momentum states for the transitions in table D.22.

Daughter state	ℓ^π
Ground	3^-
1st	0^+
2nd	2^+
3rd	2^+

Conservation of angular momentum and parity allows for the calculation of the minimum ℓ^π for the α-particle for each transition as given in table D.23.

Since the energy differences are relatively small, the branching ratios are determined from quantum mechanical consideration. The values of f shown follow directly from a qualitative consideration of ℓ^π.

8.10.

(a) For the decay

$$^8\text{Be} \rightarrow 2\alpha$$

the Q is given by

$$Q = \left[m\left(^8\text{Be}\right) - 2m\left(^4\text{He}\right) \right] c^2 = 0.095 \text{ MeV}.$$

This occurs with a lifetime of 2.8×10^{-16} s.

(b) For the decay

$$^{12}\text{C} \rightarrow 3\alpha$$

the Q is given by

$$Q = \left[m\left(^{12}\text{C}\right) - 3m\left(^4\text{He}\right) \right] c^2 = -7.27 \text{ MeV}.$$

The energy levels of ^{12}C are fairly high energies—as is typical for light nuclei. The second excited state of ^{13}C is at 7.66 MeV and is known to decay by this method, as do higher energy excited states. If the decay

$$^{12}\text{C} \rightarrow {}^8\text{Be} + \alpha$$

occurs, the ^{8}Be will spontaneously decay on a short time scale.

(c) The decay

$$^{16}\text{O} \rightarrow 4\alpha$$

has

$$Q = \left[m(^{16}O) - 4m(^4He) \right] c^2 = -14.44 \text{ MeV}.$$

This is not known to occur from ^{16}O excited states. However, some excited states are known to decay by

$$^{16}O \rightarrow ^{12}C + \alpha.$$

This has

$$Q = \left[m(^{16}O) - m(^{12}C) - m(^4He) \right] c^2 = -2.42 \text{ MeV}$$

In general, the ^{16}O excited state energy will be insufficient to leave the ^{12}C in a state above the α-decay threshold.

9.2. Tritium consists of a nucleus with 1 proton and 2 neutrons with 1 atomic electron. Possible decays are

$$^3H \rightarrow ^3He + e^- + \bar{\nu}_e$$

and

$$^3H \rightarrow ^2H + n.$$

β^+ (ec) cannot occur as it would produce a daughter nucleus with 3 neutrons. Proton separation cannot occur as the daughter nucleus would consist of two neutrons. For the above processes the Q values are

$$\beta^- \ Q = \left[m(^3H) - m(^3He) \right] c^2$$

$$\text{n separation } Q = [m(^3H) - m(^2H) - m_n] c^2.$$

From the measured masses this gives

$$\beta^- \ Q = 0.018 \text{ MeV}$$

$$\text{n separation } Q = -6.25 \text{ MeV}.$$

Thus β^- decay is energetically favorable while neutron emission is not. This situation is observed experimentally.

9.4. For the following decays

$$^{10}Be \rightarrow ^{10}Be + e^- + \bar{\nu}_e$$
$$^{21}F \rightarrow ^{21}Ne + e^- + \bar{\nu}_e$$
$$^{37}S \rightarrow ^{37}Cl + e^- + \bar{\nu}_e$$
$$^{60}Co \rightarrow ^{60}Ne + e^- + \bar{\nu}_e$$

we can tabulate the I^π for the parent (P) and daughter (D) as well as the change in parity. From these, the allowed values of ΔI are determined as shown in table D.24.

A comparison with the tabulated information from the text allows for a determination of the degree to which these transitions are forbidden. The lowest degree to which the transition is forbidden dominates as given in table D.25.

9.6. The decay of the $3/2^+$ ground state of ^{43}K has the branching ratios as shown in table D.26.

We must consider the change in I and π for each transition as shown in table D.27.

We can also determine the β end point energy from a calculation of the Q for the decay (1.82 MeV). The degree to which each transition is forbidden is determined by comparison of the above table with table 9.1. We summarize this information in table D.28.

The most probable decay is the allowed transition with the greatest energy.

Table D.24. Allowed ΔI for some β-decays.

Parent	$I_P{}^\pi$	$I_D{}^\pi$	π change	Allowed ΔI
^{10}Be	0^+	3^+	N	3
^{21}F	$5/2^+$	$3/2^+$	N	1, 2, 3, 4
^{37}S	$7/2^+$	$3/2^+$	Y	2, 3, 4, 5
^{60}Co	5^+	0^+	N	5

Table D.25. Forbiddenness for some β-decays.

Parent	Decay
^{10}Be	2nd forbidden
^{21}F	Allowed
^{37}S	1st forbidden
^{60}Co	4th forbidden

Table D.26. Branching ratios of the decay of ^{43}K to different ^{43}Ca states.

^{43}Ca state (MeV)	I^π	f
0	$7/2^-$	0.015
0.373	$5/2^-$	0.03
0.593	$3/2^-$	0.034
0.990	$5/2^+$	0.89
1.395	$3/2^+$	0.03

Table D.27. Spin and parity changes for the decay of ^{43}K to different ^{43}Ca states.

^{43}Ca state	ΔI_{min}	ΔI_{max}	Parity change
0	2	5	Y
0.373	1	4	Y
0.593	0	3	Y
0.990	1	4	N
1.395	0	3	N

Table D.28. β-decay end point energy and forbiddenness.

^{43}Ca state (MeV)	β end point (MeV)	Forbidden
0	1.82	1st
0.373	1.45	1st
0.593	1.23	1st
0.990	0.83	Allowed
1.395	0.42	Allowed

9.8. For each of the processes mentioned we need to calculate the values of Q;

(i) Proton emission,

$$^9\text{B} \rightarrow {}^8\text{Be} + \text{p}$$

$$Q = \left[m\left({}^9\text{B}\right) - m\left({}^8\text{Be}\right) - m\left({}^1\text{H}\right) \right]c^2 = +1.85 \text{ MeV}.$$

(ii) Electron capture,

$$\text{e}^- + {}^9\text{B} \rightarrow {}^9\text{Be} + \nu_e$$

$$Q = \left[m\left({}^9\text{B}\right) - m\left({}^9\text{Be}\right) \right]c^2 = +1.068 \text{ MeV}$$

where the electronic binding energy has been ignored.

(iii) β^+ decay,

$$^9\text{B} \rightarrow {}^9\text{Be} + \text{e}^+ + \nu_e$$

$$Q = \left[m\left({}^9\text{B}\right) - m\left({}^9\text{Be}\right) - 2m_e \right]c^2 = +0.046 \text{ MeV}.$$

(iv) Neutron emission,

$$^9\text{B} \rightarrow {}^8\text{B} + \text{n}$$

$$Q = \left[m\left({}^9\text{B}\right) - m\left({}^8\text{B}\right) - m_\text{n} \right] c^2 = -18.6 \text{ MeV.}$$

(v) β^- decay,

$$^9\text{B} \rightarrow {}^9\text{C} + \text{e}^- + \bar{\nu}_\text{e}$$

$$Q = \left[m\left({}^9\text{B}\right) - m\left({}^9\text{C}\right) \right] c^2 = -16.5 \text{ MeV.}$$

This shows that the first three processes are exothermic and will occur. The last two processes are endothermic and will not occur.

9.10. Referring to the energy level diagram as shown in figure D.3, we need to calculate E_B and Q.
  E_B is given in terms of nuclear mass as

$$E_\text{B} = \left[m_N\left({}_{Z+1}^{A}\text{Y}\right) - m_N\left({}_{Z}^{A}\text{X}\right) \right] c^2.$$

Adding appropriate electron masses

$$E_\text{B} = \left[m_N\left({}_{Z+1}^{A}\text{Y}\right) + (Z + 1)m_\text{e} - m_N\left({}_{Z}^{A}\text{X}\right) - Zm_\text{e} - m_\text{e} \right] c^2$$

gives an expression in terms of atomic masses

$$E_\text{B} = \left[m\left({}_{Z+1}^{A}\text{Y}\right) - m\left({}_{Z}^{A}\text{X}\right) \right] c^2.$$

The energy, Q, of the process

$$^A_Z\text{X} \rightarrow {}_{Z+2}^{A}\text{Z} + 2\text{e} + 2\bar{\nu}_\text{e}$$

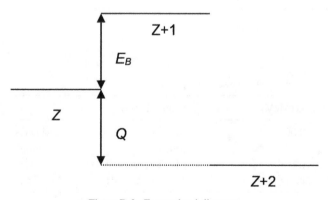

Figure D.3. Energy level diagram.

is

$$Q = \left[m_N\left(_Z^A X\right) - m_N\left(_{Z+2}^A Z\right) - 2m_e \right] c^2.$$

Adding appropriate electron masses

$$Q = [m_N(_Z{}^A X) + Zm_e - m_N(_{Z+2}{}^A Z) - (Z+2)m_e + 2m_e - 2m_e]c^2.$$

In terms of atomic masses this is

$$Q = \left[m\left(_Z^A X\right) - m\left(_{Z+2}^A Z\right) \right] c^2.$$

10.2. The four transitions to the $9/2^+$ ground state have the properties shown in table D.29, where $|I_i - I_f| \leqslant \Delta I \leqslant |I_i + I_f|$.

Based on the allowed ΔI and the parity change the allowed multiplicities are given in table D.30.

The leading term dominates with, in some cases, an admixture of the second term.

10.4. For ^{58}Co, the ground state is 2^+ so the transition has $\Delta I_{min} = 3$ with no parity change. This is an M3 transition and from the expression in table 10.2, the lifetime is found to be 5×10^7 s (or about a year and a half). This is not very good agreement with the actual value of 12 h. For ^{58}Fe, the ground state is 0^+ so the transition has $\Delta I_{min} = 2$ with no parity change. This is an E2 transition and from the expression in table 10.2, the lifetime is found to be 1.7×10^{-10} s. This is in much better agreement with the actual value than the Weisskopf estimate for ^{58}Co. These results may be explained in terms of the contribution from internal conversion for the two decays.

Table D.29. Properties of γ-transitions to the ground state of ^{83}Kr.

E (MeV)	I^π	ΔI (Allowed)	π Change
0.009	$7/2^+$	1, 2, 3, 4, 5, 6, 7, 8	N
0.042	$1/2^-$	4, 5	Y
0.56	$5/2^-$	2, 3, 4, 5, 6, 7	Y
0.57	$3/2^-$	3, 4, 5, 6	Y

Table D.30. Allowed multiplicities.

E (MeV)	Multiplicities
0.009	M1, E2, M3, E4, M5, E6, M7, E8
0.042	M4, E5
0.56	M2, E3, M4, E5, M6, E7
0.57	E3, M4, E5, M6,

Table D.31. Recoil energies for some γ-transitions.

Nuclide	E_γ (MeV)	E_R (MeV)
^{15}O	5.183	9.6×10^{-4}
^{19}O	0.096	2.6×10^{-7}
^{57}Fe	0.0144	1.9×10^{-9}
^{70}Ge	1.0396	8.3×10^{-6}
^{227}Th	0.0093	2.0×10^{-10}
^{228}Th	0.0578	7.9×10^{-9}

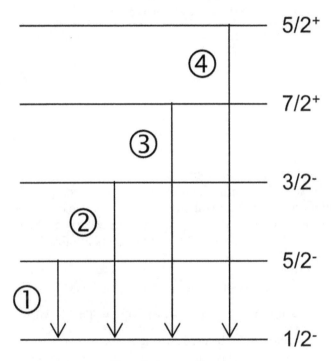

Figure D.4. γ-transitions.

10.6. (a) The recoil energy is given in equation (10.1) as

$$E_R = \frac{E_\gamma^2}{2m_f c^2}.$$

Using m_f as the ground state mass, results are given in table D.31.

E_R decreases as either E_γ decreases or m_f increases. Heavier nuclei typically have more closely spaced energy levels. Both these factors lead to a decrease in the importance of the recoil energy for heavier nuclei.

10.8. The expected transitions are shown below in figure D.4. These transitions have properties given in table D.32. Transition '1' has an energy E_1, etc.

Table D.32. Properties of γ-transitions in figure D.4.

Transition	$\sim E$	π Change	ΔI	Multiplicity
1	E_1	N	2, 3	E2, M3
2	$2E_1$	N	1, 2	M1, E2
3	$3E_1$	Y	3, 4	E3, M4
4	$4E_1$	Y	2, 3	M2, E3

Table D.33. Calculated transition rates.

Transition	$\lambda \ (s^{-1})$
1	5×10^8
2	7×10^{12}
3	8×10^2
4	2×10^7

We make the following assumptions and consider the relevant Weisskopf estimates for the leading term in the multiple radiation field: $A = 55$ and $E_1 = 0.5$ MeV. This latter estimate is based on an estimate of the typical energy spacing of the low-lying levels of a nucleus with $A \approx 55$ from an inspection of the *Table of Isotopes*. The relevant expressions in table 10.2 give the results in table D.33.

11.2.

(a) The excess energy available is

$$Q = [m(^{13}C) + m(^4He) - m(^{17}O)]c^2 = +6.359 \text{ MeV}.$$

The atomic masses can be used as long as the mass of ^{4}He is used in place of the α-particle mass. This energy is above the neutron separation energy of ^{17}O which is given by equation (4.9) to be

$$S_n = [m_n + m(^{16}O) - m(^{17}O)]c^2 = 4.143 \text{ MeV}.$$

Hence, the ^{17}O will be formed in an energy level above its neutron separation energy.

(b) The resonances that appear in the figure correspond to the population of excited ^{17}O* states that form prior to neutron separation. The lowest energy peak shown in the figure corresponds to an energy of 1.05 MeV. This is above the energy that is available from the reaction, or an energy of 1.05 + 6.36 = 7.41 MeV.

11.4.

(a) The missing particle/nucleus can be determined by equating neutrons and protons on the two sides of the reaction.

    (i) ^{29}Si$(\alpha, n)^{32}$S

    (ii) ^{60}Ni$(p, n)^{60}$Cu

    (iii) ^{111}Cd$(n, \gamma)^{112}$Cd

    (iv) ^{189}Os$(p, d)^{188}$Os

    (v) ^{156}Gd$(d, n)^{157}$Tb

(b) The reactions involve nuclei, so when using atomic masses, it is important to properly account for all electrons.

    (i) $Q = [m(^{29}\text{Si}) + m(^{4}\text{He}) - m_n - m(^{32}\text{S})]c^2 = -1.525$ MeV

    (ii) $Q = [m(^{60}\text{Ni}) + m(^{1}\text{H}) - m_n - m(^{60}\text{Cu})]c^2 = -6.909$ MeV

    (iii) $Q = [m(^{111}\text{Cd}) + m_n - m(^{112}\text{Cd})]c^2 = +9.398$ MeV

    (iv) $Q = [m(^{189}\text{Os}) + m(^{1}\text{H}) - m(^{188}\text{Os}) - m(^{2}\text{H})]c^2 = -3.696$ MeV

    (v) $Q = [m(^{156}\text{Gd}) + m(^{2}\text{H}) - m_n - m(^{157}\text{Tb})]c^2 = +3.293$ MeV

11.6.

(a) Conservation of the number of neutrons and protons requires that the reaction is

$$^{28}\text{Si}(d, p)^{29}\text{Si}.$$

(b) The Q for this process is

$$Q = \left[m\left(^{28}Si\right) + m\left(^{2}H\right) - m\left(^{29}Si\right) - m\left(^{1}H\right) \right]c^2 = +6.249 \text{ MeV}.$$

(c) Using equation (11.9) we relate ΔE, the change in the ^{29}Si energy and E_f. Here we use $E_i = 10$ MeV, $Q = 6.249$ MeV, $\theta = 30°$ and the masses: $m_B = 28.968$ u, $m_b = 1.0072$ u and $m_a = 2.0136$ u. Equation (11.9) reduces to

$$\Delta E = 15.554 - 1.035E_f + 0.269\sqrt{E_f}.$$

From the *Table of Isotopes* we find the energies of the first three excited states of ^{29}Si are 1.273, 2.028 and 2.426 MeV. Using these values for ΔE in the above equation, E_f is most easily determined by iterative methods as shown in table D.34.

Table D.34. Calculated final proton energies.

State	ΔE (MeV)	E_f (MeV)
Ground	0	16.07
1st	1.273	14.80
2nd	2.028	14.04
3rd	2.426	13.64

11.8. The reaction

$$n + {}^{16}O \rightarrow {}^{17}O^* + \gamma$$

has a Q of 4.15 MeV. In order to populate the 4.56 MeV state of ^{17}O, an additional 0.41 MeV of kinetic energy (in the cm frame) is needed. The excess energy width due to Doppler broadening is

$$\Delta E = 2[mE_{cm}k_BT/m_A]^{1/2}$$

where $m = m_n$, $m_A = m({}^{16}O)$ and $k_BT = 0.04$ eV at room temperature. Thus,

$$\Delta E = 2\left[0.41 \times 10^6 \times \frac{0.04}{16}\right]^{1/2} = 64 \text{ eV}$$

or a relative increase of $64/0.41 \times 10^6 = 1.6 \times 10^{-4}$. For the data in figure 11.9 the width at 400 K is about 0.1 eV out of 6.65 eV or a relative width of $0.1/6.65 = 1.5 \times 10^{-2}$. In absolute terms the broadening of the resonance in ^{17}O is large, but compared with the energy scales involved it is much smaller than the low energy resonance in heavier nuclei.

12.2.

(a) In this case $m_n \approx 1$ u and $M \approx 1$ u for the ^{1}H nucleus then

$$T_f = \frac{M^2 + m_n{}^2}{(M + m_n)^2}T_i = \frac{1}{2}T_i.$$

Thus, after N collisions the relative energy will be

$$\frac{T_N}{T_i} = \left(\frac{1}{2}\right)^N.$$

So, if $T_N = 0.01$ eV and $T_i = 2$ MeV we solve for N as $27.6 \approx 28$.

(b) For a particle of cross section σ and velocity v the volume swept out per unit time is σv. In a medium of number density ρ the mean time between collisions is

$$\Delta t = \frac{1}{\sigma v \rho} = \frac{1}{\sigma \rho}\sqrt{\frac{m_n}{2T}}.$$

The energy change per collision is

$$\Delta T = -\left[1 - \left(\frac{M^2 + m_n^2}{(M + m_n)^2}\right)\right]T.$$

We write the rate of energy loss as

$$\frac{dT}{dt} = \frac{\Delta T}{\Delta t} = -\left[1 - \left(\frac{M^2 + m_n^2}{(M + m_n)^2}\right)\right]T\sigma\rho\sqrt{\frac{2T}{m_n}} \approx -\frac{\sigma\rho T^{3/2}}{\sqrt{2m_n}}.$$

Integrating from T_i to T_f gives the time as

$$t = \int dt = -\int_{T_i}^{T_f}\left(\frac{dT}{dt}\right)^{-1}dT.$$

Substituting from above and integrating gives

$$t \approx -\frac{\sqrt{2m_n}}{\sigma\rho}\int_{T_i}^{T_f}T^{-3/2}dT \approx \frac{2\sqrt{2m_n}}{\sigma\rho}\left[\frac{1}{\sqrt{T_f}} - \frac{1}{\sqrt{T_i}}\right] \approx \frac{2\sqrt{2m_n}}{\sigma\rho}\frac{1}{\sqrt{T_f}}$$

using $\sigma \approx 0.33$ b $= 3.3 \times 10^{-29}$ m^2 and ρ as the number of ^{1}H nuclei per unit volume. Then

$$\rho = \frac{2N_A\rho_m}{W}$$

where ρ_m is the mass density 10^3 kg·m^{-3} and W is the molecular weight 18×10^{-3} kg·mole^{-1}. Therefore, $\rho = 6.7 \times 10^{28}$ m^{-3}. Substituting in the above gives $t \approx 1.3$ ms.

12.4. For the fission process

$$^A\text{X} \rightarrow {}^{A/2}\text{Y} + {}^{A/2}\text{Y}$$

the fission energy is

$$E = [m(^A\text{X}) - 2m(^{A/2}\text{Y})]c^2$$

where we have assumed no prompt neutrons. This may be written in terms of binding energies as

$$E = [2B(^{A/2}\text{Y}) - B(^A\text{X})].$$

Ignoring the paring term, the binding energy is given by the liquid drop model as

$$B = a_V A - a_S A^{2/3} - \frac{a_C Z(Z - 1)}{A^{1/3}} - \frac{a_{\text{sym}}(A - 2Z)^2}{A}.$$

The fission energy is, therefore, written as

$$E = a_S A^{2/3} - 2a_S \left(\frac{A}{2}\right)^{2/3} + \frac{a_C Z(Z-1)}{A^{1/3}} - \frac{2a_C \frac{Z}{2}\left(\frac{Z}{2} - 1\right)}{(A/2)^{1/3}}.$$

Combining terms in powers of A gives

$$E = a_S(1 - 2^{1/3})A^{2/3} + a_C \left(\frac{Z^2}{2}\right)A^{-1/3}.$$

Using $a_S = 16.8$ MeV and $a_C = 0.72$ MeV we obtain the energy in MeV as

$$E = -4.36A^{2/3} + 0.36Z^2 A^{-1/3}.$$

It is now necessary to express Z as a function of A. Using figure 3.1 as a guide, we may draw a straight line through all the data points and find $N \approx 1.5Z$ or $Z \approx 0.4A$. The energy is, therefore, given as

$$E = -4.36A^{2/3} + 0.058A^{5/3}.$$

12.6. The number of ^{256}Fm nuclei present in a 1 μg sample is

$$N = \frac{(10^{-6}\text{g}) \times (6.02 \times 10^{23} \text{ nuclei mol}^{-1})}{256 \text{ g mol}^{-1}} = 2.35 \times 10^{15} \text{ nuclei}.$$

Since we are concerned with a time period that is much less than the lifetime, we can write the decay rate as

$$\left|\frac{dN}{dt}\right| = \lambda N = \frac{1}{3.8 \text{ h} \times 3600 \text{ s h}^{-1}} \times 2.35 \times 10^{15} = 1.72 \times 10^{11} \text{ s}^{-1}.$$

Assuming an energy release of 200 MeV per fission the energy release per unit time is

$$P = (200 \text{ MeV}) \times (1.72 \times 10^{11} \text{ s}^{-1}) \times (1.6 \times 10^{-13} \text{ J MeV}^{-1}) = 5.5 \text{ W}.$$

Thus, in one second the energy release would be 5.5 J. The temperature increase will be

$$\Delta T = \frac{E}{MC}$$

where M is the number of moles and C is the molar specific heat. Substituting appropriate values gives

$$\Delta T = \frac{5.5 \times 256}{25 \times 10^{-6}} = 5.6 \times 10^7 \text{ K}.$$

12.8. The possible nuclides are ^{236}Pu, ^{237}Pu, ^{238}Pu, ^{239}Pu, ^{240}Pu, ^{241}Pu, ^{242}Pu and ^{244}Pu. Those that are fissile have Q for a thermal neutron absorption reaction that is greater than the Coulomb fission barrier. For $A = 235$ this is 6.2 MeV, so we will use this as a guide and calculate Q for the reaction.

$$\text{n} + {}^A\text{Pu} \rightarrow {}^{A+1}\text{Pu} + \gamma.$$

Table D.35. Fission energy of Pu isotopes.

A	Q (MeV)
236	5.88
237	7.00
238	5.65
239	6.53
240	5.24
241	6.31
242	5.03
244	4.77

Table D.36. Calculated Coulomb barrier heights for some fission processes.

Nucleus	A	Z	R (fm)	V (MeV)
d	2	1	1.51	0.48
^{6}Li	6	3	2.18	2.97
^{20}Ne	20	10	3.26	22.1

This is

$$Q = [m(^A\text{Pu}) + m_n - m(^{A+1}\text{Pu})]c^2.$$

Results are given in table D.35 below.

It is obvious, as we expected, that odd A isotopes of Pu are fissile while even A isotopes are not.

13.2. The nuclear radii are given (in fm) by

$$R = 1.2A^{1/3}$$

and the distance between the centers of the two nuclei when their surfaces are just in contact is $2R$. The Coulomb potential is

$$V = \frac{Z^2e^2}{4\pi\varepsilon_0 \cdot 2R}.$$

Results are shown in table D.36.

13.4.

(a) The process of fusing 4 protons into a ^{4}He nucleus must be considered in the context of atomic masses. We add 4 m_e to each side of equation (13.10) to obtain

$$Q = \left[4m(^1\text{H}) - m(^4\text{He}) - 4m_e\right]c^2 = 24.69 \text{ MeV}.$$

To this we can add the energy gained from the annihilation of the 2 positrons with electrons or, $4m_ec^2 = 2.044$ MeV, for a total of $Q = 26.73$ MeV.

(b) The energy in each case is calculated from atomic masses. Being careful with electron masses, we obtain

$$Q = \left[m\left(^{12}C\right) + m\left(^1H\right) - m\left(^{13}N\right) \right]c^2 = 1.94 \text{ MeV}$$

$$Q = \left[m\left(^{13}N\right) - m\left(^{13}C\right) \right]c^2 = 1.20 \text{ MeV}$$

$$Q = \left[m\left(^{13}C\right) + m\left(^1H\right) - m\left(^{14}N\right) \right]c^2 = 7.55 \text{ MeV}$$

$$Q = \left[m\left(^{14}N\right) + m\left(^1H\right) - m\left(^{15}O\right) \right]c^2 = 7.29 \text{ MeV}$$

$$Q = \left[m\left(^{15}O\right) - m\left(^{15}N\right) \right]c^2 = 1.74 \text{ MeV2}$$

$$Q = \left[m\left(^{15}N\right) + m\left(^1H\right) - m\left(^{12}C\right) - m\left(^4He\right) \right]c^2 = 4.96 \text{ MeV}$$

This gives a total energy of 24.63 MeV, and including the annihilation energy from the positions emitted in the two β^+ decays, the energy is 26.67 MeV. The same, within the accuracy of the measured masses, as obtained in part (a).

13.6. Each p–p cycle produces 26.7 MeV of which 0.52 MeV is lost as neutrino energy giving a net energy which can be observed as 26.18 MeV per p–p cycle. This represents the fission of four hydrogen nuclei. The observable energy from the sun is 3.86×10^{26} W or

$$\frac{3.86 \times 10^{26} \text{ W}}{1.6 \times 10^{-13} \text{ J MeV}^{-1}} = 2.41 \times 10^{39} \text{ MeV} \cdot \text{s}^{-1}.$$

This corresponds to

$$\frac{2.41 \times 10^{39}}{26.7} \times 4 = 3.61 \times 10^{38} \text{ hydrogen nuclei per second.}$$

and represents a mass of $3.61 \times 10^{38} \times 1.66 \times 10^{-27} = 6 \times 10^{11}$ kg s^{-1}.

14.2. Feynman diagrams are shown in figure D.5.

14.4. The lifetime of a particle, τ, is the lifetime in its own reference frame. Thus, the lifetimes given for particles are, strictly speaking, the lifetimes for particles at rest in the frame in which the measurement is made. This applies to moving particles as long as their velocities are non-relativistic. For relativistic particles the measured

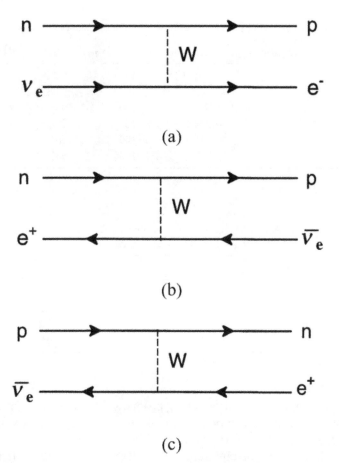

Figure D.5. Feynman diagrams for the processes (a) $\nu_e + n \rightarrow p + e^-$, (b) $e^+ + n \rightarrow p + \bar{\nu}_e$ and (c) $\bar{\nu}_e + p \rightarrow n + e^+$.

(laboratory frame) lifetime, τ_{lab}, will be longer than the rest lifetime by a relativistic correction factor. For a particle with a total energy E and a rest mass energy mc^2, the laboratory lifetime is given in terms of the rest lifetime as

$$\tau_{\text{lab}} = \frac{\tau E}{mc^2}.$$

(For a more complete discussion see e.g., Fraunfelder H and Henley E M 1991 *Subatomic Physics* 2nd edn, Prentice-Hall pp 4–5.) For a relativistic particle $v \approx c$ and the range in the rest laboratory frame will be given by $c\tau_{\text{lab}}$. Thus, the distance traveled in the laboratory frame will be

$$d = \frac{c\tau E}{mc^2}.$$

For the μ and τ-lepton we consider some typical relativistic energies and calculate the relative lifetime in the laboratory frame and the subsequent distance traveled. Results are shown in table D.37.

Table D.37. Calculated distances traveled for some relativistic particles.

Particle	mc^2 (MeV)	τ (s)	E (GeV)	τ_{lab}/τ	d (m)
μ	105.7	3.3×10^{-6}	1	9.46	9.4×10^3
			10	94.6	9.4×10^4
			100	946	9.4×10^5
τ	1784	4.9×10^{-13}	10	5.61	9.4×10^{-4}
			100	56.1	9.4×10^{-3}
			1000	561	9.4×10^{-2}

Table D.38. Values of T_3 for some quarks. The value of T_3 for all other quarks and antiquarks is 0.

Quark	T_3
u	+1/2
$\bar{u}$	−1/2
d	−1/2
$\bar{d}$	+1/2

14.6. The decays of π^+ and π^- produce by-products that are lepton–antilepton pairs. In order for a hadron (the pion) to couple to leptons, the weak interaction is required. The decay of π^0 is to photons, thus the electromagnetic interaction is responsible. Since the weak interaction is weaker than the electromagnetic interaction the decay proceeds more slowly. In general, the rate of a decay or reaction is determined by the weakest interaction that is necessary for the process to proceed.

15.2. The values of T_3 for quarks are given in table D.38.
The value of $\sum T_3$ for some mesons is given in table D.39.

15.4. The diagram for the exchange of a positive pion is shown in figure D.6.
For the exchange of a negative pion, the diagram is shown in Figure D.7.

16.2. For the first decay we write the quark relations as

$$\overline{D}^0 \rightarrow K^- + \mu^+ + \nu_\mu$$

$$u\bar{c} \rightarrow s\bar{u} + \mu^+ + \nu_\mu$$

and the Feynman diagram is shown in figure D.8.
For the second decay we have

$$\overline{D}^0 \rightarrow K^+ + \mu^- + \bar{\nu}_\mu$$

Table D.39. Values of the sum of T_3 for some mesons.

Particle	Quarks	$\sum T_3$
π^-	$\bar{u}d$	-1
K^+	$u\bar{s}$	$+1/2$
D^0	$\bar{u}c$	$-1/2$
J	$c\bar{c}$	0
Ξ	dss	$-1/2$
n	udd	-1
Δ^{++}	uuu	$+3/2$

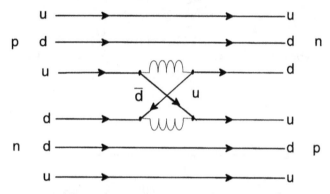

Figure D.6. Feynman diagram for positive pion exchange.

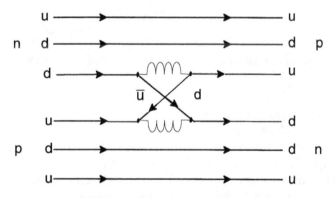

Figure D.7. Feynman diagram for negative pion exchange.

$$u\bar{c} \rightarrow u\bar{s} + \mu^+ + \bar{\nu}_\mu$$

and the Feynman diagram is shown in figure D.9.

In the first instance two weak bosons are necessary while in the second case only one is required. As well, the first process requires two quark generation changes while the second conserves quark generation. Thus, the first process is suppressed.

D-36

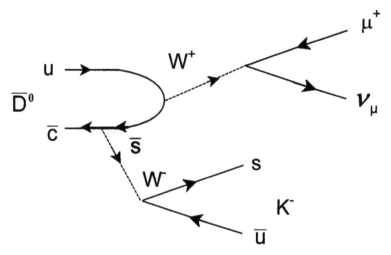

Figure D.8. Feynman diagram for the decay $\bar{D}^0 \to K^- + \mu^+ + \nu_\mu$.

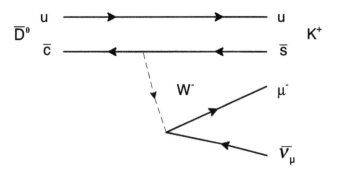

Figure D.9. Feynman diagram for the decay $\bar{D}^0 \to K^+ + \mu^- + \bar{\nu}_\mu$.

16.4. The branching ratios for the decays of the W^+, as shown in table 16.1, add up to $\approx 100\%$. We, therefore, expect any other decays to be either nonexistent or very unlikely. We consider the decay modes as suggested in the problem as these all conserve charge. These may be classified into three categories as follows:

$u\bar{s}$, $c\bar{d}$, $c\bar{b}$ These modes all involve one quark generation change and are, therefore, suppressed relative to the decays shown in table 16.1.

$u\bar{b}$ This mode involves two quark generation changes and is therefore strongly suppressed.

$t\bar{d}$, $t\bar{s}$, $t\bar{b}$ These modes are strictly forbidden as they violate conservation of mass/energy.

16.6. One possible combination of colors is shown in figure D.10.

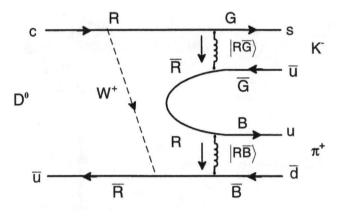

Figure D.10. Possible quark coloring scheme for the decay in figure 16.2.

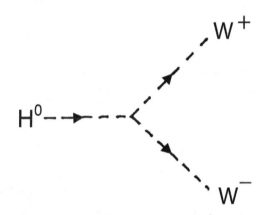

Figure D.11. Feynman diagram for the decay $H^0 \to W^+ + W^-$.

16.8.
   (a) $\pi^+ \to \pi^0 + e^+ + \nu_e$ allowed, weak interaction;
   (b) $\tau^+ \to \pi^+ + \bar{\nu}_\tau$ allowed, weak interaction;
   (c) $\Xi^- \to \Lambda^0 + \pi^-$ allowed, weak interaction;
   (d) $\mu^- \to \pi^- + \nu_\mu$ violates mass/energy conservation;
   (e) $\pi^+ \to e^+ + \nu_e$ allowed, weak interaction;
   (f) $\pi^0 \to e^+ + e^- + \gamma$ allowed, weak/electromagnetic interaction;
   (g) $\phi \to K^+ + K^- + \gamma$ allowed, strong/electromagnetic interaction;
   (h) $\Omega^- \to \Xi^0 + K^-$ violates mass/energy conservation;
   (i) $K^- \to e^- + \nu_e$ violates lepton number conservation;
   (j) $D^0 \to \Delta^- + e^+ + \nu_e$ violates baryon number conservation.

17.2 The Higgs boson can decay, as shown below, to two vector bosons. These decays include,

$$H^0 \to W^+ + W^-.$$

A Feynman diagram for this decay is shown in figure D.11.

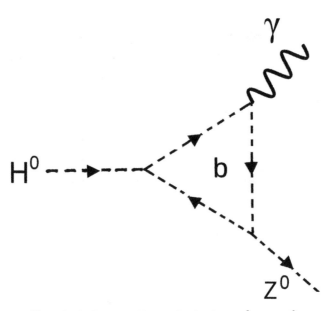

Figure D.12. Feynman diagram for the decay $H^0 \rightarrow \gamma + Z^0$.

The weak charged bosons decay as shown in table 16.1 to fermion–antifermion pairs

$$e + \nu_e, \ \mu^+\nu_\mu, \ \tau^+\nu_\tau, \ u\bar{d}, \ \bar{s}c.$$

The decay products of the W^- will be the charge conjugates of these.

The Higgs boson can also decay through a virtual weak boson loop by the process

$$H^0 \rightarrow \gamma + Z^0$$

and the Z^0 can then decay to a lepton–antilepton pair or a quark–antiquark pair (except top quarks) where the decay particles are of the same flavor. A diagram of this decay is shown in figure D.12.

18.2. From equation (18.14) the threshold energy is given as

$$E = \frac{m_e c^2}{\left[1 - \dfrac{v^2}{c^2}\right]^{1/2}}$$

table D.40 shows the result of this calculation for some charged particles.

19.2. For the process

$$\nu_e + {}^A X \rightarrow {}^A Y + e^-$$

the threshold energy is the same as the Q for the reverse process, i.e. electron capture,

$$e^- + {}^A Y \rightarrow {}^A X + \nu_e.$$

Table D.40. Cherenkov threshold energy for some charged particles.

Particle	Mass (MeV/c^2)	Cherenkov threshold (MeV)
p	938.27	1419
μ^+	105.7	160
K^+	494	147
π^+	139.57	211

Table D.41. Threshold energy for electron neutrino absorption by some nuclei.

^AX	E_{th} (MeV)
^{55}Mn	0.231
^{40}Ca	14.32
^{125}Te	0.178
^{146}Nd	1.482
^{136}Xe	0.067

This is given in terms of atomic masses as

$$E_{th} = [m(^AY) - m(^AX)]c^2.$$

Using tabulated masses, we obtain the results in table D.41.

19.4. From equation (19.14) the energy given up to the γ-ray in the (n,γ) reaction is

$$E\gamma = \left[m(^{35}Cl) + m_n - m(^{36}Cl) \right]c^2 = [34.96885271 \text{ u} + 1.008664904 \text{ u} - 35.96830695 \text{ u}]$$
$$\times 931.494 \text{ MeV/u} = 8.58 \text{ MeV}.$$

The maximum energy given up to an electron in a Compton scattering event is

$$E_e = \frac{E_\gamma^2}{m_e c^2} \frac{1}{1 + \frac{E_\gamma}{m_e c^2}}.$$

When $E_\gamma \gg m_e c^2$ then $E_e \approx E_\gamma$, so, $E_e \approx 8.58$ MeV, well above the Cherenkov threshold of 0.77 MeV given in the text.

Printed in the USA
CPSIA information can be obtained
at www.ICGtesting.com
JSHW061341241223
54197JS00004B/71